International Symposium on
Metabolism, Physiology, and Clinical Use of

Pentoses and Pentitols

Hakone, Japan, August 27th—29th, 1967

Edited by

B. L. Horecker · K. Lang · Y. Takagi

With 254 Figures

Springer-Verlag Berlin · Heidelberg · New York 1969

Professor Dr. BERNARD L. HORECKER
Department of Molecular Biology, Albert Einstein College of Medicine
Bronx, New York, U.S.A.

Professor Dr. KONRAD LANG
Physiologisch-Chemisches Institut der Johannes Gutenberg-Universität
Mainz, West Germany

Professor Dr. YASUYUKI TAKAGI
Department of Biochemistry, Kyushu University School of Medicine
Fukuoka, Japan

ISBN-13: 978-3-642-46193-4 e-ISBN-13: 978-3-642-46191-0
DOI: 10.1007/978-3-642-46191-0

Softcover reprint of the hardcover 1st edition 1969
Library of Congress Card Number 75-97994.

 Title-No. 1631

Preface

In recent times it has come to be realized that pentoses are important as building blocks in the organism and as intermediate products in metabolism. The finding that ribose and desoxyribose are nucleic acid constituents and the discovery of the pentose-phosphate cycle have put these substances in a central position in biochemical research. The concurrent realization that the pentitol, xylitol, is involved in an important metabolic cycle and that the body can handle it in quantities comparable in magnitude to the well-known carbohydrates has resulted in extensive experimental and clinical studies, especially in Germany and Japan. Some of the properties of xylitol, e.g. its independence of insulin and also its utilization in other disturbances of the glucose metabolism have aroused considerable interest among clinicians. The initial experiments have shown that xylitol may be of significant therapeutic value, and more extensive research work into the possible clinical applications of xylitol is being undertaken. Xylitol research is a typical example of the interrelation of the clinical and experimental approaches to a problem. These rapid advances in knowledge made it necessary to bring together workers in both the research and clinical fields in order that they might discuss their findings in critical fashion and advance our understanding of this subject.

Through the generosity and cooperation of two pharmaceutical companies, Eisai Co., Ltd. (Japan) and J. Pfrimmer & Co., Erlangen (Germany), it was possible to meet this need. The excellent papers and the profitable and fruitful discussions which followed made it obvious, both to the organizers of the meeting and to the participants, that these findings should be made available to a wider audience. The editors would like to express their thanks to the publishers, Springer-Verlag, and to the sponsors of the Symposium for making this possible.

November 1969

B. L. Horecker, K. Lang, Y. Takagi

Table of Contents

Session III: Clinical Use of Pentoses and Pentitols

Opening Remarks

It is a great honour and pleasure for me to open the

"International Symposium on Metabolism, Physiology, and Clinical Use of Pentoses and Pentitols"

and to welcome you here at this wonderful place.

The problems which shall be discussed here have no long history. It is even the first time that they are discussed on an international basis.

Biological, physiological and clinical aspects of these substances have been obtained in different parts of the world which are far away from each other. The biochemical basis was found in the U.S.A., above all by the good work done by Touster *et al.* The discovery that the pentoses and the pentitol xylitol are normal metabolites of the organism which are connected by the metabolic cycle, i.e. the glucuronic acid/xylulose-cycle, has at first drawn the attention of more and more experts to these substances.

Then, in Europe, the physiological, pharmacological and toxicological effects and properties of these substances have been elucidated especially by the experiments of my laboratory and of other German laboratories and hospitals closely connected with us.

Because of our results, we supposed that these substances, especially xylitol, may be of therapeutical value in disturbances of metabolism, above all in diabetes mellitus but also in catabolism after trauma and surgical interventions.

Moreover, we could show that the participation in metabolism of the physiological pentoses and of xylitol is much higher than was at first expected and that a significant percentage of the energy requirements can be met by them. In Japan, in a relatively short time, important clinical data for the therapeutical use have been obtained. Numerous investigations and experiments confirmed the good therapeutical effect of xylitol. The field of indications for its use has been determined as well.

Most of the physiological, pharmacological and clinical investigations have been carried out with xylitol because this is the very substance within this field that can be obtained in the easiest way and to a great extent. Therefore, it is also comprehensible that our knowledge of the

Session I

Metabolism of Pentoses and Pentitols

The Role of Pentitols and other Polyols in Evolutionary Development*

B. L. HORECKER

Department of Molecular Biology, Albert Einstein College of Medicine, Bronx, New York, U.S.A.

I am very pleased to have the opportunity to participate in this Symposium and to present the opening paper. Since this will serve as a general introduction to the discussions that follow, I will take the opportunity to indulge in some frank speculation regarding the evolutionary origin of the pathways of carbohydrate metabolism and the possible role of pentitols and other polyols. I have been interested in the metabolism of pentoses and polyols for some time and I believe that this is a good time to review the subject and to examine the present state of our knowledge with respect to the biochemistry and physiology of the pentoses and pentitols, as well as the general role of polyols in carbohydrate metabolism.

Dr. Lang has stated the general purpose of this Symposium in a very elegant manner. Obviously, of greatest interest to most of us here will be a discussion of the metabolism and physiology of these substances in man, as well as their possible clinical application. However, the evolutionary origin of these pathways is of great interest and significance. As J. B. Bernal wrote recently in the New Scientist (January 5, 1967): "The major practical reason for studying the origin of life is that we cannot understand current life without this study, and if we cannot understand it we cannot control it." It was the opinion of many scientists that a better awareness of the origin of life and its evolutionary development will help us to reach a more complete understanding of the significance and function of existing metabolic processes.

In this general review of carbohydrate metabolism I will attempt not only to relate the pentoses and pentitols to the more important metabolic pathways, but I will also consider these substances from the point of view of comparative biochemistry, particularly as they occur in primitive metabolic pathways. Polyols are frequently encountered as intermediates

* This work was supported by grants from the National Institutes of Health (GM 11301) and the National Science Foundation (GB 1465). This is Communication No. 106 from the Joan and Lester Avnet Institute of Molecular Biology.

in the metabolism of certain classes of microorganisms, and perhaps an understanding of their function in these primitive organisms will provide us with a better insight into their importance in mammalian metabolism.

An examination of the specific enzymatic reactions in which pentitols and other polyols are formed or utilized reveals that these are directly related to the metabolism of the pyridine nucleotides, since the polyols are produced from the aldoses and ketoses by reactions involving DPN and TPN, or their reduced forms. A general indication of how these polyols may arise is given in Fig. 1, using as an example the reactions

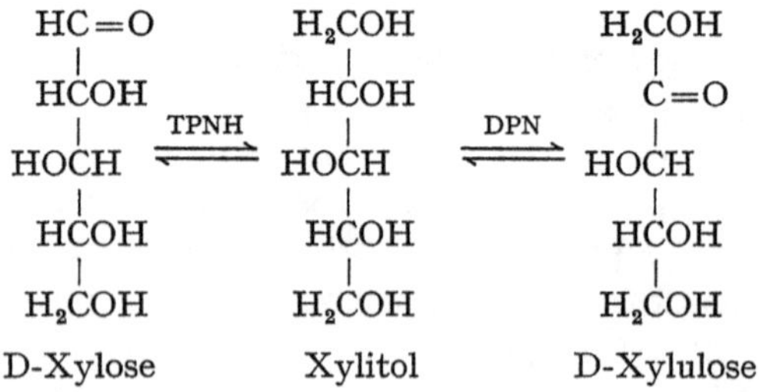

Fig. 1. Formation of ketopentose in Candida

which occur in the aerobic yeast, *Candida utilis*. This yeast may be regarded as a primitive organism which makes extensive use of the polyol pathway for the metabolism of pentoses. In this and related species the pentose xylose is reduced by DPNH to xylitol, which is in turn oxidized by DPN to D-xylulose. This pathway thus provides two mechanisms which are frequently encountered for the formation of pentitol; one is the reduction of a ketopentose, such as D-xylulose, by DPNH, and the other is the reduction of an aldopentose, such as D-xylose, by TPNH. Both of these reactions are readily reversible, but together they provide a mechanism for the conversion of D-xylose to D-xylulose. In Candida D-xylulose is phosphorylated to yield D-xylulose 5-phosphate, which is metabolized by the pentose phosphate pathway.

Although similar mechanisms for the conversion of aldoses to ketoses are encountered in mammals, historically this was not the first type of reaction described for the formation of pentitols. The earliest was the discovery by Hollmann and Touster [1] of the reduction of L-xylulose to xylitol by a TPN-linked system in mammalian liver. It is of interest that TPN systems are involved in the formation of polyols from aldoses, as well as ketoses, and I will present evidence later which suggests that a similar aldose reductase, which can be linked to TPNH-generating reactions in many tissues, may have considerable physiological importance in mammalian metabolism.

Since the formation and utilization of pentitols is directly linked to the state of oxidation of the pyridine nucleotides, it is obvious that we must first understand the metabolic functions of these coenzymes. Although this subject may be quite familiar to most of you, it is worth reviewing briefly in order to provide a basis for further discussion. It is useful to consider the functions of the pyridine nucleotides from a historical point of view. Cozymase (DPN or NAD) was first discovered by Harden and Young in 1906 [2]. They found this coenzyme to be

$$\text{Glucose} \xrightarrow{\text{DPN}} \text{Ethanol} + CO_2$$

Harden and Young (1906)

$$\text{Glucose} \xrightarrow{\text{DPN}} \text{Lactic acid}$$

Meyerhof (1918)

Fig. 2. Discovery of pyridine nucleotides—DPN

essential for the conversion of glucose to ethanol in yeast extracts (Fig. 2). It, therefore, served as the coferment of fermentation and, in fact, the name cozymase derives from this function in alcoholic fermentation in yeast. A number of years later Meyerhof [3] found that the same coenzyme was needed for the conversion of glucose to lactic acid in muscle extract. On the basis of these early observations, it was generally accepted that this coenzyme (DPN) was required for fermentation mechanisms. With the discovery by Warburg and his coworkers [4] of a second

$$\text{Glucose 6-phosphate} \xrightarrow{\text{TPN}} \text{6-Phosphogluconate} \xrightarrow{\text{TPN}} CO_2$$

Warburg and Christian (1935), Lipmann (1936)

$$\text{Substrate} \xrightarrow{2\text{H}} \text{TPN} \xrightarrow{2\text{H}} \text{2-Cytochrome C} \xrightarrow{2e} O_2$$

Haas *et al.* (1941)

Fig. 3. Discovery of pyridine nucleotides—TPN

coenzyme (Fig. 3), which appeared to be concerned more directly with oxidative mechanisms, the distinction was made between the coferment of fermentation (coenzyme I or DPN) and the coferment of respiration (coenzyme II or TPN). Warburg and his coworkers found that the second coenzyme was involved in the oxidation of glucose 6-phosphate to 6-phosphogluconate, and also that the next step in this pathway, which yields pentose phosphate and carbon dioxide, also required the same coenzyme [5]. Thus, the direct oxidation of glucose 6-phosphate appeared to constitute a different pathway for the metabolism of carbohydrate from the glycolytic pathway which had been studied by Embden, Meyerhof, Parnas, and others. The conclusion that this was an oxidative pathway rested on the fact that aerobic conditions were required [6] and that carbon dioxide was produced [5, 7]. All of these considerations supported the view that TPN was the coenzyme of respiration, in contrast to DPN, the coenzyme of fermentation. It was generally agreed that if the link between reduced TPN and cytochrome C could be found, then the remaining gap in our knowledge of electron transport would have been closed, and we would understand respiration as well as fermentation. For this reason, in 1939 Haas and I, in the laboratory of T. R. Hogness, began a search for the enzyme which would catalyze the oxidation of TPNH by cytochrome C. The enzyme was isolated from yeast and named cytochrome C reductase [8], and later a similar enzyme was isolated from mammalian liver [9]. At that time we thought that we had isolated an important enzyme in the respiratory chain. However, it soon became apparent, especially from the work of Lehninger [10], and others [11], that TPNH was not readily oxidized by the respiratory enzyme and that DPNH was the true coenzyme for the cytochrome system and oxidative phosphorylation. DPN, therefore, has an important function not only as the coenzyme for the utilization of carbohydrate by the Embden-Meyerhof pathway, but also as the primary coenzyme for energy production. On the other hand, it is now clearly established that the function for the second coenzyme, TPN, is quite different from that which had originally been proposed. This coenzyme appears to be required for reductive, rather than oxidative, processes.

The hexose monophosphate pathway then provides a mechanism for generating TPNH for reductive requirements of the cell. A good example of the function of these pathways is provided by the red cell. In this cell glucose is converted to lactic acid by the Embden-Meyerhof pathway. This pathway requires DPN and yields ATP as the principal product. The red cell also possesses enzymes which catalyze the oxidation of glucose 6-phosphate (G6P) to 6-phosphogluconate (6PG) and then to pentose phosphate, producing TPNH. Both ATP, produced by the DPN system, and TPNH, produced by the TPN system, are essential for the healthy state of the red cell. In mutants lacking the enzyme for G6P oxidation [12] the red cells tend to be fragile and hemolytic anemias result. One function of TPNH is to maintain the integrity of the cell membrane, perhaps through reduction of disulfide bonds of protein. Thus, in the red cell we have the clearest demonstration of the distinct functions of the two coenzymes. The special function of TPNH is as a source of reducing power, or what Wald proposes to call "metabolic hydrogen" [13]. We will return to this concept later.

Any discussion of the pathways of carbohydrate metabolism must begin with a consideration of the Embden-Meyerhof pathway, which is quantitatively the most important mechanism in mammalian cells (Fig. 4). Two facts are pertinent to the present discussion. One is that this pathway utilizes DPN, and the second is that there are two major branch points which link this pathway to the pentose phosphate pathway.

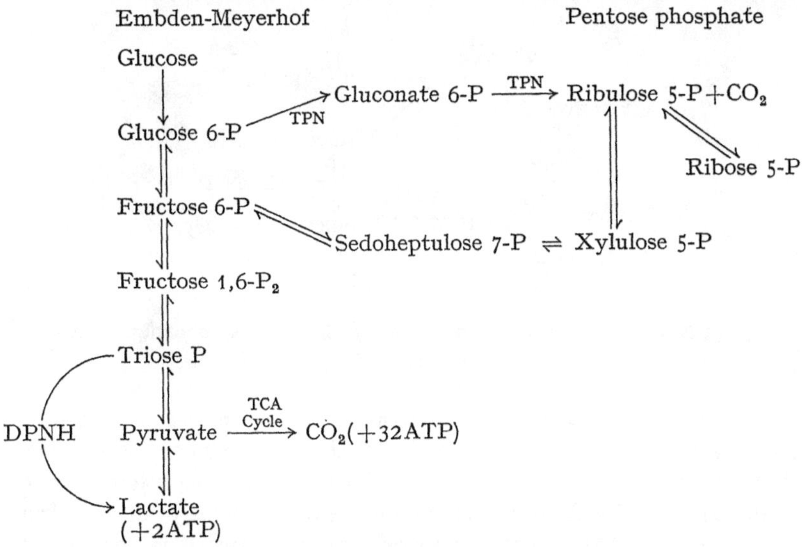

Fig. 4. Pathways of carbohydrate metabolism

These occur at the levels of glucose 6-phosphate and fructose 6-phosphate, respectively. There is now considerable evidence for the hypothesis that major control of the Embden-Meyerhof pathway in the direction of glycolysis occurs at the level of phosphofructokinase. This enzyme is inhibited by ATP and citrate, which may be regarded as products of the glycolytic pathway and the citric acid cycle. If the conversion of fructose 6-phosphate to fructose diphosphate is blocked, then an accumulation of hexose monophosphate would occur which might be expected to result in an increased activity of the pentose phosphate pathway. This was proposed many years ago by Englehardt and Barkash [14] as an explanation of the Pasteur effect. Although definite evidence that a block at the level of phosphofructokinase necessarily increases the activity of the oxidative or pentose phosphate pathway is lacking, this mechanism has not been eliminated.

The reactions of the pentose phosphate pathway are outlined in Fig. 5. This pathway provides two distinct mechanisms for the formation of pentose phosphate. In the first, glucose 6-phosphate is oxidized to 6-phosphogluconate and then to ribulose phosphate and CO_2. Both reactions produce TPNH, and ribulose phosphate is in equilibrium with ribose phosphate [15]. If an excess of pentose phosphate is produced by these oxidative reactions, this excess can be converted back to hexose monophosphate, as was first shown by Dische [16] with red cell hemolysates. This process has been studied in many laboratories and it is

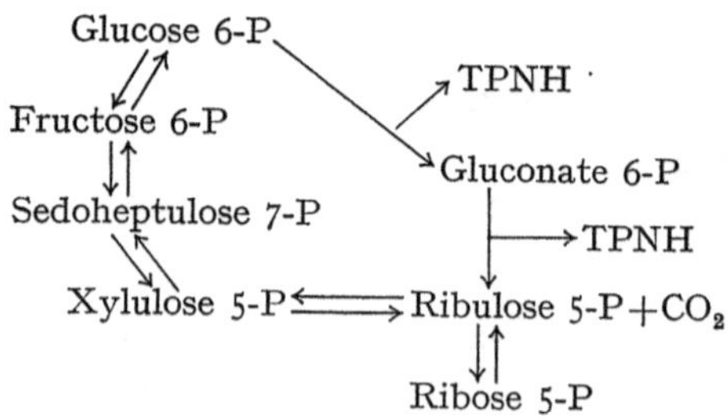

Fig. 5. TPNH and pentose P formation by the pentose phosphate pathway

now well established that most tissues catalyze a very rapid interconversion of hexose and pentose phosphate, with sedoheptulose 7-phosphate as an intermediate [17, 18]. Since pentose phosphate can be produced either by the oxidation of glucose 6-phosphate or directly from fructose 6-phosphate by the non-oxidative reactions, it is pertinent to ask which mechanism is more important in mammalian cells. Evidence based on the use of radioactive tracers indicates that much of the pentose phosphate required for nucleic acid synthesis is produced directly from fructose 6-phosphate by the non-oxidative reactions. The precise quantitative contributions of the two pathways to pentose phosphate reduction is difficult to judge because of the exchange reactions which occur in the non-oxidative pathway [19]. As a result, labeled carbons from positions 1 and 2 of hexose are rapidly equilibrated with the carbon atoms in positions 1 and 2 of pentose phosphate and the results of isotopic experiments are difficult to interpret. There is some indication, however, that the relative contribution of the oxidative reactions to pentose phosphate production depends on the requirement of the tissues for TPNH. If large quantities of the reduced coenzyme are required for lipid synthesis and the formation of other reduced cell constituents, then the amount of pentose phosphate produced by the TPN-linked reactions will be in excess of that required for nucleic acid and nucleotide synthesis; this excess will be converted back to hexose monophosphate and the pathway will operate as a true cycle. This cycle is encountered in certain microorganisms, such as *Candida utilis* [20] and also in actively-synthesizing mammalian tissues, such as the lactating mammary gland [21]. In most cells, however, it appears that the amount of pentose phosphate produced by the oxidative reactions is insufficient to satisfy the needs for nucleotide synthesis and additional pentose phosphate is synthesized by the non-oxidative steps. In these tissues, rather than a complete cycle, we find a flux from hexose monophosphate to pentose phosphate in both arms of the pathway. There is evidence that production of TPNH is linked to polyol formation; this subject will be considered later.

A third important pathway of carbohydrate metabolism in mammalian cells is the uronic acid pathway (Fig. 6). This pathway will be considered in detail in a later chapter by Touster, who will review more specifically the metabolism of L- and D-xylulose. However, I should like to emphasize one important feature of the pathway because it provides an interesting generalization for all of the metabolic pathways which are under consideration at this Symposium. This generalization states that reactions which proceed in the direction of oxidation usually utilize the coenzyme DPN, while reactions which proceed in the direction of reduction tend to utilize TPNH. This permits these reactions to proceed with greatest efficiency, since in the intact cell DPN is maintained largely in the oxidized form by virtue of its very rapid interaction with the electron transport mechanism. On the other hand, TPNH which is not rapidly oxidized by molecular oxygen is more likely to be available for reduction reactions. Thus, in the uronic acid pathway the oxidation of UDPG to UDP-glucuronic acid, of L-gulonate to L-xylulose, and of xylitol to D-xylulose all require DPN, whereas the reduction of glucuronate to L-gulonate and of L-xylulose to xylitol both require TPNH.

The uronic acid pathway is also of interest to the present discussion because it provides for the synthesis of another pentose, D-xylose, which is one of the most abundant monosaccharides in nature, second only to D-glucose. Most of this xylose is found in plant tissues in the form of hemicellulose. It is produced as the nucleotide sugar, UDP-xylose, by decarboxylation of UDP-glucuronic acid [22]. UDP-Xylose is then the immediate precursor of xylose containing polysaccharides. Since D-xylose has recently been shown to be a component of glycoproteins, it is also of interest in mammalian metabolism; this subject will be reviewed by Rodèn. There is reason to believe that in mammalian cells UDP-xylose is also formed by decarboxylation of UDP-glucuronic acid. The occurrence of D-xylose and L-arabinose in mammalian tissues is a very recent development and the full significance of these pentoses in mammalian metabolism remains to be evaluated.

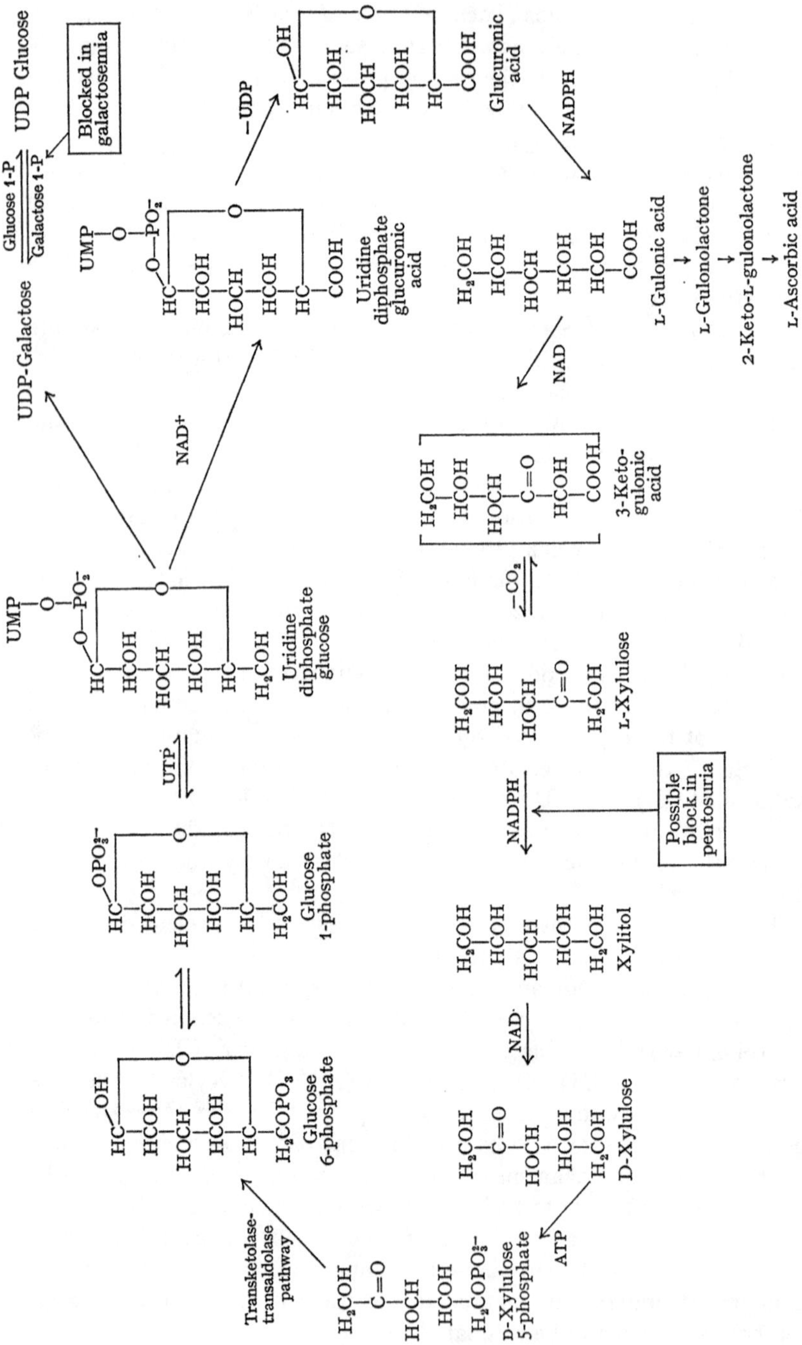

Fig. 6. Uronic acid pathway

As indicated earlier, there is much to learn from a consideration of the metabolism of primitive forms, particularly with respect to their utilization of the pentitol pathways. In these microorganisms, where metabolic processes may be very much simpler, we can more readily reach an understanding of the metabolic function of the pentitols. We can then attempt to relate these mechanisms to those found in mammalian tissues.

It is generally agreed that when life first appeared on the earth there was little or no oxygen in the atmosphere, and both atmosphere and hydrosphere were very different in composition from what they are today (Fig. 7). The atmosphere is thought to have contained water,

Atmosphere	Hydrosphere
H_2O	Carbohydrates (from CH_2O)
CH_4	Purines
H_2	Pyrimidines
N_2, NH_3	Amino acids, porphyrins

Fig. 7. Composition of primitive earth

methane, hydrogen, nitrogen, ammonia, and other reducing substances [23]. It is generally believed that the primordial seas were rich in complex organic compounds which were probably the precursors of primitive life. Wald [13] has compiled an alphabet of 27 such compounds which are essential for life and it is probable that all of these were present in the primitive seas. They arose from the substances present in the atmosphere under the conditions of temperature, electrical discharge, and ultraviolet radiation which then prevailed. Under these conditions, carbohydrates were probably produced by the polymerization of formaldehyde. However, the strongly reducing conditions which prevailed would be inimical to the existence of free aldoses and ketoses. These would either become stabilized by formation of glycosidic linkages with purines and pyrimidines, or by polymerization to form polysaccharides. Those sugars which remained free would eventually undergo reduction to the corresponding polyols. This would be particularly true of the ketoses. Thus, one of the first metabolic requirements for the primitive living forms would be to derive energy from polyols and to convert these to aldoses and ketoses. If we assume that all the essential precursors were present in the primitive seas and that life originated by a single spontaneous event in which the right combination of these precursors were assembled, then the chief metabolic function required to sustain life would be to provide energy for the continued assembly of the materials available into living matter. This energy was derived from fermentation reactions, since oxygen was absent, and in these fermentation reactions the original coenzyme, DPN, was utilized. Thus, energy for synthetic processes was generated.

Unfortunately, we cannot test these concepts directly, since the atmospheric conditions of the primitive earth cannot be known precisely, but can only be inferred from the composition of ancient rocks; it is even less likely that these conditions can be duplicated in the laboratory. Unfortunately, truly primitive forms of life no longer exist, although we may hope eventually to learn which existing forms most resemble those which were present when life began. One important fact, however, has been adequately documented. This is the absence of oxygen in the early atmosphere [24]. Since ozone and oxides of nitrogen were also absent, ultraviolet radiation from the sun would have been highly lethal to life as we now know it, and it is therefore probable that life originated and existed only in deep pools where it would have been adequately protected from the destructive action of this radiation.

Oxygen appeared in the atmosphere with the development of photosynthesis. Early photosynthetic processes were probably based on the decomposition of inorganic substances, including hydrogen or hydrogen disulfide [Eq. (1)] rather than of water [Eq. (2)].

$$2\,H_2S + CO_2 \xrightarrow{h\nu} (CH_2O) + H_2O + 2\,S \tag{1}$$

$$2\,H_2O + CO_2 \xrightarrow{h\nu} (CH_2O) + H_2O + O_2. \tag{2}$$

As the reducing substances became depleted, the mechanisms for utilization of water characteristic of present-day higher plant photosynthesis appeared, and oxygen was produced. With the accumulation of this oxygen in the atmosphere, living forms were induced to undergo several significant changes in their metabolic patterns. In the first place, they were able to produce far greater quantities of energy by utilizing oxygen (respiration) for the oxidation of reduced TPN than they could gain in simple fermentation. This led to the development of the highly efficient process of oxidative phosphorylation. A second consequence of the accumulation of oxygen in the atmosphere was an accelerated oxidation of pre-existing substrates in the hydrosphere (Fig. 8). These substances would have been relatively stable in the highly-reducing atmosphere, but would tend to become oxidized spontaneously by oxygen to forms which were no longer directly useful for the cell. Thus, pre-existing substrates which could be utilized without change were now present in a more highly oxidized state, and it was necessary to reduce them back again to the forms with which the early forms of life were familiar. However, since DPNH was now being oxidized efficiently by oxygen, this coenzyme was not available for reduction reactions, and it was necessary to develop a second coenzyme, stable in the reduced form, which could be utilized to reduce the oxidized substances. This is the "metabolic hydrogen" of Wald [13].

We may summarize the effect of the appearance of oxygen in the atmosphere as follows: 1. The development of oxidative phosphorylation; 2. the disappearance from the primitive oceans of the reduced substrates, and 3. the development of a second coenzyme which could be utilized as a source of metabolic hydrogen for reduction reactions. This was achieved by the phosphorylation of DPN to TPN (Fig. 8).

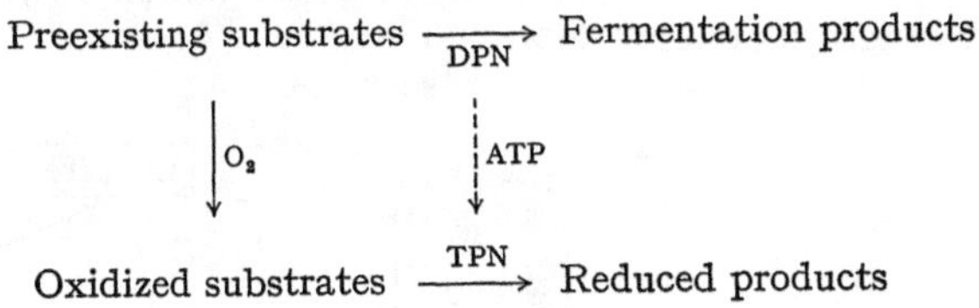

Fig. 8. Origin of coenzymes in evolutionary development

The reduced form of TPN, which was a poor substrate for the electron transport mechanism, could be utilized for reduction reactions. Finally 4., thanks to the screening of ultraviolet radiation by ozone, life was able to move out onto the dry land and develop its present sophisticated forms.

In order to evaluate these very speculative hypotheses and to understand metabolism in higher forms, we must search for living fossils which may still possess the metabolic mechanisms which existed at very early stages of evolution. I would like to review several fermentation mechanisms which appear to have been primitive types. The organisms to which I refer appear to possess fermentation mechanisms which are related to the pentose phosphate pathway, rather than to the Embden-Meyerhof pathway. While there is no proof that the pentose phosphate pathway is a more primitive mechanism than the Embden-Meyerhof pathway, there is some fragmentary evidence which suggests that this may be the case. For example, the critical enzymes of the pentose phosphate pathway, transketolase and transaldolase, are found in all cells which have been examined. In contrast, many species of organisms lack the Embden-Meyerhof pathway, and it is common to find organisms devoid of fructose diphosphate aldolase. Furthermore, on the basis of current studies at the molecular level [25], there is reason to believe that aldolase developed from transaldolase. On the basis of these and other considerations, Wald [13] has also reached the conclusion that the reactions of the pentose phosphate pathway antedate those of the Embden-Meyerhof pathway.

The fermentation pathways which are related to hexose monophosphate oxidation provide some very interesting models for reduction mechanisms. The pathway of glucose metabolism in *Leuconostoc mesenteroides* is shown in Fig. 9. Glucose 6-phosphate is oxidized to 6-phosphogluconate, utilizing either DPN or TPN [26]. 6-Phosphogluconate is oxidized to CO_2 and pentose phosphate and the latter is cleaved by inorganic phosphate to yield acetyl phosphate and triose phosphate [27]. The latter is converted to lactate by the familiar reactions involving oxidation and reduction of DPN, but acetyl phosphate utilizes the two equivalents of DPNH which were formed in the oxidative steps and is thereby converted to ethanol.

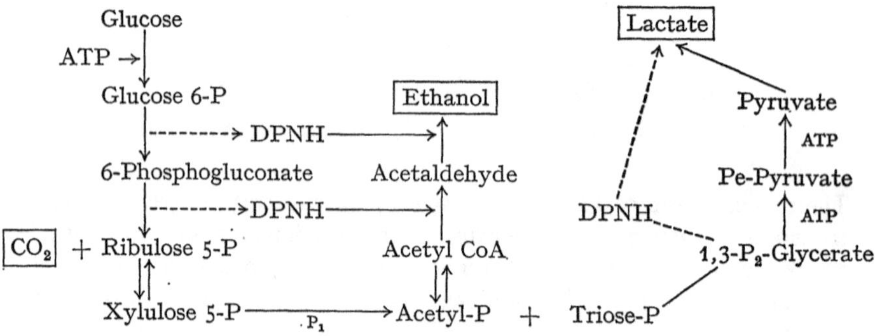

Fig. 9. Hexosemonophosphate oxidation pathway in Leuconostoc

It is important to note that in the fermentation mechanism just described the number of reactions in which pyridine nucleotide is reduced is precisely matched by an equal number in which it becomes oxidized. This balance is not so simply achieved in certain related organisms which are thus forced to rely on polyol formation (Fig. 10). One such organism is *Lactobacillus brevis*, which also ferments fructose by the oxidative pathway (Fig. 10). Glucose 6-phosphate, formed from fructose by a kinase and isomerase, is oxidized in two steps to ribulose 5-phosphate, which is converted to xylulose 5-phosphate before being split to acetyl phosphate and triose phosphate. Thus far the mechanism is identical with that found in Leuconostoc. However, in *L. brevis* acetyl phosphate is not reduced, but is utilized to produce energy in the form of ATP, and the final fermentation products are lactic acid and acetic acid [28]. However, this leaves two equivalents of reduced pyridine nucleotide which must be reoxidized before the fermentation can proceed. *L. brevis* does not contain a physiological mechanism for the oxidation of DPNH by oxygen. Instead, DPN is regenerated from DPNH by a specific polyol dehydrogenase [29]. This enzyme catalyzes the reduction of fructose

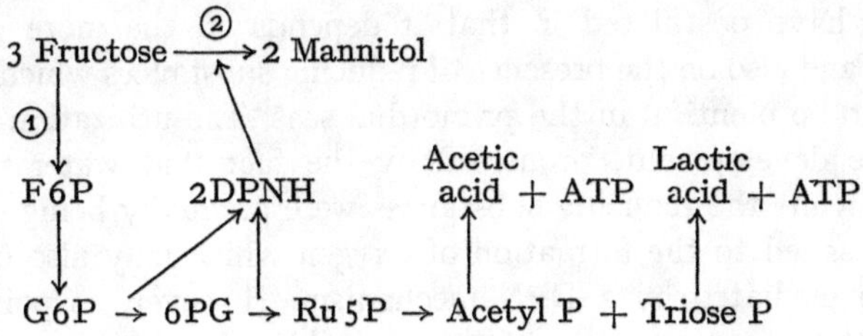

Fig. 10. Fermentation pathway in Lactobacilli

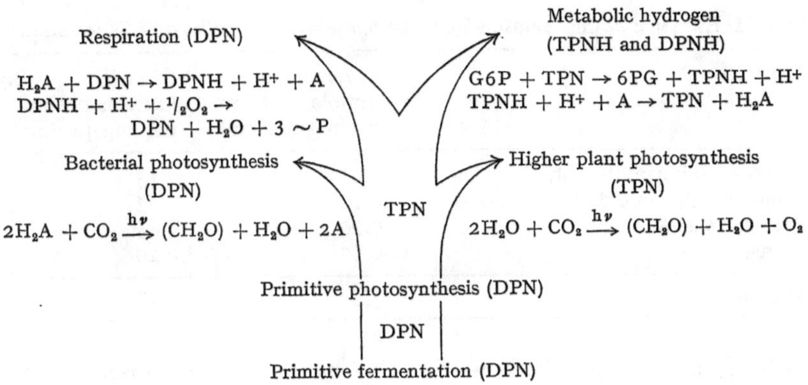

Fig. 11. Proposed evolutionary development of metabolic pathways

to mannitol, which is the normal fermentation product. In the fermentation balance, three equivalents of fructose are converted to one each of acetic acid, lactic acid, carbon dioxide, and two equivalents of mannitol. This is an excellent example of polyol formation linked to the dehydrogenation steps in the pentose phosphate pathway; similar mechanisms, to be discussed below, provide an important source of polyols in certain animal cells.

The evolution of the major metabolic pathways may be summarized as illustrated in Fig. 11. First we have the primitive fermentation mechanisms which depend on DPN as a coenzyme for fermentation and the production of energy. Early photosynthesis was probably also dependent on DPN; this is continued in present-day photosynthetic bacteria where the action of light brings about the reduction of DPN to DPNH. Furthermore, the bacterial photosynthetic mechanisms uniformly require a reducing substance, listed in the figure as H_2A, which becomes oxidized to A during photosynthesis. This type of photosynthetic mechanism thus utilizes an organic reducing substance rather than water, and does not produce oxygen. It resembles the primitive photosynthetic mechanism which we have postulated in that it depends on the more primitive coenzyme and also on the presence of reducing substances which we have assumed to be plentiful in the primordial seas. The utilization of water was a late development, promoted by the fact that water was very abundant while the reducing substances were gradually being depleted. This process led to the formation of oxygen which may also originally have been mediated by a DPN mechanism. However, as pointed out earlier, with the accumulation of oxygen and the development of respiration, it was necessary to convert to a system based on another coenzyme,

Table I. *Enzymes of the pentose phosphate pathway in C. utilis and S. cerevisiae*

Enzyme	*S. cerevisiae* (µmoles per min per mg protein)	*C. utilis* (µmoles per min per mg protein)
Glucose 6-phosphate dehydrogenase	0.06	2.15
6-phosphogluconic dehydrogenase	0.12	0.35
Transketolase	0.20	1.45
Transaldolase	0.02	0.20
Aldolase	3.0	1.1

TPN. Thus, in higher plants we find that light energy is utilized for the reduction of TPN to TPNH, rather than of DPN to DPNH (Fig. 11). With the appearance of respiration, based on the oxidation of DPNH, TPNH could be utilized as a source of metabolic hydrogen. On the right side of the figure are the TPN functions and on the left side the DPN functions. This is, to be sure, a very speculative and incomplete model. For one thing, it is lacking the uronic acid pathway which is of primary interest to this Symposium.

Returning to the metabolic function of the polyols, there is still another organism which possesses a pattern of polyol metabolism similar to that of higher forms. This is the aerobic yeast, *Candida utilis*, which shows a number of interesting differences from ordinary yeast, *Saccharomyces cerevisiae*, in its utilization of carbohydrates. There is, for example, a striking difference in the relative activities of the Embden-Meyerhof and pentose phosphate pathways (Table I). Candida is very much richer in glucose 6-phosphate dehydrogenase, which is about 40 times as active as in Saccharomyces, and in 6-phosphogluconic dehydrogenase, transketolase, and transaldolase, which are 3—10 times as active as they are in Candida. On the other hand, the activity of the glycolytic enzyme, aldolase, is much smaller in Candida than in Saccharomyces [30]. Thus, while Saccharomyces utilizes a typical Embden-Meyerhof pathway, the metabolism of *Candida utilis* is based very strongly on the hexose monophosphate pathway. The mechanism for the utilization of D-xylulose by this organism is shown in Fig. 12. It is interesting to note that this organism, which is so dependent on the hexose monophosphate pathway, also appears to retain the information for the utilization of pentitols. This would be consistent with a more primitive evolutionary state. In *Candida utilis* the first step in pentose metabolism is reduction to xylitol, followed by oxidation to D-xylulose. The first step requires TPNH and the second DPN, according to the familiar pattern which was mentioned earlier. The source of TPNH for the first reduction step is the pentose phosphate pathway; the enzymes in this pathway have already been

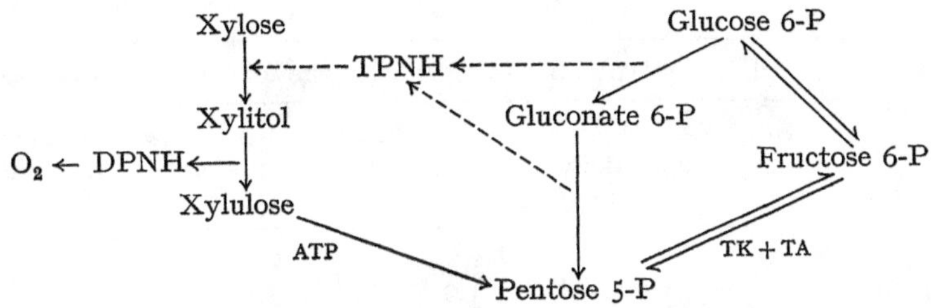

Fig. 12. Carbohydrate metabolism in Candida

shown to be very active in this organism. It is significant that although *Candida utilis* contains very active transketolase and transaldolase, which together catalyze the interconversion of fructose 6-phosphate and pentose phosphate, it does not appear to form any pentose from hexose by this mechanism. Sowden *et al.* [20] have shown that all of the ribose present in this organism originates from glucose by the oxidative mechanism, with the loss of carbon-1 of hexose. This is one of the few examples in nature where pentose formed from glucose-1-^{14}C is completely unlabeled, and it is probably related to the fact that this organism possesses a very high requirement for TPNH. This requirement for TPNH may be related to the fact that Candida is very rich in lipid, whose synthesis depends on TPNH. However, even the mechanism of triglyceride formation is not fully understood, since a DPN-linked α-glycerophosphate dehydrogenase, which is required for the formation of α-glycerophosphate from dihydroxyacetone phosphate, cannot be detected in extracts of this organism. An alternate mechanism for glycerol production may be the direct reduction of glyceraldehyde (Table II). It is noteworthy that the best substrate for the TPN-linked polyol dehydrogenase is D-glyceraldehyde. The polyol dehydrogenase activity in cells grown on various substrates is shown in Table II. The reduction of xylose is catalyzed by an inducible enzyme, present in highest concentration in cells grown on xylose. Very little of this activity is detected in cells grown on glucose,

Table II. *Activity of the TPN-linked dehydrogenase in cels of the C. utilitis grown on four carbon sources*

Carbon source	Specific activity with the following substrates				
	D-xylose (units/mg)	D-glycer-aldehyde (units/mg)	L-arabinose (units/mg)	D-erythrose (units/mg)	D-glucose (units/mg)
D-xylose	2.4	9.2	3.1	7.0	0.22
D-glucose	0.18	2.1	0.34	1.0	0.05
Glycerol	0.17	1.5	0.45	1.7	0.039
Acetate	0.14	0.46	—	—	0.019

Carbon source	Activity relative to D-xylose with the following substrates				
	D-xylose (units/mg)	D-glycer-aldehyde (units/mg)	L-arabinose (units/mg)	D-erythrose units/mg)	D-glucose (units/mg)
D-xylose	1.0	3.8	1.3	2.9	0.09
D-glucose	1.0	6.7	1.9	5.5	0.28
Glycerol	1.0	8.8	2.6	10.0	0.23
Acetat	1.0	3.3	—	—	0.13

glycerol, or acetate. It will be noted, however, that in cells grown on glucose there is significant activity for the reduction of glyceraldehyde and the ratios of the two activities in the cells grown under various conditions suggest the presence of two different enzymes, one induced by D-xylose, which is also active with D-glyceraldehyde, and perhaps a second enzyme which is specific for glyceraldehyde and/or erythrose.

It is hardly necessary to point out that glycerol is also a polyol, perhaps the most important polyol present in mammalian cells. The pathway for the formation of glycerol, based on the reduction of glyceraldehyde, may occur in higher forms; the presence of this enzyme in mammalian cells will be discussed presently.

It is of interest to compare the polyol pathway in *Candida utilis* with some similar reactions found in mammalian cells (Fig. 13). Hers [31] has shown that the formation of fructose in the seminal vesicle is the result of a two-step reaction in which glucose is reduced to sorbitol by TPNH, and sorbitol oxidized to fructose by DPN. This pathway bears a striking resemblance to the pathway of xylose utilization in Candida; indeed it is difficult to distinguish between the polyol dehydrogenases of Candida, whose natural substrates are D-xylose and xylitol, and the enzymes in seminal vesicle, whose natural substrates are glucose and sorbitol. As we have seen (Table II), the Candida enzyme will also utilize

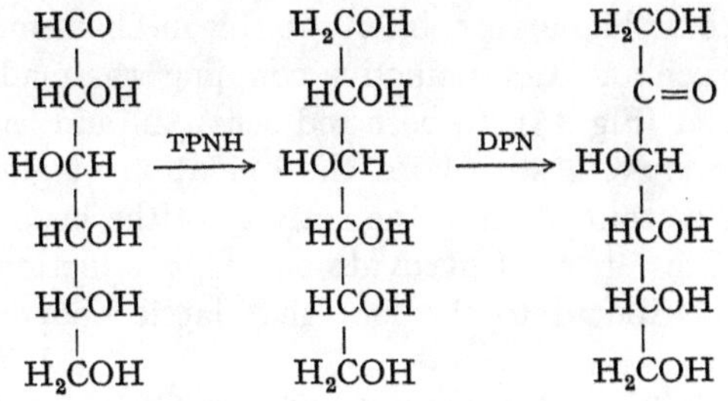

Glucose Sorbitol Fructose

Fig. 13. Conversion of glucose to fructose in the seminal vesicle

D-Glucose	$\xrightarrow{\text{TPNH}}$	D-Sorbitol	$\xrightarrow{\text{DPN}}$	D-Fructose
D-Mannose	$\xrightarrow{\text{TPNH}}$	D-Mannitol	$\xrightarrow{\text{DPN}}$	D-Fructose
D-Ribose	$\xrightarrow{\text{TPNH}}$	Ribitol	$\xrightarrow{\text{DPN}}$	D-Ribulose
D-Xylose	$\xrightarrow{\text{TPNH}}$	Xylitol	$\xrightarrow{\text{DPN}}$	D-Xylulose
L-Arabinose	$\xrightarrow{\text{TPNH}}$	L-Arabitol		
		L-Iditol	$\xrightarrow{\text{DPN}}$	L-Sorbose

Fig. 14. Common substrates for seminal vesicle and Candida enzymes

glucose, and it has been demonstrated that xylose is an excellent substrate for the corresponding dehydrogenase in seminal vesicle. The two enzymes are indeed active with a very similar range of substrates (Fig. 14). All of the aldoses listed in the figure are reduced by both seminal vesicle extracts and Candida TPN polyol dehydrogenase, while the corresponding DPN polyol dehydrogenases will oxidize the same set of polyols.

In addition to the substrates shown in the table, the best substrates for the TPN enzyme from either source are D-erythrose and glycerol. In view of the similar substrate specificities of the polyol dehydrogenases from these very different sources, it is of considerable interest to isolate the proteins and compare their primary structures.

The polyol pathway of carbohydrate metabolism is also encountered in certain species of ungulates [31]. In the placenta of these species fructose is reduced to sorbitol which passes into the fetal blood and is carried to the fetal liver, where it is oxidized to fructose. In all of the cases cited the polyol pathway occurs in reproductive or embryonic tissue, suggesting again that it may be a vestige of a primitive mechanism.

In other mammalian tissues, polyol production is linked to the hexose monophosphate pathway in a manner which resembles the mannitol fermentation mechanism in Lactobacillus. This will be discussed in later chapters, but I would like to refer briefly to this mechanism. The original observation for such an oxido-reductive coupling was made with pyruvate as the oxidant (Fig. 15). In both red cells [32] and lens tissue [33] the addition of pyruvate stimulates the oxidation of glucose-1-^{14}C to $^{14}CO_2$, indicating a stimulation of the activity of the hexose monophosphate pathway. This effect of pyruvate on CO_2 production from C-1 of glucose can be attributed to the fact that lactic dehydrogenase will

$$\begin{array}{rl} & \text{Glucose 6-P} + 2\,\text{TPN} \rightarrow \text{Pentose P} + CO_2 + 2\,\text{TPNH} + 2\,H^+ \\ & 2\,\text{Pyruvate} + 2\,\text{TPNH} + H^+ \rightarrow 2\,\text{Lactate} + 2\,\text{TPN} \\ \hline \text{Sum} & \text{Glucose 6-P} + 2\,\text{Pyruvate} \rightarrow \text{Pentose P} + CO_2 + 2\,\text{Lactate} \end{array}$$

Fig. 15. Glucose 6-phosphate dehydrogenase coupling to pyruvate

$$\begin{array}{rl} & \text{Glucose 6-P} + 2\,\text{TPN} \rightarrow \text{Pentose P} + CO_2 + 2\,\text{TPNH} + 2\,H^+ \\ & 2\,\text{Aldose}\begin{pmatrix}\text{Glucose}\\ \text{Galactose}\\ \text{Xylose}\end{pmatrix} + 2\,\text{TPNH} + H^+ \rightarrow 2\,\text{Polyol}\begin{pmatrix}\text{Sorbitol}\\ \text{Dulcitol}\\ \text{Xylitol}\end{pmatrix} + 2\,\text{TPN} \\ \hline \text{Sum} & \text{Glucose 6-P} + 2\,\text{Aldose} \rightarrow \text{Pentose P} + CO_2 + 2\,\text{Polyol} \end{array}$$

Fig. 16. Glucose 6-phosphate dehydrogenase coupling to hexoses

utilize pyruvate for the oxidation of TPNH; under normal conditions in the red cell TPNH oxidation is the rate-limiting step. By converting TPNH back to TPN pyruvate permits the reaction to continue. The same result is achieved by polyol dehydrogenase in lens and nerve tissue (Fig. 16) [33, 34]. It has been demonstrated that the primary pathway for carbohydrate metabolism in the lens is the oxidation of hexose monophosphate [35]. In the presence of any one of a number of hexoses or pentoses the TPNH produced will be utilized to form the corresponding polyol: sorbitol from D-glucose, dulcitol from D-galactose, and xylitol from D-xylose. There is some evidence which suggests that polyol formation is related to the production of cataracts. It is not clear whether this is a direct consequence of the accumulation of polyol and the osmotic changes which result, or whether it is related to the depletion of TPNH, which is thus no longer available to maintain lens protein in the reduced (sulfhydryl) form. Dische and his coworkers [36] have presented evidence suggesting that the formation of cataracts is associated with the tendency of lens proteins to become oxidized to the disulfide form. This question will probably be discussed by others.

In this brief review I have tried to summarize the major pathways of carbohydrate metabolism and their relation to the formation and utilization of polyols. As Professor Lang has already pointed out, our knowledge of pentitol and pentose metabolism in mammalian cells is quite recent. Only a few years ago the work of Touster provided the sole indication for an important role of pentitols in mammalian metabolism. New reactions involving these substances are being reported constantly and I think we may assume that there are still many aspects of their metabolism which remain to be elucidated. It is likely that the pentitols play a much more important role than we suspect even now.

References

1. Hollmann, S., and O. Touster: An enzymatic pathway from L-xylulose to D-xylulose. J. Amer. chem. Soc. **78**, 3544 (1956).
2. Harden, A., and W. J. Young: The alcoholic ferment of yeast juice. Proc. roy. Soc. B **77**, 405 (1906).
3. Meyerhof, O.: Über das Vorkommen des Coferments der alkoholischen Hefegärung im Muskelgewebe und seine mutmaßliche Bedeutung im Atmungsmechanismus. Hoppe-Seylers Z. physiol. Chem. **101**, 165 (1918).
4. Warburg, O., W. Christian u. A. Griese: Wasserstoffübertragendes Co-Ferment, seine Zusammensetzung und Wirkungsweise. Biochem. Z. **282**, 157 (1935).
5. — — Abbau von Robisonester durch Triphospho-Pyridin-Nucleotid. Biochem. Z. **292**, 287 (1937).
6. Lipmann, F.: Fermentation of phosphogluconic acid. Nature (Lond.) **138**, 588 (1936).
7. Dickens, F.: CCXI. Oxidation of phosphohexonate and pentose phosphoric acids by yeast enzymes. I. Oxidation of phosphohexonate. II. Oxidation of pentose phosphoric acids. Biochem. J. **32**, 1626 (1938).
8. Haas, E., B. L. Horecker, and T. R. Hogness: The enzymatic reduction of cytochrome C. J. biol. Chem. **136**, 747 (1940).
9. Horecker, B. L.: Triphosphopyridine nucleotide-cytochrome C reductase in liver. J. biol. Chem. **183**, 593 (1950).
10. Lehninger, A., in: Phosphorus metabolism (W. D. McElroy and B. Glass, editors), p. 344. Baltimore, Maryland: Johns Hopkins Press 1951.
11. Jacobson, K. B., and N. O. Kaplan: Pyridine coenzymes of subcellular tissue fractions. J. biol. Chem. **226**, 603 (1957).
12. Carson, P. E., C. L. Flanagan, C. E. Ickes, and A. L. Alving: Enzymatic deficiency in primaquine-sensitive erythrocytes. Science **124**, 484 (1956).
13. Wald, G.: The origins of life. Proc. nat. Acad. Sci. (Wash.) **52**, 595 (1964).
14. Englehardt, V. A., and A. P. Barkash: Oxidative breakdown of phosphogluconic acid. Biokhimiya **3**, 500 (1938).
15. Horecker, B. L., P. Z. Smyrniotis, and J. E. Seegmiller: The enzymatic conversion of 6-phosphogluconate to ribulose 5-phosphate and ribose 5-phosphate. J. biol. Chem. **193**, 383 (1951).
16. Dische, Z.: Phosphorylierung der im Adenosin enthaltenen d-Ribose und nachfolgender Zerfall des Esters unter Triosephosphatbildung im Blute. Naturwissenschaften **26**, 252 (1938).

17. Horecker, B. L., M. Gibbs, H. Klenow, and P. Z. Smyrniotis: The mechanism of pentose phosphate conversion to hexose monophosphate. I. With a liver enzyme preparation. J. biol. Chem. **207**, 393 (1954).
18. Pontremoli, S., A. Bonsignore, E. Grazi, and B. L. Horecker: A coupled reaction catalyzed by the enzymes transketolase and transaldolase. J. biol. Chem. **235**, 1881 (1960).
19. Katz, J., and H. G. Wood: The use of glucose-C^{14} for the evaluation of the Pathways of glucose metabolism. J. biol. Chem. **235**, 2165 (1960).
20. Sowden, J. C., S. Frankel, B. H. Moore, and J. E. McClary: Utilization of 1-C^{14}-D-glucose by *Torula utilis* yeast. J. biol. Chem. **206**, 547 (1954).
21. Wood, H. G., P. Schambye, and E. J. Peeters: Lactose synthesis. II. The distribution of C^{14} in lactose of milk from the perfused isolated cow udder. J. biol. Chem. **226**, 1023 (1957).
22. Feingold, D. S., E. F. Neufeld, and W. Z. Hassid: The 4-epimerization and decarboxylation of uridine diphosphate D-glucuronic acid by extracts from *Phaseolus aureus* seedlings. J. biol. Chem. **235**, 910 (1960).
23. Oparin, I.: Life, its nature, origin and development. London: Oliver and Boyd 1961.
24. Berkner, L. V., and L. C. Marshall: Discussions Faraday Soc. 122 (1964).
25. Morse, D. E., and B. L. Horecker: The mechanism of action of aldolases. Advanc. Enzymol. **30** (in press).
26. De Moss, R. D.: Routes of ethanol formation in bacteria. J. cell. comp. Physiol. **41** (Suppl. 1), 207 (1953).
27. Hurwitz, J.: Pentose phosphate cleavage by *Leuconostoc mesenteroides*. Biochim. biophys. Acta (Amst.) **28**, 599 (1958).
28. Eltz, R. W., and P. J. Vandermark: Fructose dissimilation in *Lactobacillus brevis*. J. Bact. **79**, 763 (1967).
29. Martinez, G., H. A. Barker, and B. L. Horecker: A specific mannitol dehydrogenase from *Lactobacillus brevis*. J. biol. Chem. **238**, 1598 (1963).
30. Scher, B. M., and B. L. Horecker: Pentose metabolism in Candida. III. The TPN-specific polyol dehydrogenase of *C. utilis*. Arch. Biochem. **116**, 117 (1966).
31. Hers, H. G.: Le metabolisme du fructose. Brussels: Arscia 1957.
32. Szeinberg, A., and P. Marks: Substances stimulating glucose catabolism by the oxidative reactions of the pentose phosphate pathway in human erythrocytes. J. clin. Invest. **40**, 914 (1961).
33. Kinoshita, J. H., S. Futterman, K. Satoh, and L. O. Merola: Factors affecting the formation of sugar alcohols in ocular lens. Biochim. biophys. Acta (Amst.) **74**, 340 (1963).
34. Gabbay, K. H., L. O. Merola, and R. A. Field: Sorbitol pathway: Presence in nerve and cord with substrate accumulation in diabetes. Science **151**, 209 (1966).
35. Kinoshita, J. H.: The stimulation of the phosphogluconate oxidation pathway by pyruvate in bovine corneal epithelium. J. biol. Chem. **228**, 247 (1957).
36. Dische, Z., E. Borenfreund, and G. Zelmenis: Changes in lens proteins of rats during aging. Arch. Ophthal. **55**, 471 (1956).

Discussion

Dr. Touster: Will you tell us about the intracellular distribution of enzymes of the pathways you have discussed. Such information is often lacking in publications and is important understanding metabolic interrelationships.

Dr. Horecker: Most of the enzymes of the glycolytic and pentose phosphate pathways are in the so-called soluble fraction, although I do not believe that any enzymes are really just floating in the cytoplasm. Very likely, we will find them to be organized as enzyme complexes, much like the pyruvate and α-ketoglutarate oxidases. With respect to the polyol dehydrogenases in animal tissues, you will tell us later more about the liver enzymes, but I believe that the aldose reductase of seminal vesicle has not yet been obtained in soluble form.

Dr. Minakami: In conncetion with the comparison of metabolism in *Saccharomyces* and *Candida,* I would like to ask Prof. Horecker about metabolic control in pentose phosphate cycle and also I would like to know how he thinks about the parallelism between enzyme pattern and actual metabolic pattern.

Dr. Horecker: There is good reason to believe that control of the oxidation pathway in higher forms is through changes not only in the level of the enzymes, glucose 6-phosphate dehydrogenase and 6-phosphogluconate dehydrogenase, but also through the formation of TPN from DPN. For example, in the *Arbacia* egg, fertilization results in increased levels of dehydrogenase, and also a rapid formation of TPN from DPN. There is reason to believe that reoxidation of TPNH determines the rate of glucose 6-phosphate oxidation; an increase in the level of TPN would also increase the activity of the pathway. It may be controlled by biosynthetic processes which utilize TPNH.

Vitamin B_{12}-Dependent Conversion of Ribonucleotides to Deoxyribonucleotides

William S. Beck

Department of Medicine, Harvard Medical School, and the Hematology Unit of the Medical Service of the Massachusetts General Hospital, Boston, Massachusetts, U.S.A.

Introduction

I should like this morning to describe some of our studies on the biosynthetic pathway of deoxyribose, a pentose of some biological importance. At least two pathways of deoxyribose synthesis occur in nature. One requires the participation of a vitamin B_{12} derivative; the other does not. This essay deals with the vitamin B_{12}-dependent pathway. The next speaker, Dr. Larsson, will consider the vitamin B_{12}-independent pathway.

The studies to be described began with an inquiry into the mechanism of megaloblastic erythropoiesis. Megaloblastic anemias are macrocytic anemias in which the bone marrow erythroid precursors display characteristic morphological abnormalities: 1. large size; 2. increased ratio of cytoplasmic to nuclear area; 3. cytoplasmic basophilia (in the more immature forms); and 4. a curious and distinctive fine-grained chromatin texture. Although megaloblastic anemia occasionally appears in ill-defined circumstances, it is most often due to deficiency of vitamin B_{12} or folic acid (or their active forms). Ordinarily repletion of the lacking nutrient is the only treatment needed to replace megaloblastic erythrocyte precursors with normal normoblastic ones.

Although bone marrow cells are difficult to study biochemically, it is well established that megaloblasts contain abnormally high quantities of RNA[1] — hence the cytoplasmic basophilia — and normal amounts of DNA [1, 2]. As I shall presently mention, the primary difficulty in megaloblasts is not an abnormal excess of RNA synthesis but a defect in the duplication of DNA. Since these cells readily incorporate radioactive deoxyribosyl compounds such as thymidine into their DNA [3, 4], it may be assumed that megaloblasts are not incapable of synthesizing DNA from its deoxyribonucleotide precursors. Rather, the problem appears to be an incapacity to synthesize the precursors.

1. The following abbreviations have been used: DBCC, 5,6-dimethylbenzimidazolyl-cobamide coenzyme; FH_2, 7,8-dihydrofolate; FH_4, 5,6,7,8-tetrahydrofolate; standard abbreviations for nucleotides, nucleic acids, and other biochemicals in accordance with the useage approved by *The Journal of Biological Chemistry*.

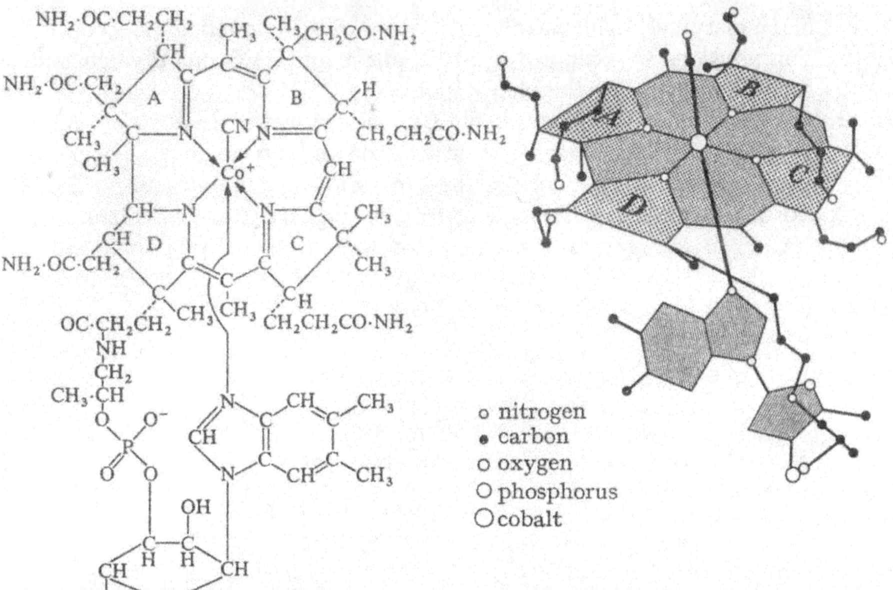

I II

Fig. 1. Chemical formula of vitamin B_{12} (formula I) and semidiagrammatic representation of three-dimensional structure (formula II), showing relations of planar and nucleotide moieties. Hydrogen atoms and several oxygen atoms are omitted

At the time we began our work, it was known why folic acid deficiency might impair DNA synthesis. Although folic acid deficiency has a number of metabolic sequelae [5], the biochemical defect responsible for megaloblast formation appeared to be a block in the synthesis of one DNA precursor, thymidylate. Thymidylate synthetase, the folate-linked enzyme that catalyzes thymine methyl synthesis, had been purified [6] and the evidence was unequivocal that the following reaction takes place:

$$\text{dUMP} + \text{N}^5,\text{N}^{10}\text{-methylene FH}_4 \rightarrow \text{dTMP} + \text{FH}_2 \quad (1)$$

The reaction has no vitamin B_{12} requirement, though conceivably a vitamin B_{12} deficiency *in vivo* could embarrass the supply of N^5,N^{10}-methylene FH_4. The question, then, to which we sought an answer concerned the role of vitamin B_{12} in DNA synthesis.

For purposes of review, a brief word might be in order on the chemical properties of vitamin B_{12} and certain of its derivatives. Cyanocobalamin, the most familiar form of vitamin B_{12}, contains two major portions: a planar group, which bears a close but imperfect resemblance to a porphyrin structure, and a nucleotide that lies nearly perpendicular to the planar group (Fig. 1). The four reduced pyrrole rings of the porphyrin-like moiety link to a central cobalt atom whose two remaining coordination positions are occupied by —CN above and a 5,6-dimethylbenzimidazolyl moiety below the planar group.

A coenzyme form of vitamin B_{12} was discovered in 1959 by Weissbach, Toohey, and Barker in studies of an enzyme from *Clostridium tetanomorphum* that catalyzes the conversion of glutamate to β-methylaspartate [7]. This derivative of vitamin B_{12}, the so-called cobamide or vitamin B_{12} coenzyme, was shown by Lenhert and Hodgkin [8] to be

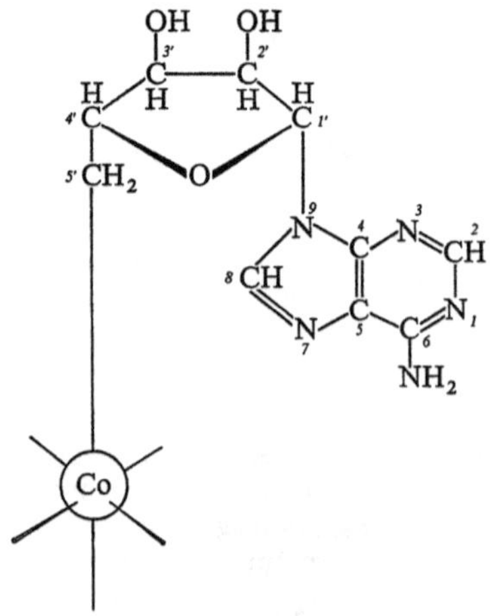

Fig. 2. The 5′-deoxyadenosyl ligand of cobalt in DBCC

5,6-dimethylbenzimidazolyl 5′-deoxyadenosyl-cobamide coenzyme, also known as deoxyadenosylcobalamin and DBC coenzyme. We shall refer to it as DBCC. It is characterized by the presence above the plane of a 5′-deoxyadenosyl moiety that is attached by a bond between the 5′-methylene carbon atom and the cobalt, a novel Co–C bond (Fig. 2). This group, which occupies the same coordinate position on the cobalt as the —CN in cyanocobalamin, is rapidly detached by treatment with light or cyanide.

Our attempts to identify the locus of vitamin B_{12} participation in deoxyribonucleotide synthesis began with a survey of the biosynthetic pathways as they were understood in the early 1960's (Fig. 3). The *de novo* pathways of purine and pyrimidine synthesis were known to yield ribonucleotides [9, 10], a large portion of which are incorporated into RNA. Reichard and coworkers had just discovered that an enzyme system in *Escherichia coli* catalyzes the reduction of ribonucleotides to deoxyribonucleotides [11]. The enzyme, ribonucleotide reductase, was later shown to attack only ribonucleoside diphosphates [12]. The resulting deoxyribonucleoside diphosphates are converted to triphosphates which are the substrates of DNA polymerase. Fig. 3 also shows the shunt pathway in which thymidylate synthetase catalyzes the conversion of uracil in deoxyuridylate to thymine, the characteristic pyrimidine of DNA.

That vitamin B_{12} may participate in the reductive conversion of ribonucleotides to deoxyribonucleotides was suggested by four lines of

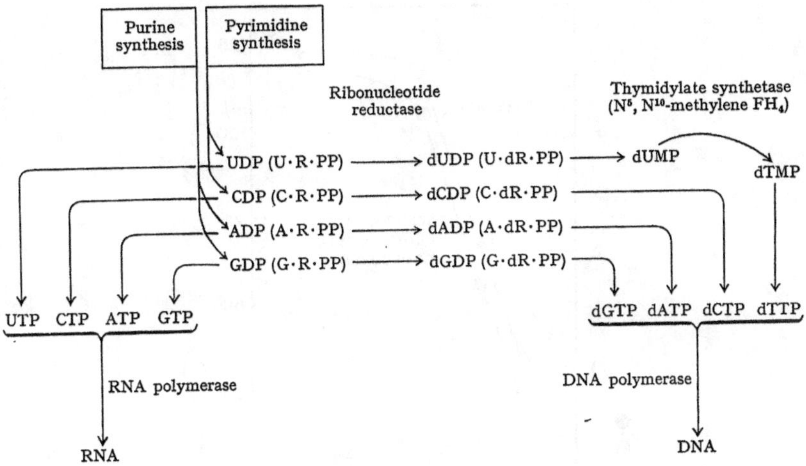

Fig. 3. Schematic diagram of biosynthetic pathways of RNA and DNA precursors in *E. coli*

indirect evidence: 1. the well known fact that the vitamin B_{12} requirement of the vitamin B_{12} auxotroph *Lactobacillus leichmannii* is eliminated by any one of several deoxyribonucleosides but not by ribonucleosides or free deoxyribose [13]; 2. isotopic studies in *L. leichmannii* [14—16] and in intact human bone marrow cells [17] indicating poor conversion of ribosyl groups to DNA deoxyribose in vitamin B_{12} deficiency; 3. decreased levels of acid-soluble deoxyribosyl compounds in vitamin B_{12}-deficient bacteria with abrupt increases following repletion with excess vitamin B_{12} [18]; and 4. typical unbalanced growth behavior (i.e., filamentous growth, elevated RNA/DNA ratios, and impaired cell division) in vitamin B_{12}-starved *L. leichmannii* [19] and other vitamin B_{12}-dependent organisms [20]. We have considered that the megaloblastic erythroid cell is yet another example of unbalanced growth due to impairment of DNA synthesis by vitamin B_{12} or folate deficiency. Indeed, we have looked upon the elongated *L. leichmannii* cell as a "model megaloblast" [21].

Ribonucleotide Reductase Activity in Crude Extracts

In an attempt to obtain more decisive evidence of vitamin B_{12} participation in ribonucleotide reduction, we undertook enzymological studies which I should now like to summarize.

It was the goal initially to determine whether ribonucleotide reductase activity requires the presence of vitamin B_{12} or one of its derivatives. Although Reichard had elegantly demonstrated ribonucleotide reductase

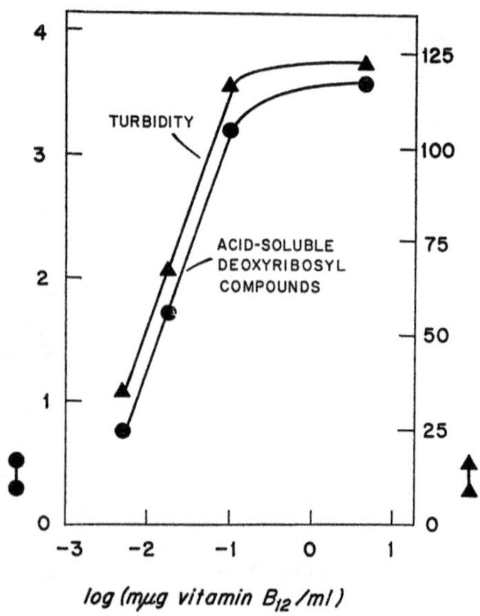

Fig. 4. Effect of vitamin B_{12} concentration in the medium on growth kinetics and acid-soluble deoxyribosyl pool of *L. leichmannii* after cultivation for nine hours. Turbidity is expressed in Klett units (66 filter). Acid-soluble deoxyribosyl pools (expressed as mμmole deoxyribonucleoside per mg protein) were assayed microbiologically after snake venom treatment as described elsewhere [18]

in *E. coli*, we rejected this organism because *E. coli* is not known to require vitamin B_{12} as a nutrient, and turned instead to *L. leichmannii*, an organism which does require vitamin B_{12} as a nutrient (except when offered a deoxyribonucleoside). We hoped to demonstrate that reductase activity is present in extracts of vitamin B_{12}-grown cells and absent in extracts of deoxyribonucleoside-grown cells. Unfortunately, crude extracts of *L. leichmannii* displayed feeble and inconstant reductase activity.

Repressor Control. It seemed that if synthesis of the enzyme were under repressor control — a possibility implied by its strategic position in the pathway of deoxyribonucleotide synthesis — more active preparations might be obtained from derepressed cells. An approach to the study of a postulated repressor system was suggested by data on the effects of vitamin B_{12} and folate deficiency on the size and composition of the acid-soluble deoxyribosyl pool in bacteria [22, 23]. When *L. leichmannii* is starved of vitamin B_{12}, this pool is strikingly decreased (Fig. 4). However, when *L. leichmannii* (Fig. 5 A) or *L. casei* (Fig. 5 B) are starved

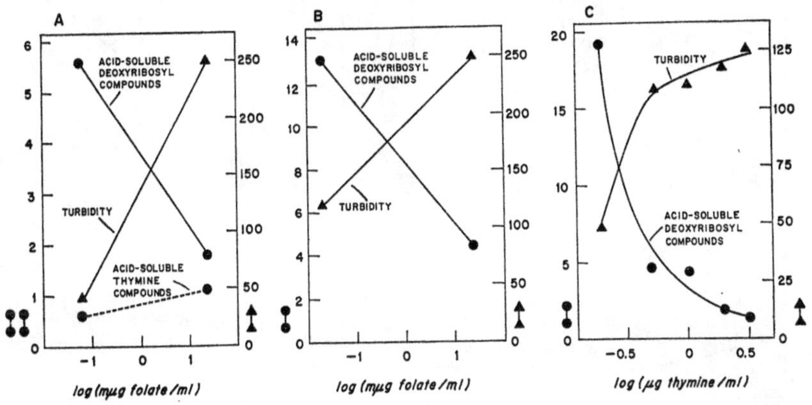

Fig. 5 A—C. Effects of decreased availability of folate or thymine on growth kinetics and acid-soluble deoxyribosyl pool. A, *L. leichmannii* limited in folate; B, *L. casei* limited in folate; C, *E. coli*$_{15T^-}$ limited in thymine. The thymineless state was produced in *L. leichmannii* and *L. casei* by limiting exogenous folate in cultures containing inocula from cell lines serially cultivated in thymine rather than folate (to avoid folate storage [26]). In the experiment summarized above, *L. leichmannii* were cultivated 15 hours, *L. casei* 30 hours, and *E. coli*$_{15T^-}$ 7 hours. Acid-soluble thymine compounds were assayed microbiologically by a microbiological method described elsewhere [23]

of folate and thus of endogenously synthesized thymidylate, or when the thymine-requiring 15T⁻ mutant of *E. coli* is starved of thymine (Fig. 5 C), acid-soluble non-thymine-containing deoxyribonucleotides accumulate in large quantities [22—24].

If the increased deoxyribosyl pool in thymine starvation is attributable to derepressed reductase activity, it seemed doubtful that the hypothetical repressor system would be governed by the intracellular concentration of one or more of the deoxyribonucleotides, which like dATP is present in increased amounts. Rather, a decreased intracellular concentration of a thymine derivative seemed a more likely stimulus to increased reductase synthesis. This supposition was confirmed first in experiments with the thymine-requiring mutant *E. coli*$_{15T^-}$ [22, 25].

When these cells were cultivated in thymine-rich media and then transferred to media containing no thymine, reductase activity in crude extracts (with CDP as substrate) increased 12.5-fold in 2 hours, reaching specific activities of approximately 2.0 (mμmoles of CDP reduced per mg of protein under standard conditions) (Fig. 6A).

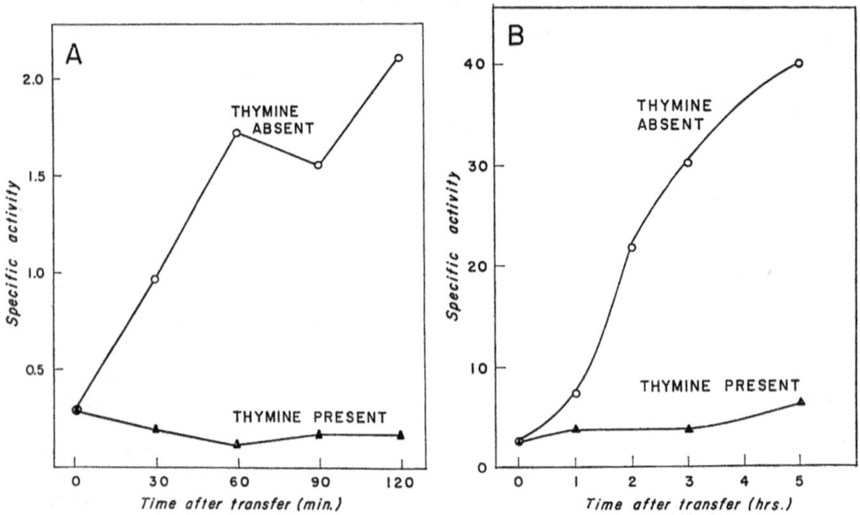

Fig. 6A and B. Time-courses of specific activity of ribonucleotide reductase in crude bacterial extracts following transfer from thymine-rich primary cultures to secondary cultures containing no thymine (○), or thymine (▲). A, *E. coli*$_{15T^-}$; secondary cultures contained 8 mμmoles per ml of thymine; B, *L. leichmannii*; secondary cultures contained 50 mμmoles per ml of thymine. Folate was absent in all cultures. Specific activity is defined as mμmoles of CDP reduced per mg of protein under standard conditions

We then tried to deplete *L. leichmannii* of endogenously synthesized thymine by transferring optimally grown cells to media lacking folate, but folate is stored in substantial quantities by these cells, and stored folate was found to be metabolically available to cells deprived of exogenous folate [26]. If, however, folate in the original medium was replaced by thymine (50 mμmoles/ml), the thymineless state was readily achieved when cells were grown in a thymine-containing primary culture, harvested, washed, and resuspended in a secondary culture containing thymine-free medium. Secondary cultures were further incubated for various time intervals, cells were collected, washed, and sonicated, and ribonucleotide reductase was assayed in sonicates.

Results of a typical experiment [27] are shown in Fig. 6B. Increased reductase activity was observed, specific activities rising to 40—50.

That the enzyme increase was due to derepression of *de novo* enzyme synthesis and not to release of feedback inhibition was indicated by experiments showing that chloramphenicol inhibits the increase in enzyme specific activity following thymine withdrawal [27].

Requirements for Ribonucleotide Reduction. Extracts containing active derepressed enzyme displayed an absolute requirement for added cyanocobalamin or DBCC (Fig. 7) [27]. These results confirmed and extended

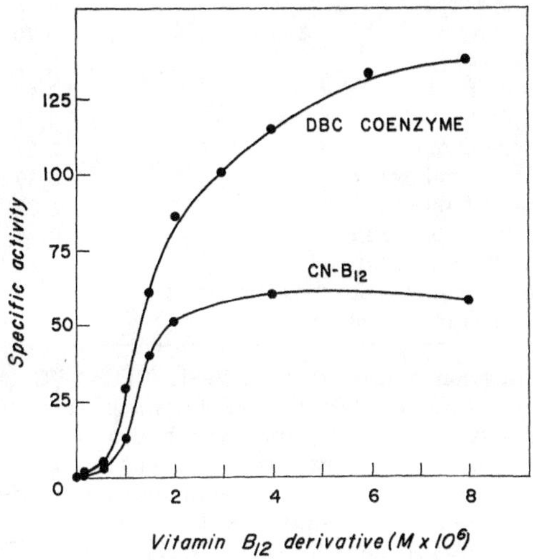

Fig. 7. Effect upon reductase activity of varying concentrations in the incubation medium of cyanocobalamin and DBCC. Incubations were of the standard type [27] and included 2 mg of whole sonicate prepared from cells that had been cultivated 4 hours in thymine-free secondary cultures

the earlier demonstration by Blakley and Barker [28] that conversion of CMP to dCMP in a crude extract of *L. leichmannii* is stimulated by DBCC and a reducing system. Activity in our system was doubled when cyanocobalamin was replaced by DBCC. For the reduction of CDP, there were absolute requirements for ATP and a dithiol reductant (i.e., dihydrolipoate) and a relative requirement for Mg^{++} (Table I) When CDP reducing activity in crude extracts was tested with various cobamide derivatives, activity was highest with DBCC. Hydroxocobalamin was slightly less active than cyanocobalamin and methylcobalamin; Factor B and vitamin B_{12} monocarboxylic acid and monoanilide were relatively inert.

We had earlier observed that *L. leichmannii* and certain other vitamin B_{12}-requiring lactobacilli are capable of binding far more vitamin B_{12} than is needed to satisfy nutritional requirements [28]. The

Table I. *Requirements of ribonucleotide reductase in crude extracts*

Incubation mixture	ΔdCDP formed (mμmoles/mg/min)
Complete	3.19
— DBCC	0.00
— dihydrolipoate	0.01
— dihydrolipoate, + 2-mercaptoethanol	0.05
— ATP	0.00
— Mg^{++}	0.79
+ TPNH	3.17
+ glucose 6-phosphate	3.15
+ glucose 6-phosphate, + TPN	3.21
— DBCC, + BC coenzyme	3.21
— DBCC, + methylcobalamin	0.57
— DBCC, + cyanocobalamin	1.02
— DBCC, + hydroxocobalamin	0.76
— DBCC, + vitamin B_{12} anilide	0.11
— DBCC, + vitamin B_{12} monocarboxylic acid	0.02
— DBCC, + Factor B (cobinamide)	0.00

Complete incubation mixtures contained: CDP-2-^{14}C, 1.25 millimoles (100,000 cpm); ATP, 4 μmoles; DBCC, 2 μmoles; magnesium chloride, 10 μmoles; Tris-chloride buffer, pH 7.3, 25 μmoles; dihydrolipoate, 15 μmoles; and 1.2 mg of L. leichmannii sonicate prepared from cells that had been cultivated in a thymineless secondary culture for 5 hours and were suspended in Mg^{++}-free buffer prior to sonication. Volume of the incubation mixture, 0.5 ml. The amount of other additions were: TPNH, 1 μmole; TPN, 1 μmole; glucose 6-phosphate, 1 μmole; 2-mercaptoethanol, 15 μmoles; various vitamin B_{12} derivatives, 2 μmoles. Incubation and other operations were conducted in the dark [34].

bulk of the added vitamin B_{12} appeared to be bound in the ribosomes. Purified ribosomes actively bind vitamin B_{12} *in vitro* [30]. High molecular weight vitamin B_{12}-binding material is reversibly released in concentrated salt solutions [31], and it appeared from their relative resistance to ribonuclease digestion that vitamin B_{12}-binding ribosomes may constitute a specific class within the ribosome population [32]. In light of these facts, it was of interest to determine the effects of removing ribosomes on reductase activity in crude extracts. This experiment had two interesting results (Fig. 8): 1. the reductase system of ribosome-free 144,000 $\times g$ supernatant was active only with DBCC, cyanocobalamin being virtually inert as a cofactor; and 2. activity of the supernatant fraction with DBCC was considerably higher than that of whole sonicate. These results suggested that ribosomes are the locus of the conversion of cyano- or hydroxocobalamin to DBCC, that coenzyme synthesized in or on the ribosome is held there until released in accordance with undetermined desorption equilibria or other regulatory influences, and that ribosomes also bind exogenously added coenzyme, thereby making it unavailable to reductase apoenzyme. These conclusions were supported by later studies showing that purified *L. leichmannii* ribosomes, naturally or adventitiously, contain 1. a vitamin B_{12} binding material and 2. an active coenzyme synthetase system

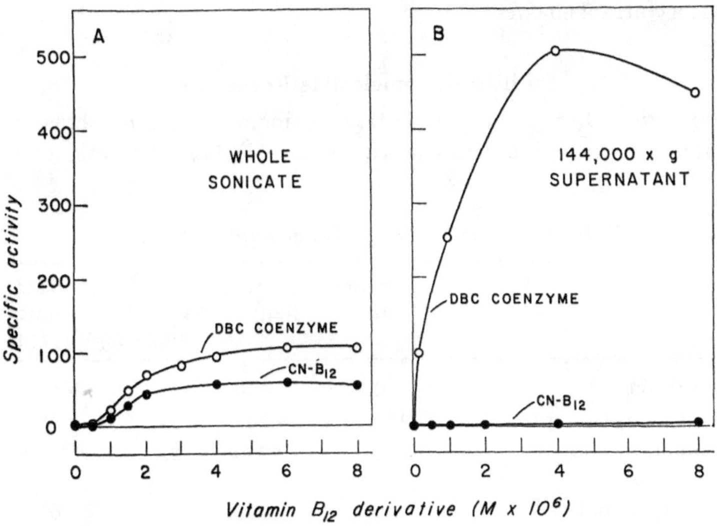

Fig. 8A and B. Effect of ribosome removal upon requirement of reductase system of cyanocobalamin and DBC coenzyme. Incubations were of the standard type [27]. Abscissas, concentrations of cyanocobalamin and DBCC; ordinates, specific reductase activity (mμmoles of CDP reduced per mg). A, Reductase activity in whole sonicate. B, Activity in 144,000 $\times g$ supernatant fraction

[33] consisting of a reductase, which catalyzes the conversion of vitamin B_{12a} (cyano- or hydroxocobalamin containing tervalent cobalt) to vitamin B_{12s} (univalent cobalt), and a vitamin B_{12s} adenosylating enzyme, which catalyzes the ADP-dependent adenosylation step with the formation of tripolyphosphate.

We have been unable to find such components in or on the ribosomes of *E. coli*, *L. acidophilus*, *Cl. tetanomorphum*, rat liver, or indeed any species other than a vitamin B_{12}-requiring lactobacillus (i.e., *L. leichmannii* and *L. lactis*). If these are in fact natural ribosomal components in these species, there would be good reason to look upon them as useful biological adaptations. Even under saturating conditions, only 10,500 molecules of vitamin B_{12} are bound per cell [29]; moreover, the organism often finds itself in an environment incapable of saturating it with vitamin B_{12}. Hence, the binding of vitamin B_{12} onto a particulate surface may be a device for ensuring that the relatively small number of intracellular vitamin B_{12} molecules are efficiently gathered to an activation site, much as intrinsic factor gathers them to an absorption site. Activated vitamin could then be stored on the ribosome or released to serve as a cofactor in accordance with equilibrium kinetics or other regulatory arrangements.

Purified Ribonucleotide Reductase

Properties. During the purification of ribonucleotide reductase from partially derepressed *L. leichmannii* (Table II) [34] the ratio of CTP

Table II. *Purification of ribonucleotide reductase*

Step	Volume (ml)	Protein (mg)	Specific activity (units/mg)	Total units ($\times 10^{-3}$)
Crude extract	380	10,400	1.0	10.4
78,000 × *g* supernatant	310	4,780	11	53.1
Ammonium sulfate fractionation	105	2,050	28	57.4
Acetone fractionation	66	1,090	57	62.1
Ammonium sulfate gradient fractionation	320	322	94	30.3
DEAE-cellulose chromatography	112	36	225	8.1 [a]
Hydroxylapatite chromatography	80	18	576	11.2

[a] It is not known why in this preparation the total units recovered in the DEAE-cellulose step were exceeded by total recovered in the following step. In other preparations, the DEAE-cellulose fraction contained approximately 13% more units than the hydroxylapatite fraction.

reducing activity to CDP reducing activity rose with each step. Indeed, as shown also by Abrams [35] and Blakley *et al.* [36] the reductase of *L. leichmannii* turned out to be a ribonucleoside *tri*phosphate reductase. In this respect, among others, it differs from the ribonucleoside *di*-phosphate reductase of *E. coli*. Table II indicates that inhibitors were present in the crude extract; hence, it is not possible to calculate precisely the amount of enzyme present. Nevertheless, the data indicate that at least 1.6% of the total protein in the crude extract was enzyme.

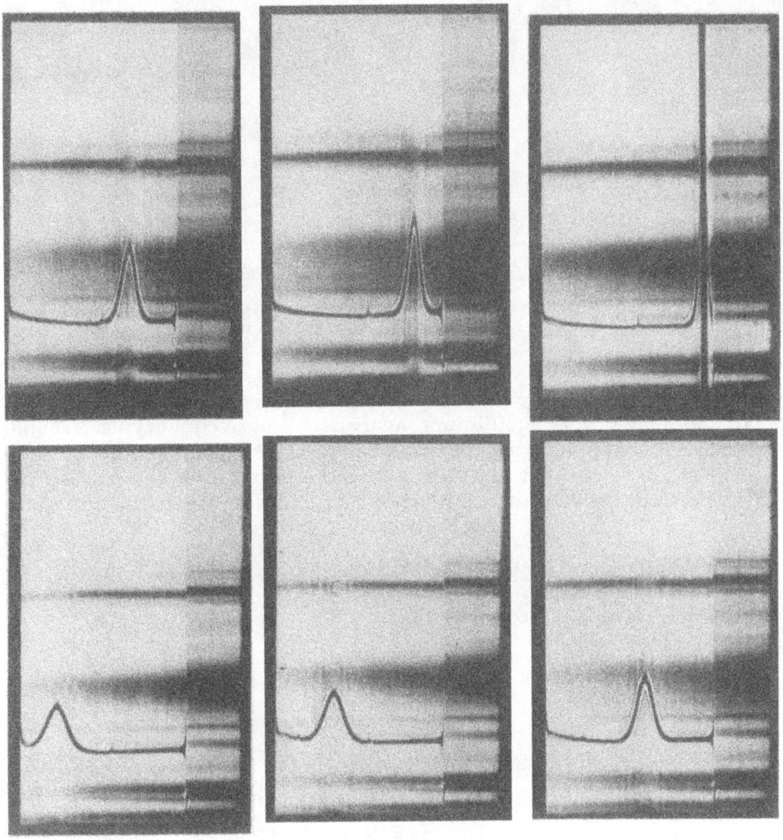

Fig. 9. Schlieren patterns showing sedimentation of purified enzyme in a single sector Kel-F cell at time intervals of 0 to 80 min (from *right* to *left*) after attainment of top speed. Rotor speed, 68,000 rpm; protein concentration, 8.7 mg per ml; temperature, 9.0°; buffer, 0.1 M KCl, 0.01 M potassium phosphate (pH 7), 0.001 M 2-mercaptoethanol

The product of purification was homogeneous by electrophoresis in cellulose acetate at several pH's. Sedimentation velocity analysis showed a single symmetrical peak (Fig. 9) with an $s_{20,w}$ of 5.8. High speed equilibrium sedimentation according to Yphantis [37] suggested a small amount of heterogeneity, possibly attributable to aggregation (Fig. 10). Molecular weight, calculated from the slope near the beginning of the plot to minimize the effects of inhomogeneity and by the use of a value for the partial specific volume (0.73) determined from the amino acid composition, is 115,000. Surprisingly, polyacrylamide gel electrophoresis at pH 9.5 revealed two principal bands of equal intensity (Fig. 11). I shall comment later on this interesting result.

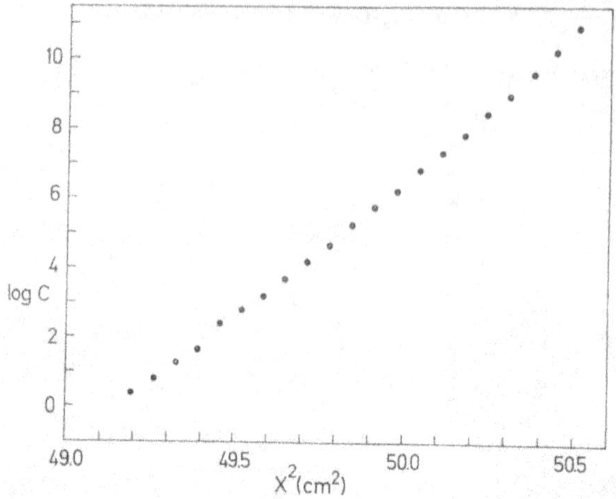

Fig. 10. Sedimentation equilibrium analysis of purified enzyme. Double sector cell with sapphire windows; rotor speed, 15,220 rpm; protein concentration 0.26 mg per ml; temperature, 6.0°; buffer, same as in Fig. 9. *Abscissa*, square of distance from center of rotation (X); *ordinate*, logarithm of protein concentration (C)

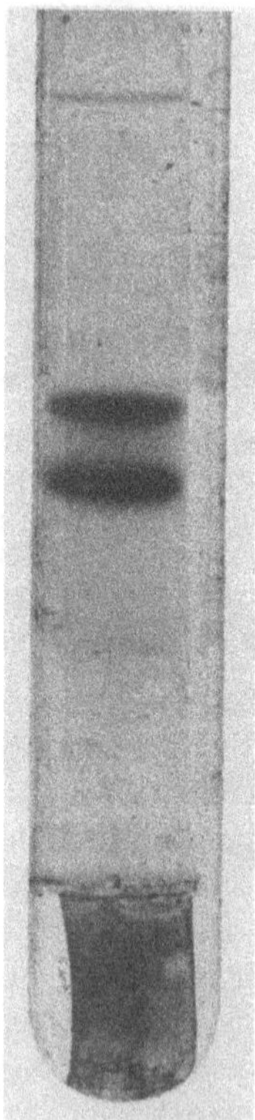

Fig. 11. Polyacrylamide gel electrophoresis of purified enzyme at pH 9.5

Table III. *Requirements of purified ribonucleotide reductase* .

Incubation mixture	dCTP formed (mμmoles/mg/min)
Experiment 1	
Complete	362.7
– Enzyme	4.7
– DBCC	1.6
– Dihydrolipoate	0.0
– ATP	9.4
– Mg^{++}	211.4
– Mg^{++}, + Mn^{++}	354.1
– Mg^{++}, + Ca^{++}	274.6
– DBCC, + BC coenzyme	197.3
– DBCC, + methylcobalamin	2.3
– DBCC, + hydroxocobalamin	0.0
– DBCC, + cyanocobalamin	1.6
– Dihydrolipoate, + TPNH	3.9
– Dihydrolipoate, + DPNH	3.9
Experiment 2	
Complete	450.1
– ATP, + dATP	579.5
– ATP, + dTTP	55.4
– ATP, + GTP	27.3
– ATP, + UTP	23.4
Experiment 3	
Complete	580.3
– Mg^{++}	300.3
– ATP	11.7
– Mg^{++}, – ATP	138.1
– Mg^{++}, – DBCC	2.3
– Mg^{++}, – dihydrolipoate	3.1
– Mg^{++}, – ATP, + dATP	140.4

Standard incubation conditions have been described elsewhere [34]. The substrate was CTP-2-C^{14}; the enzyme, 0.64 μg hydroxylapatite fraction. Concentrations of other additions were equal to the concentrations of the replaced ingredients. Mixtures were incubated 20 min.

In a standard incubation system, purified reductase had the same requirements for CTP reduction (Table III) as crude enzyme had for CDP reduction (Table I): DBCC, ATP, and dihydrolipoic acid with a relative requirement for Mg^{++} (or Mn^{++} or Ca^{++}). Addition of EDTA in the absence of added Mg^{++} had a negligible effect.

Reichard's early demonstration that reductase activity in crude *E. coli* requires a dithiol such as dihydrolipoate [38] led us and other investigators to observe that *L. leichmannii* has a similar requirement [27] and that reduction of the substrate is related stoichiometrically to oxidation of the dithiol [39]. The very high concentration of dihydrolipoate required (0.03 M or higher) made it unlikely that this compound is the natural reductant in both systems. Dihydrolipoate can be replaced by compounds containing 6,8-, 1,4-, or 1,3-dithiol groups, but not by 2,3-dithiols, TPNH, DPNH, or monothiols such as 2-mercaptoethanol [39]. Following Reichard's discovery that the natural reductant in *E. coli* is thioredoxin, a low molecular weight protein (viz. 12,000) bearing two thiol groups which are kept reduced by the TPNH-linked flavoprotein, thioredoxin reductase, we found that *E. coli* thioredoxin and thioredoxin reductase can serve as hydrogen donor with *L. leichmannii* reductase [40]. Orr and Vitols subsequently obtained evidence of a thioredoxin-like protein in *L. leichmannii* extracts [41].

The requirement of pure enzyme for DBCC coenzyme is virtually absolute (Table III). DBCC is partially replaceable by an analogue, BC (benzimidazolylcobamide) coenzyme, but not by methylcobalamin, cyanocobalamin, or hydroxocobalamin [27, 34]. Maximum activity occurs at a DBCC concentration of $2 \times 10^{+6}$ M. The apparent K_m is 5.8×10^{-7}.

Nucleotide and Divalent Cation Effects. The ATP requirement observed in early studies of CTP reductase activity was at first believed to indicate the occurrence of a phosphorylation, perhaps of the 2′ carbon. However, a number of considerations made it seem unlikely that ATP participates stoichiometrically: 1. ATP is itself a substrate, which is converted to dATP; 2. the ATP requirement of CTP reduction can be met by dATP — indeed, the participation of ATP in CTP reduction might well require its prior conversion to dATP; 3. if purified reductase is a single protein, participation of ATP as a substrate-level cofactor would increase the number of required cofactors from two (DBCC and dihydrolipoic acid) to three; and 4. in the presence of Mg^{++}, GTP reduction requires no ATP, and in the absence of Mg^{++} the ATP requirement for CTP reduction is only a relative one. This supposition was supported by studies of the fate of ATP during CTP reduction in incubations containing non-radioactive substrate CTP and radioactive ATP. In one experiment, reduction of 18.3 mμmoles of CTP was accompanied by the conversion of 2.61 mμmoles of ATP to dATP. Conversions of ATP to ADP, dADP, AMP, or dAMP could not be detected. It appeared, therefore, that ATP participates catalytically. Kinetic studies revealed that it decreases the K_m for CTP about 6-fold. These results suggested that ATP (or dATP) functions as an allosteric effector (in accordance with the terminology of Monod, Wyman, and Changeux [42]).

Although Mg^{++} stimulates CTP reduction in the presence of ATP or dATP, it is strongly inhibitory in their absence. Omission of Mg^{++} also decreases the requirement for ATP or dATP. Further studies on the role of dATP and Mg^{++} in CTP reduction were then performed.

Allosteric Regulation of Ribonucleotide Reductase

Effects of "Prime Effectors" and Mg^{++}. The indications that dATP (and perhaps ATP) stimulate CTP reduction catalytically led to studies of the effects of various nucleotides on the capacity of the enzyme to attack various substrates (Table IV). Reduction of each ribonucleoside triphosphate is maximally stimulated by a different deoxyribonucleoside triphosphate. These are conveniently designated "prime" effectors. Thus, dATP is the prime effector for CTP reduction, dCTP for UTP reduction, dGTP for ATP reduction, and dTTP for GTP reduction. ATP and GTP are actively reduced in the absence of an effector nucleotide; in both cases, however, reduction is enhanced by the prime effector.

Table IV. *Influence of various deoxyribonucleoside triphosphates on reduction of ribonucleoside triphosphates*

Substrate	Effector nucleotide added				
	none (mμmoles)	dATP (mμmoles)	dCTP (mμmoles)	dGTP (mμmoles)	dTTP (mμmoles)
CTP	0.21	6.95	0.51	0.80	0.76
UTP	0.25	0.40	2.32	0.48	0.54
ATP	2.48	1.15	0.80	7.00	0.73
GTP	2.91	3.30	3.35	1.42	7.01

Complete incubation mixtures contained: substrate (CTP-2-^{14}C, UTP-2-^{14}C, ATP-8-^{14}C, or GTP-8-^{14}C), 50 mμmoles (30,000 cpm); effector nucleotide, 8 mμmoles; magnesium acetate, 320 mμmoles; DBCC, 0.08 mμmoles; dihydrolipoate, 600 mμmoles; Tris-succinate, pH 7.5, 1 μmole; and pure reductase, 0.64 μg, in a volume of 20 μl. The figures (indicating mμmoles of deoxyribonucleotide formed per 20 min per 0.64 μg of enzyme) are means of separate determinations performed with at least three different enzyme preparations.

Various nucleotides were then tested in the presence and absence of Mg^{++} for their effects on the reduction of the four ribonucleoside triphosphates. The results (Fig. 12) yielded curves of characteristic shape when reductase activity with each substrate (in the presence of Mg^{++}) was plotted as a function of prime effector concentration. With rising effector concentration, activity rises abruptly, reaches a maximum at a low effector concentration (viz. 1 to $3 \times 10^{+4}$ M) and then levels off or drops slightly. Apparent Michaelis constants for prime effectors ranged from 1 to 3×10^{-5}.

In the absence of Mg^{++}, three of the four nucleotides that are prime effectors in the presence of Mg^{++} (i.e., dATP, dCTP, and dGTP) maximally stimulate the reduction of the corresponding substrate (CTP, UTP, and ATP, respectively). In the absence of an effector nucleotide, however, reduction of all four ribonucleotides, especially of GTP, is higher

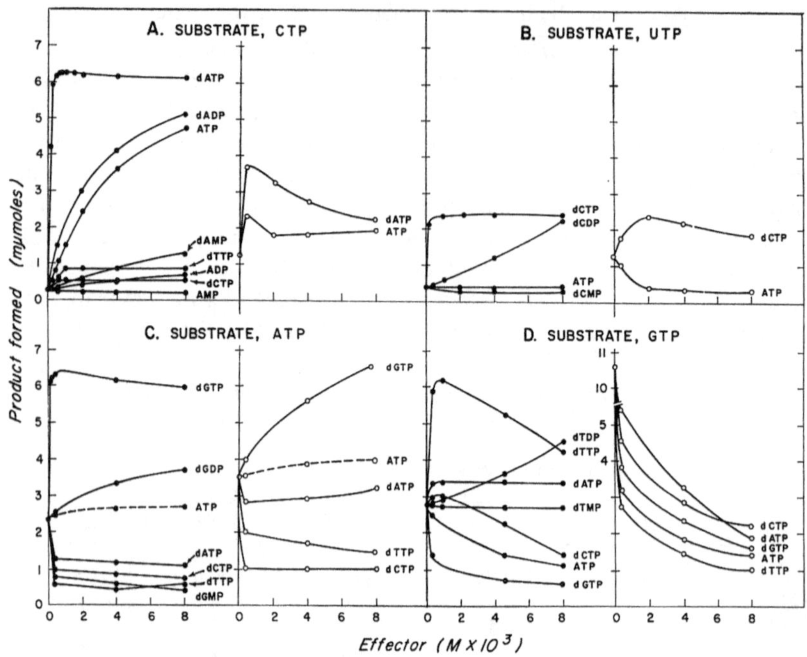

Fig. 12A—D. Effects on ribonucleotide reduction of various effector nucleotides and of Mg^{++}. Incubation mixtures were variations of the standard system described in the text. *Abscissas*, concentrations of indicated effector nucleotides; *ordinates*, mμmoles of products formed in 20 min. ●, Mg^{++} (1.6×10^{-2} M) present; ○, Mg^{++} absent. Substrates were: A, CTP; B, UTP; C, ATP; and D, GTP. *Dashed lines* in C denote special case in which "substrate ATP" (2.6×10^{-3} M) was supplemented by "effector ATP". In this case, the abscissa scales refer only to ATP added as effector; thus, total ATP concentration equals 2.6×10^{-3} M plus the abscissa value

in the absence than in the presence of Mg^{++}. It is seen in Fig. 12 that in the absence of Mg^{++} these three prime effectors produce lower or slower rises of activity with rising effector concentrations, in contrast with the abrupt rises produced in the presence of Mg^{++}.

The highest reduction rate observed was that of GTP in the absence of Mg^{++} and effector. It is noteworthy that dTTP, prime effector for GTP reduction in the presence of Mg^{++}, is strongly inhibitory when Mg^{++} is absent (Fig. 12D). All nucleotides tested inhibited GTP reduction in the absence of Mg^{++}.

Fig. 12 reveals other features of interest. In the presence of Mg^{++}, reducing activities are lower when prime effectors are replaced by the corresponding deoxyribonucleoside diphosphates and monophosphates; the activities decrease progressively as the phosphorylation level decreases from 3 to 1. At high concentrations (8×10^{-3} M), diphosphates support moderate reduction rates — indeed, the second most active effector nucleotide in all four cases is the diphosphate analogue of the prime effector — but their velocity-concentration curves lack abrupt early rises. Monophosphate analogues of the prime effectors are virtually inert or, as in the case of ATP reduction (Fig. 12C), inhibitory. ADP and AMP do not stimulate CTP reduction.

Of those nucleotides that are not prime effectors in the presence of Mg^{++}, some are weak positive effectors (e.g., dTTP in CTP reduction; dATP in GTP reduction) and some are negative effectors (e.g., dATP, dCTP, and dTTP in ATP reduction; dGTP in GTP reduction). Evidence will be presented later that some nucleotides, having no influence on the enzyme *per se*, do influence the activity of the prime effector, and that different nucleotides in this category affect prime effector activity differently.

The Mg^{++}-containing incubations summarized in Fig. 12 were $1.6 \times 10^{+2}$ M in Mg^{++}, a concentration that had earlier been found optimal in incubations in which substrate was CTP and effector nucleotide was $8 \times 10^{+3}$ M ATP [34]. In light of the Mg^{++} effects shown in Fig. 12, observations were made of the effects of variations in Mg^{++} concentration on the reduction of each ribonucleotide in the presence of different concentrations of prime effector, and also in the case of CTP reduction, of ATP. The results, shown in Fig. 13, revealed several striking phenomena. In the absence of effector and Mg^{++}, the rates of reduction of the four substrates were as shown in Fig. 12. Small amounts (approximately 1.2 mμmoles) of CTP and UTP were reduced; a moderate amount (3.1 mμmoles) of ATP was reduced; and a large amount (9.2 mμmoles) of GTP was reduced. Despite these differences, increasing additions of Mg^{++} (in the absence of effector) brought all four rates to zero. Patterns were different, however, in the presence of effector. With Mg^{++} absent, the rates of CTP, UTP, and ATP reduction were higher in the presence of prime effector than in its absence; the rate of GTP reduction was lower. In all four cases, the addition of Mg^{++} in increasing concentrations increased reductase activities; maximal reduction rates occurred at Mg^{++} concentrations of 6 to 8×10^{-3} M. The largest stimulation by Mg^{++} was of CTP reduction in the presence of dATP.

Curious results were obtained when the substrate was CTP and the effector nucleotide ATP (Fig. 13A). The optimal Mg^{++} concentration varied with the concentration of ATP, and maximal reduction rates seemed to occur when the concentration of ATP was approximately twice that of Mg^{++}. Similar effects were not observed on CTP reduction

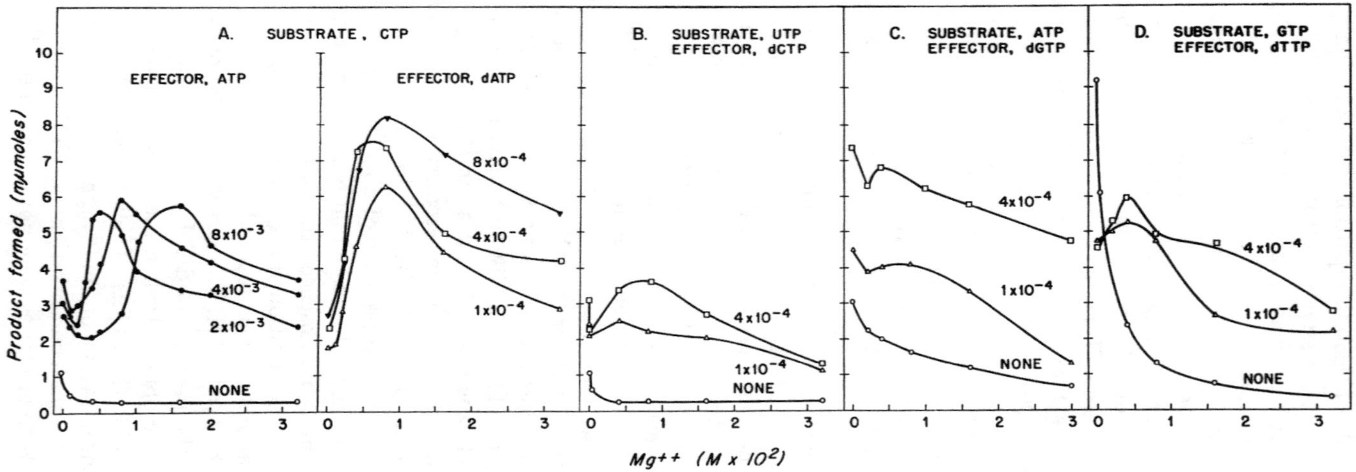

Fig. 13 A—D. Effects on ribonucleotide reduction of changing Mg^{++} concentration at various concentrations of prime effectors. Incubation mixtures were variations of the standard system described in the text. *Abscissas*, Mg^{++} concentration; *ordinate*, mμmoles of product formed in 20 min. Substrates and prime effectors were, respectively: A, CTP and ATP or dATP; B, UTP and dCTP; C, ATP and dGTP; D, GTP and dTTP. Effector concentrations were: ○, zero; △, 1×10^{-4} M; □, 4×10^{-4} M; ▼, 8×10^{-4} M; ●, 2, 4, or 8×10^{-3} M, as indicated

at different dATP concentrations (although the curves in Fig. 13A suggest possible small minor shifting of the optimal Mg^{++} concentration), on UTP reduction at different dCTP concentrations (Fig. 13B), on ATP reduction at different dGTP concentrations (Fig. 13C), or on GTP reduction at different dTTP concentrations (Fig. 13D).

Experiments described elsewhere [43] showed an interesting effect of the sequence of effector and Mg^{++} addition. When reactions were initiated by the addition of substrate to an otherwise complete incubation mixture — in contrast to the usual procedure in which reactions are initiated by the addition of enzyme — it was found that when Mg^{++} and effector nucleotide were both absent during the period preceding the addition of substrate (which we will designate the "background incubation") moderate rates of reduction occurred when substrate was added; however, subsequent addition of Mg^{++} abruptly stopped the reaction. When Mg^{++} (but not effector) was present in the background incubation mixture, addition of substrate produced no reaction, but later addition of prime effector stimulated brisk reactions. With prime effector (but not Mg^{++}) in a background incubation mixture, active reduction followed addition of substrate, but complete inhibition followed subsequent addition of Mg^{++}. Finally, reaction velocity was rapid when substrate was added to background incubation mixtures containing Mg^{++} *and* prime effector. These results suggest that the physical state of the enzyme is affected by Mg^{++} and effector nucleotide, that the effect varies with the addition sequence, and that optimal reduction of each substrate (or group of substrates) is dependent upon a distinctive physical state of the enzyme.

Substrate-velocity curves for the four ribonucleotide substrates were determined under four conditions: (a) with prime effector and Mg^{++} present; (b) with effector absent and Mg^{++} present; (c) with effector present and Mg^{++} absent; and (d) with effector and Mg^{++} absent. With prime effector absent and Mg^{++} present, all four substrate-velocity curves are sigmoid-shaped (Fig. 14). Addition of prime effector increases velocities and decreases the sigmoidal nature of the curves. With Mg^{++} absent, however, different effects occur with different substrates. With effector absent, the CTP and UTP curves rapidly reach a low maximum velocity and level off (Fig. 14A and B); with effector present, velocities are only slightly higher. Of the four ATP curves (Fig. 14C), that obtained with effector present and Mg^{++} absent has the highest maximum velocity. Of the four GTP curves (Fig. 14D), that obtained with effector and Mg^{++} both absent has the highest maximum velocity.

A sigmoidal shape was unambigously present in all four curves obtained with effector absent and Mg^{++} present. Its presence was also suggested in the UTP curve obtained with effector and Mg^{++} both

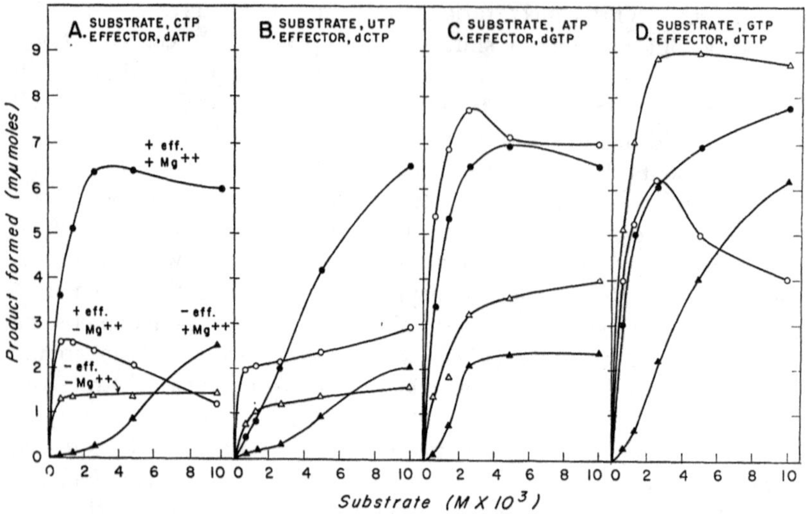

Fig. 14A—D. Effects of substrate concentration on rates of reduction. Incubation mixtures were variations of the standard system. *Abscissas*, substrate concentration; ordinates, mμmoles of product formed in 20 min. *Closed symbols*, Mg^{++} (1.6×10^{-2} M) present; *open symbols*, Mg^{++} absent; *circles*, prime effector present; *triangles*, prime effector absent. Substrates and prime effectors were, respectively: A, CTP and dATP; B, UTP and dCTP; C, ATP and dGTP; and D, GTP and dTTP. *eff*, effector

present (Fig. 14B). This curve indicates that the enzyme has a lower affinity for UTP than for the other three substrates under comparable conditions.

Relationships between Simultaneously Added Effectors and Substrates. When the effectors, dATP and dTTP, and the substrates, CTP and GTP, are all present in the same incubation mixture, the rates of CTP and GTP reduction are determined by the relative proportions of the two effectors. As shown in Fig. 15, a stepwise increase in the molar ratio of dATP to dTTP transforms the enzyme from one that reduces GTP in preference to CTP to one that reduces CTP in preference to GTP. Similar though less striking effects occur in the absence of Mg^{++}.

It appears from the results in Table IV that those nucleotides incapable of strongly stimulating a given reduction are simply weak positive effectors. It is not clear, however, whether such nucleotides would be capable of competing with a strong positive effector for a binding site. To obtain evidence on this question, studies were performed of the relationship of dTTP and dATP in CTP reduction. The results in Fig. 16 show that although dTTP is a weak positive effector for CTP reduction,

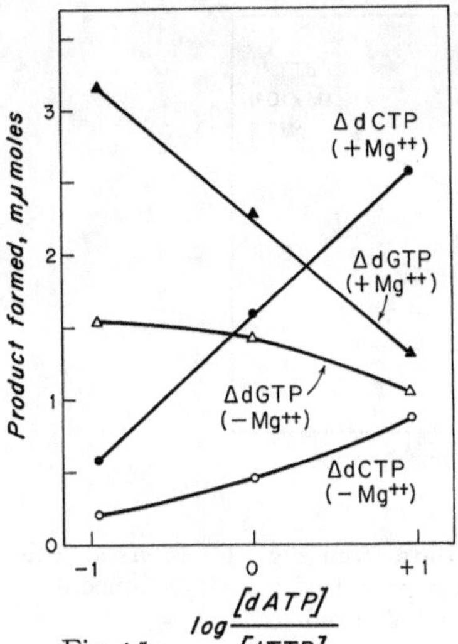

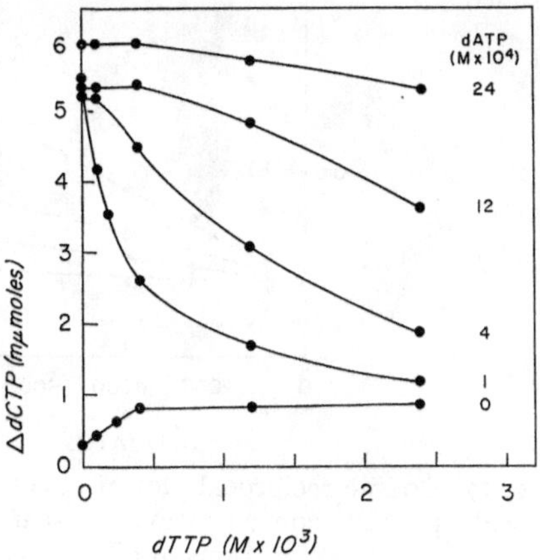

Fig. 15 Fig. 16

Fig. 15. Effects on reduction of two simultaneously present substrates of variations in the concentrations of their simultaneously present prime effectors. Incubations were of the standard type except that each contained the two substrates, CTP and GTP (each 2.6×10^{-3} M), and the two effectors, dATP and dTTP, in a combined concentration of 2×10^{-4} M, but in individual concentrations, respectively, of 2×10^{-5} M and 1.8×10^{-4} M (ratio, 1:9), 1×10^{-4} M and 1×10^{-4} M (ratio, 1:1), and 1.8×10^{-4} M and 2×10^{-5} M (ratio, 1:9). *Abscissa,* log [dATP]/[dTTP]; *ordinate,* mμmoles of product formed in 20 min. Incubation mixtures were run in pairs, one with CTP-2-^{14}C and unlabeled GTP, the other with unlabeled CTP and GTP-8-^{14}C. CTP reduction was measured in the former, GTP reduction in the latter. *Closed symbols,* Mg^{++} (1.6×10^{-2} M) present; *open symbols,* Mg^{++} absent; *circles,* dCTP formation; *triangles,* dGTP formation

Fig. 16. Effects of variations in the relative concentration of dATP and dTTP on CTP reduction. *Abscissa,* dTTP concentration; *ordinate,* mμmoles of product formed in 20 min. Concentrations of dATP were as indicated

it inhibits dATP-stimulated CTP reduction. A double reciprocal plot of the curves plotting velocity as a function of dATP concentration (Fig. 17) suggests that dTTP inhibits by competing with dATP for a binding site on the enzyme. When the reciprocal of velocity is plotted as a function of the reciprocal of substrate concentration (Fig. 18), it is evident that negative effector dTTP does not compete with substrate CTP for a binding site on the enzyme. The results suggest that the enzyme has a binding site for effector nucleotides that is separate and distinct from the catalytic site.

Extensive studies were performed to determine the extent to which a competitive relationship exists between pairs of substrates present

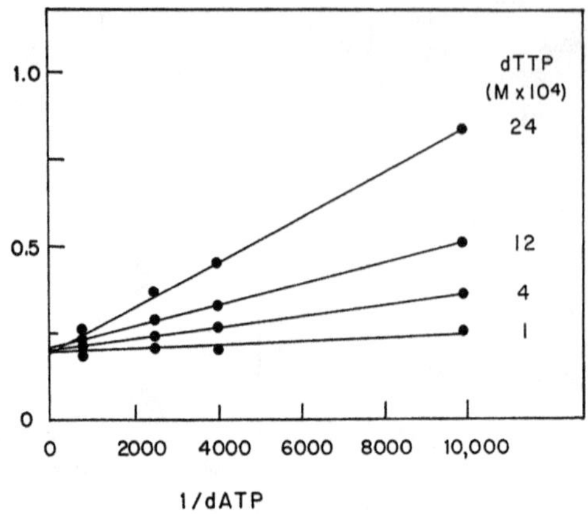

Fig. 17. Double reciprocal plot of selected data from Fig. 16. *Abscissa*, reciprocal of dATP concentration; *ordinate*, reciprocal of velocity. Concentrations of dTTP were as indicated

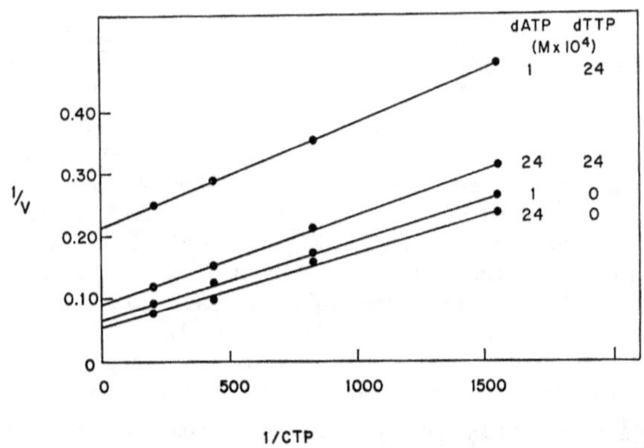

Fig. 18. Double reciprocal plot of substrate-velocity curves (substrate, CTP) determined in the indicated concentration of dATP and dTTP. *Abscissa*, reciprocal of CTP concentration; *ordinate*, reciprocal of velocity (expressed as mμmoles per 20 min)

together in a mixture. Pairs of incubations containing Substrates A and B were prepared so that compositions were identical, but only A was radioactive in one mixture, and only B in the other. Thus, reduction of a substrate could be specifically assayed in the presence of another substrate.

Results of these studies have been described elsewhere [43]; space will permit only a brief summary here. Although indications were obtained that substrates probably do compete for a single catalytic site, it was concluded that kinetic studies cannot resolve this question. This conclusion derives from the complexity of the observed network of stimulatory and inhibitory effects by reaction products and substrates. It was found, for example, that a given nucleotide can serve both as a substrate and an effector or as a product and an effector. Thus, when CTP and GTP were present together, the ability of each to compete with the other for a position on the catalytic site depend on which other effectors, if any, were present; also, GTP (or the product of its reduction, dGTP) is a weak positive effector for CTP reduction in the absence of an added effector and CTP (or dCTP) is a weak positive effector for GTP reduction (see Fig. 11 D). Clearly, direct binding studies will be needed to determine if the enzyme possesses one or more catalytic site.

Regulation of Reductase Activity in vivo. The results may be examined for their possible implications concerning the regulation *in vivo* of deoxyribonucleotide synthesis in *L. leichmannii.* As in other species, the pool of intracellular low molecular weight deoxyribosyl compounds is small relative to the pool of ribosyl compounds and to the pools of deoxyribosyl and ribosyl moieties in DNA and RNA. Thus, we have found [18, 23] that optimally nourished cells of *L. leichmannii* in log phase contain approximately 0.6 μmole of acid-soluble deoxyribosyl compounds (exclusive of a dTDP-sugar complex), 78 μmoles of acid-soluble purine ribosyl compounds, 280 μmoles of acid-insoluble (i.e., DNA) deoxyribonucleotides, and 4,000 μmoles of acid-insoluble (i.e., RNA) ribonucleotides per g of protein. Evidence has been presented which suggests that intracellular dTTP represses reductase synthesis. Therefore, synthesis of deoxyribonucleotides in *L. leichmannii* is subject to dual regulation: (a) by a repressor control system, and (b) by the system of positive and negative feedback effects described above.

The feedback control system has several distinctive properties: (a) the enzyme attacks four substrates, the common ribonucleoside triphosphates; (b) the reduction of each is maximally stimulated by a different deoxyribonucleoside triphosphate (the prime effector) that is presumably bound to an allosteric regulatory site; (c) divalent cations (Mg^{++}, Ca^{++}, or Mn^{++}) and spermidine stimulate or inhibit the enzyme, the effect depending on the substrate and effector present and the sequence of cation and effector addition; and (d) the reduction of each substrate is stimulated and inhibited by a variety of nucleoside triphosphates (and to some extent diphosphates) other than the prime effector of that reduction, including the substrates and products of other reductions. Hence, the situation *in vivo* may be too complex for precise analysis.

Nevertheless, if it is assumed that regulation *in vivo* results mainly from the basic network of interrelations among the prime effectors and the four reductions, as outlined in Fig. 19, a number of interesting surmises become possible. In considering the pattern in Fig. 19, it is useful to recall that GTP and ATP reduction (*Reactions 1* and *2* in Fig. 19) are less dependent on effectors than CTP and UTP reduction (*Reactions 3* and *4*). Thus, the sequence could begin with the reduction of GTP or of ATP in the absence of the prime effector. Revolutions around the square would be unlikely to exceed a certain maximum velocity since the reactions led to an accumulation of deoxyribonucleotides; those that are negative effectors would limit reaction velocities. The low affinity of the enzyme for UTP would also have a damping effect on revolution velocity.

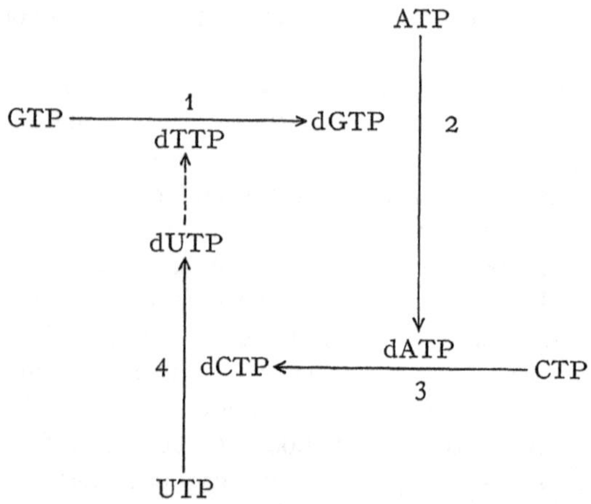

Fig. 19. Interrelations of the products and prime effectors of the four major ribonucleotide reductions. The product of *Reaction 1* is prime effector for *Reaction 2*, the product of *Reaction 2* is prime effector for *Reaction 3*, and so on. It is assumed (dashed arrow) that dUTP, the product of *Reaction 4*, is rapidly converted via several intermediary reactions to dTTP, prime effector for *Reaction 1*

It is of interest to note that in optimally nourished *L. leichmannii*, the deoxyribosyl compounds in lowest concentration are the guanine derivatives [23]. In folate-free *L. leichmannii*, wherein thymidylate synthesis (dashed arrow in Fig. 19) is blocked, the intracellular deoxyribosyl pool increases drastically [23]. Although the rise is due mainly to deoxyadenylates and deoxyguanylates, calculations based on our published data [23] show that folate deficiency causes the deoxyribosyl derivatives of uracil to increase approximately 60-fold, of guanine 25-fold, of adenine 9-fold, and of cytosine 2-fold. Presumably, the gross increases are due to derepressed reductase synthesis and to accumulations of non-thymine deoxyribosyl compounds resulting from the unavailability of dTTP for DNA synthesis. In light of the scheme in Fig. 19, the pattern of the increases may be attributable to decreased methylation of deoxyuridylate and to the ability of GTP and ATP reduction to proceed without prime effectors. The large ATP pool normally present [18] may possibly be spared undue reduction by (a) the low dGTP concentration and (b) the inhibitory effect of dATP on ATP reduction.

Although rigorous proof is not yet available, it appears from preliminary evidence that regulatory necluotides affect substrate specificity and rate behavior by inducing different states of enzyme conformation or subunit aggregation. Results to date indicate that the enzyme contains

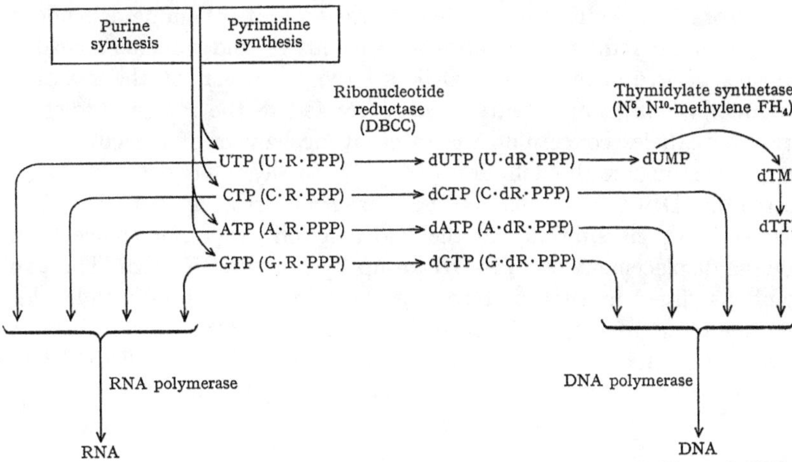

Fig. 20. Schematic diagram of biosynthetic pathways of RNA and DNA precursors in *L. leichmannii* (see Fig. 3)

2 or 4 subunits which are disaggregated or reaggregated by modifications of pH and salt concentration and by allosteric effectors and divalent cations. For example, the two bands seen on polyacrylamide gel electrophoresis at pH 9.5 (Fig. 11) are closer together but still distinguishable at pH 8.0. However, when dATP and Mg^{++} are added to the gel, a sharp single band appears. Both ingredients are necessary for this effect; dATP cannot be replaced by dADP or dAMP, but it can be replaced by ATP. Also, elevation of pH lowers the molecular weight of the main sedimenting components on sucrose density gradient centrifugation.

The above described regulatory mechanisms of the cobamide-dependent ribonucleotide reductase of *L. leichmannii* display notable differences from the regulatory mechanisms of the cobamide-independent reductase of *E. coli*, which will be described by the next essayist [44—46]. This system, it will be recalled, is a ribonucleoside diphosphate reductase (see Fig. 3); that of *L. leichmannii* is a ribonucleoside triphosphate reductase (Fig. 20), whose immediate products are ready-made substrates for DNA polymerase (if it is of the usual type)[2]. The *E. coli* reductase has an absolute requirement for divalent cations [12] and its pattern of nucleotide effects differs in certain details from that described above (e.g., CDP reduction is stimulated by ATP and inhibited by dATP). Because its effectors are deoxyribonucleotide triphosphates and its reaction products are diphosphates, it has been possible to show clearly that multiple substrates compete for a single catalytic site [45, 46].

Mechanism of Action of Cobamide Coenzyme

Introduction. Following the demonstration that cobamide-dependent ribonucleotide reductase catalyzes an exclusive and nonexchangeable transfer of hydrogen from $H_2O\text{-}^3H$ to the 2′ carbon of the reaction product [47, 48], our attention was attracted to the recent studies of Frey and Abeles concerning the reaction mechanism of the cobamide-dependent enzyme dioldehydrase [49]. These elegant experiments suggested that DBCC functions as a hydrogen-transferring agent and implied that it functions similarly in the two reactions, especially since both involve displacement of an —OH group by a hydrogen atom. The two reactions differ in that dioldehydrase catalyzes a net intramolecular transfer of hydrogen, i.e., tritium is transferred from carbon atom 1 of the substrate (1,2-propanediol-1-^{3}H) to DBCC and from DBCC-^{3}H to carbon atom 2 of the product (propionaldehyde), as follows [49]:

$$\overset{^{3}\text{H}}{\textcircled{S}} + E + \text{DBCC} \rightarrow \textcircled{S}\cdot E\cdot\text{DBCC}\cdot{}^{3}\text{H} \qquad (2)$$

$$\textcircled{S}\cdot E\cdot\text{DBCC}\cdot{}^{3}\text{H} \rightarrow E + \text{DBCC} + \overset{^{3}\text{H}}{\textcircled{P}} + H_2O \qquad (3)$$

where $\overset{^{3}\text{H}}{\textcircled{S}}$ denotes substrate with a tritium on carbon atom 1; E represents enzyme; $\textcircled{S}\cdot E\cdot\text{DBCC}\cdot\text{H}$, an enzyme-bound intermediate complex in which hydrogen has been transferred from substrate to coenzyme; and $\overset{^{3}\text{H}}{\textcircled{P}}$, product with a tritium on carbon atom 2. In contrast, ribonucleotide reductase catalyzes a net intermolecular transfer of hydrogen, i.e., tritium is transferred from $H_2O\text{-}^3H$ (or the freely

2. Since the deoxyribonucleoside triphosphates produced by reductase activity include dUTP, a non-substrate for DNA polymerase *in vivo*, we conclude that these cells contain a powerful and specific system for removing dUTP, perhaps by converting it to dUMP, the presumed substrate of thymidylate synthetase. Preliminary studies have so far failed to reveal a specific dUTPase.

tritiated —SH groups of the essential reductant, dihydrolipoate) to carbon atom 2′ of the deoxyribosyl moiety of the product [47].

Evidence was then obtained that DBCC does indeed function as an essential hydrogen-transferring agent in the cobamide-dependent ribonucleotide reductase reaction [50, 51], the reaction proceeding as follows:

$$\left[\begin{matrix} S^3H \\ \\ S^3H \end{matrix}\right. + E + \text{DBCC} \rightleftharpoons \left[\begin{matrix} S \\ | \\ S \end{matrix}\right. \cdot E\cdot \text{DBCC}\cdot {}^3\text{H} + {}^3\text{H}^+ \qquad (4)$$

$$\left[\begin{matrix} S \\ | \\ S \end{matrix}\right. \cdot E\cdot \text{DBCC}\cdot {}^3\text{H} + \text{XTP} \rightarrow \left[\begin{matrix} S \\ | \\ S \end{matrix}\right. + E + \text{DBCC} + \text{dXTP-}^3\text{H} \qquad (5)$$

where $\left[\begin{matrix} S^3H \\ \\ S^3H \end{matrix}\right.$ is the reductant, dihydrolipoate or dihydrothioredoxin; E the enzyme;

$$\left[\begin{matrix} S \\ | \\ S \end{matrix}\right. \cdot E\cdot \text{DBCC}\cdot {}^3\text{H},$$

an enzyme -bound intermediate complex containing oxidized reductant and a hydrogen that has been transferred from reductant to coenzyme; XTP, a ribonucleoside triphosphate; and dXTP-^{3}H, the corresponding deoxyribonucleoside triphosphate with a tritium at carbon atom 2′. The locus of hydrogen transport appears to be carbon atom 5′ of the DBCC deoxyadenosyl moiety. This evidence will be briefly summarized.

Transfer of Tritium from DBCC-5′-^{3}H to H_2O. Incubations were prepared (Table V) containing substrate (CTP or GTP), effector nucleotide (dATP or dTTP, respectively), reductant (dihydrolipoate or the thioredoxin system), synthetic DBCC-5′-^{3}H (prepared according to Frey and Abeles [49]), Mg^{++}, buffer, and purified ribonucleotide reductase. Assays of radioactivity in the water indicated that tritium is transferred from DBCC-5′-^{3}H to H_2O in a reaction requiring substrate, enzyme, and dithiol reductant. Results were the same with the stoichiometric reductant, dihydrolipoate, and with the cyclical TPNH-dependent dihydrothioredoxin-generating system. When CTP was the substrate, omission of its prime effector (dATP), of Mg^{++}, or of both, diminished transfer of tritium to H_2O only slightly. As we have seen (Fig. 13) and as is indicated in Table V, CTP reduction takes place in the absence of the regulatory substances Mg^{++} and dATP, but at a lower rate. Similar results were obtained when GTP was the substrate. GTP reduction, unlike CTP reduction, is stimulated by its prime allosteric effector only when Mg^{++} is present (see Fig. 13). When Mg^{++} is absent, GTP is actively reduced in the absence of dTTP.

Table V. *Requirements for transfer of tritium from DBCC-5′-³H to H_2O*

Experiment	Incubation mixture	^{3}H in water (% of total)	Product formed (mμmoles)
1	Complete system	96.2	180
	– substrate (CTP)	0.7	0
	– enzyme	0.9	0
	– dihydrolipoate	2.0	0
	– effector (dATP)	83.4	9
	– dihydrolipoate, + lipoate	0.5	0
2	Complete system	99.0	106
	– substrate (CTP)	2.4	0
	– thioredoxin	1.9	0
3	Complete system	94.3	161
	– substrate (GTP)	7.5	0
	– effector (dTTP)	93.4	44
	– Mg^{++}, – effector (dTTP)	97.0	212
	– dihydrolipoate	3.5	0

Complete incubation mixtures contained: substrate, 0.4 μmoles; effector, 40 mμmoles; DBCC-5′-^{3}H, 1 mμmole (10,100 CPM), magnesium acetate, 1.6 μmoles; Tris-succinate, pH 7.5, 5 μmoles; reductant; and pure reductase, 32 μg, in a volume of 0.1 ml. Reductant was dihydrolipoate, 30 μmoles, in experiments 1 and 3, and thioredoxin, 3.4 mμmoles, thioredoxin reductase, 18 μg, and TPNH, 75 mμmoles, in experiment 2. Substrate and effector were, respectively, CTP and dATP (ATP-free) in experiments 1 and 2, and GTP and dTTP in experiment 3. Reactions were terminated by freezing. Water of the reaction mixture, after recovery by bulb-to-bulb lyophilization, was assayed for radioactivity. Similar results were obtained when water was separated from reactants by treatment of reaction mixtures with Norit A. Product formation was assayed in parallel incubations containing radioactive substrates and unlabeled DBCC [34].

The rates of dCTP production shown in Table V indicate that, at the enzyme concentration used, the molar ratio of product formed to coenzyme added was well above unity, even in the absence of dATP. Under these conditions, coenzyme participation is catalytic, and tritium is exchanged completely. Fig. 21 shows the relation of enzyme concentration to the amount of tritium transferred from DBCC-5′-^{3}H to H_2O. When enzyme concentration is low enough, cobamide coenzyme functions stoichiometrically, and the amount of tritium released is an approximately linear function of enzyme concentration. These results show that measurement of the substrate-dependent release of tritium from DBCC-5′-^{3}H can serve as a sensitive quantitative assay for ribonucleotide reductase.

Transfer of Tritium from H_2O-^3H to DBCC. If DBCC transfers hydrogen from H_2O (or the dithiol reductant) to product, tritium should

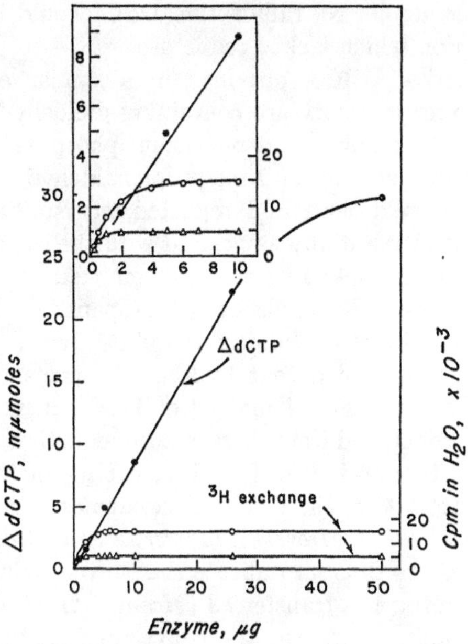

Fig. 21. Stoichiometric relation between ribonucleotide reduction and tritium transfer from DBCC-5′-^{3}H to H_2O. Incubation mixtures similar to the complete system in Table V (Experiment 1) were prepared in two series: those in one series (○) contained DBCC-5′-^{3}H, 3mμmoles (15,000 cpm); those in the second series (△) contained DBCC-5′-^{3}H, 1 mμmole (5,000 cpm). In both series, substrate was CTP-2-^{14}C, and the amount of enzyme per incubation was varied from 0.2 to 50 μg. After incubation for 20 min, H_2O-^{3}H was isolated and assayed for radioactivity (○, △), and dCTP was determined chromatographically in the resuspended residue (●). Abscissa, micrograms of emzyme per incubation; left ordinate, dCTP in mμmoles; right ordinate, ^{3}H counts recovered in H_2O; inset, enlargement of portion of graph near origin

be incorporated into coenzyme when the reaction is carried out in H_2O-^{3}H. This was observed in an incubation containing CTP, 8 μmoles; dihydrolipoate, 60 μmoles; dATP, 6.4 μmoles; DBCC, 370 mμmoles; magnesium acetate, 32 μmoles; Tris-succinate, pH 7.5, 100 μmoles; ribonucleotide reductase, 475 μg; and H_2O-^{3}H, 10 ml, 1.0 C (13×10^5 dpm per μg atom of hydrogen), in a total volume of 2 ml. After incubation for 30 minutes, the mixture was adjusted to pH 3 and DBCC was isolated in the dark by Dowex 50 (H^+) chromatography. After further purification by phenol extraction and paper electrophoresis in 0.5 M NH_4OH, the recovered DBCC had a specific radioactivity of 30×10^5 dpm per μmole. After additional purification by paper chromatography in water-saturated 2-butanol, the DBCC had a specific radioactivity of

28×10^5 dpm per μmole. No radioactive DBCC could be isolated from a control incubation which lacked reductase.

When radioactive DBCC, obtained in a similar experiment, was incubated for 6 min in a mixture containing dioldehydrase, 140 units; 1,2-propanediol, 250 μmoles; and potassium phosphate buffer, pH 8,0. 28 μmoles, in a total volume of 2 ml, propionaldehyde isolated by distillation, dimedon treatment, and repeated recrystallization contained all of the radioactivity initially associated with the coenzyme. A control experiment in which dioldehydrase was omitted and 250 μmoles of propionaldehyde were added in place of 1,2-propanediol yielded a propionaldehyde dimedon adduct containing no radioactivity.

Since it has been shown that hydrogen is transported by carbon atom 5′ of the deoxyadenosyl moiety of DBCC in the dioldehydrase reation [49], it is concluded from these experiments that, in the reduction of CTP to dCTP, tritium is transferred from H_2O (or the dithiol reductant) to carbon atom 5′ of the DBCC deoxyadenosyl moiety.

Is DBCC-mediated Hydrogen Transfer Essential in the Reductase Reaction? The results thus far indicate that in the course of ribonucleotide reduction, hydrogen is transferred 1. from H_2O-^3H (or its equivalent, R-($S^3H)_2$) to DBCC and to the 2′-deoxyribosyl carbon; and 2. from DBCC-5′-^3H, labeled synthetically or by reductase activity, to H_2O and not, in measurable amounts, to the deoxyribosyl product. Hence, the results are consistent with the conclusion that DBCC transfers hydrogen from reductant to product in the course of ribonucleotide reduction, but do not establish it conclusively.

The failure of hydrogen to be transferred from DBCC-5′-^3H to product in detectable amounts is probably due to the reversibility of the transfer from R-($S^3H)_2$ to DBCC [see Eq. (4)]. To clarify this point, the effect of increasing the concentration of unlabeled DBCC on the rate of hydrogen transfer from H_2O-^3H to product in the course of CTP reduction was studied. With increasing initial concentrations of unlabeled coenzyme, a larger pool of coenzyme would be available to accept and retain (i.e., trap) tritium, and a larger pool of nonisotopic coenzyme hydrogen would be available for transfer to product. The results indicated that when the substrate was CTP-2-^{14}C and the solvent H_2O-^3H, the ratio ^3H:^{14}C in dCTP decreased significantly, though only moderately. Presumably, it did not decrease further owing to rapid incorporation of tritium into the coenzyme.

We conclude that DBCC provides the hydrogen that displaces the —OH group of carbon atom 2′ and that DBCC functions similarly in the ribonucleotide reductase and dioldehydrase reactions. Although, according to the postulated mechanism [Eqs. (4) and (5)], the substrate does not participate in the exchange reaction, we believe it facilitates enzyme function, perhaps by playing the additional role of conformation determinant. As noted above, substrates do act as effectors and it may be that in this capacity they promote the exchange reaction. Vitols *et al.* have shown that ribonucleotides and deoxyribonucleotides increase the affinity of the enzyme for DBCC [52].

Concluding Remarks

The results which have been described delineate one of the biosynthetic pathways for what might reasonably be considered nature's most important pentose, deoxyribose. As Sable has pointed out [53], a considerable proportion of the total synthetic activity of a cell is given over to the synthesis of the nucleic acids. Thus, to study the biosynthesis of pentoses is to study a quantitatively significant cellular activity. The biosynthetic pathway holds additional interest for whatever its properties may tell us of the factors that regulate DNA synthesis.

The pathway discussed in the present essay consists of a repressible ribonucleoside triphosphate reductase that requires dimethylbenzimidazolyl-cobamide coenzyme (DBCC), the active form of vitamin B_{12}, and a dithiol reductant. The purified enzyme reduces the four major ribonucleoside triphosphates to deoxyribonucleoside triphosphates. Choice of substrate by reductase is determined by divalent cations and deoxyribonucleoside triphosphates, each an allosteric effector promoting reduction of a different ribonucleotide. Divalent cations (Mg^{++}, Ca^{++}, and Mn^{++}) also have significant stimulatory and inhibitory effects that vary with the substrate, the effector, and the sequence of cation and effector addition. Allosteric effector nucleotides and divalent cations presumably bring about modifications of the physical state of the enzyme that determine which ribonucleotides are reduced. Doubtless these effects help to maintain the balance of deoxyribonucleotide synthesis *in vivo*.

The results also show that cobamide coezyme functions as an essential hydrogen-transferring agent in the cobamide-dependent ribonucleotide reductase reaction, and that transferred hydrogen attaches to carbon atom 5′ of the coenzyme deoxyadenosyl moiety. Although hydrogen is transferred intermolecularly in the reductase reaction and intramolecularly in other DBCC-dependent reactions (e.g., the dioldehydrase reaction), the mechanism of coenzyme function is the same in both reactions.

It is of interest that the ribonucleotide reductase of *E. coli*, an organism lacking a vitamin B_{12} requirement, reduces ribonucleoside diphosphates and is cobamide-independent, yet it transfers ^3H from $R\text{-}(S^3H)_2$ to the 2′ carbon [54]. It is unclear, therefore, what agency in this system serves the hydrogen-transferring function performed by DBCC in the lactobacillus system. Which system is present in animal tissues is still unknown. A reductase of the *E. coli* type is found in Novikoff hepatoma [55, 56] but it remains to be seen whether this tumor system is a prototype for animal tissues such as bone marrow in which DNA synthesis is believed to require vitamin B_{12} [20]. We are currently engaged in the study of such tissues, but results to date do not permit useful conclusions.

Acknowledgement. The investigations described in this essay were supported by Grants CA-03728 and AM-06214 from the National Institutes of Health, United States Public Health Service, and by the John Phyffe Richardson Memorial Fund.

References

1. White, J. C., I. Leslie, and J. N. Davidson: Nucleic acids of bone marrow cells, with special reference to pernicious anemia. J. Path. Bact. **66**, 291 (1953).
2. Glazer, H. S., J. F. Mueller, T. Jarrold, K. Sakurai, J. J. Will, and R. W. Vilter: Effect of vitamin B_{12} and folic acid on nucleic acid composition of the bone marrow of patients with megaloblastic anemia. J. Lab. clin. Med. **43**, 905 (1954).
3. Lessner, H. E., and M. Friedkin: *In vitro* incorporation of deoxyuridine and thymidine into deoxyribonucleic acid of human megaloblastic and normoblastic bone marrow. Clin. Res. **7**, 207 (1959).
4. Thomas, E. C., and H. L. Lochte Jr.: Studies on biochemical defect of pernicious anemia. I. *In vitro* observations on oxygen consumption, heme synthesis, and deoxyribonucleic acid synthesis by pernicious anemia bone marrow. J. clin. Invest. **37**, 166 (1958).
5. Beck, W. S., and M. Goulian: Drugs for pernicious anemia and related megaloblastic anemias, chap. 51, p. 817—831 in: Drill's Pharmacology in Medicine, 3rd ed. (J. R. DiPalma, ed.). New York: McGraw-Hill (Blakiston Division) 1965.
6. Wahba, A. J., and M. Friedkin: The enzymatic synthesis of thymidylate. I. Early steps in the purification of thymidylate synthetase of *Escherichia coli*. J. biol. Chem. **237**, 2794 (1962).
7. Weissbach, H., J. Toohey, and H. A. Barker: Isolation and properties of B_{12} coenzymes containing benzimidazole or dimethylbenzimidazole. Proc. nat. Acad. Sci. (Wash.) **45**, 521 (1959).
8. Lenhert, P. G., and D. C. Hodgkin: Structure of the 5,6-dimethylbenzimidazolylcobamide coenzyme. Nature (Lond.) **192**, 937 (1961).
9. Buchanan, J. M., and S. C. Hartman: Enzymic reactions in the synthesis of the purines. Advanc. Enzymol. **21**, 199 (1959).
10. Reichard, P.: The enzymic synthesis of pyrimidines. Advanc. Enzymol. **21**, 263 (1959).
11. — A. Baldesten, and L. Rutberg: Formation of deoxycytidine phosphates from cytidine phosphates in extracts from *Escherichia coli*. J. biol. Chem. **236**, 1150 (1961).

12. Reichard, P.: Enzymatic synthesis of deoxyribonucleotides. I. Formation of deoxycytidine diphosphate from cytidine diphosphate with enzymes from *Escherichia coli*. J. biol. Chem. **237**, 3513 (1962).
13. Kitay, E., W. S. McNutt, and E. E. Snell: Non-specificity of thymidine as growth factor for lacticacid bacteria. J. biol. Chem. **177**, 993 (1949).
14. Downing, M., and B. S. Schweigert: Role of vitamin B_{12} in nucleic acid metabolism. IV. Metabolism of C^{14}-labeled thymidine by *Lactobacillus leichmannii*. J. biol. Chem. **220**, 521 (1956).
15. Manson, L. A.: Vitamin B_{12} and deoxyribose synthesis in *Lactobacillus leichmannii*. J. biol. Chem. **235**, 521 (1956).
16. Wacker, A., S. Kirschfeld u. L. Träger: Die Biosynthese der Desoxyribose bei Bakterien. Z. Naturforsch. **14**b, 145 (1959).
17. Williams, A. M., J. J. Chosy, and R. F. Schilling: Effect of vitamin B_{12} *in vitro* on incorporation of nucleic acid precursors by pernicious anemia bone marrow. J. clin. Invest. **42**, 670 (1963).
18. Beck, W. S., M. Goulian, and S. Hook: Metabolic functions of vitamin B_{12}. II. Participation of vitamin B_{12} in the biosynthesis of deoxyribonucleic acid and its acid-soluble precursors. Biochim. biophys. Acta (Amst.) **55**, 470 (1962).
19. — S. Hook, and B. H. Barnett: Metabolic functions of vitamin B_{12}. I. Distinctive modes of unbalanced growth behavior in *Lactobacillus leichmannii*. Biochim. biophys. Acta (Amst.) **55**, 455 (1962).
20. — The metabolic basis of megaloblastic erythropoiesis. Medicine (Baltimore) **43**, 715 (1964).
21. — M. Goulian, and S. Kashket: The role of vitamin B_{12} in deoxyribonucleic acid synthesis: studies of a model megaloblast. Trans. Ass. Amer. Phycns **78**, 343 (1965).
22. Biswas, C., M. Goulian, J. Hardy, and W. S. Beck: Factors controlling the enzymatic conversion of ribonucleotides to deoxyribonucleotides. Fed. Proc. **24**, 532 (1964).
23. Goulian, M., and W. S. Beck: Variations of intracellular deoxyribosyl compounds in deficiencies of vitamin B_{12}, folic acid, and thymine. Biochim. biophys. Acta (Amst.) **129**, 336 (1966).
24. Munch-Petersen, A., and J. Neuhard: Studies on the acid-soluble deoxynucleotide pool in thymine-requiring mutants of *Escherichia coli* during thymine starvation. I. Accumulation of deATP in *E. coli* $15T^-A^-U^-$. Biochim. biophys. Acta (Amst.) **80**, 542 (1964).
25. Biswas, C., J. Hardy, and W. S. Beck: Release of repressor control of ribonucleotide reductase by thymine starvation. J. biol. Chem. **240**, 3631 (1965).
26. Goulian, M., and W. S. Beck: Storage of folate by *Lactobacillus leichmannii*. Biochim. biophys. Acta (Amst.) **124**, 423 (1966).
27. Beck, W. S., and J. Hardy: Requirement of ribonucleotide reductase for cobamide coenzyme, a product of ribosomal activity. Proc. nat. Acad. Sci. (Wash.) **54**, 286 (1965).
28. Blakley, R. L., and H. A. Barker: Cobamide stimulation of the reduction of ribonucleotides to deoxyribotides in *Lactobacillus leichmannii*. Biochem. biophys. Res. Commun. **16**, 391 (1964).
29. Kashket, S., J. T. Kaufman, and W. S. Beck: The metabolic functions of vitamin B_{12}. III. Vitamin B_{12} binding in *Lactobacillus leichmannii* and other lactobacilli. Biochim. biophys. Acta (Amst.) **64**, 447 (1962).

30. Kashket, S., J. T. Kaufmann, and W. S. Beck: The metabolic functions of vitamin B_{12}. IV. Binding of vitamin B_{12} by ribosomes in *Lactobacillus leichmannii*. Biochim. biophys. Acta (Amst.) **64**, 458 (1962).
31. —, and W. S. Beck: Reversible release of vitamin B_{12}-binding protein from bacterial ribosomes. Biochim. biophys. Acta (Amst.) **129**, 350 (1966).
32. — — The relative ribonuclease resistance of vitamin B_{12}-binding ribosomes, a specific class within the ribosome population. Biochem. Z. **342**, 449 (1965).
33. Ohta, H., and W. S. Beck: Unpublished observations.
34. Goulian, M., and W. S. Beck: Purification and properties of cobamide-dependent ribonucleotide reductase from *Lactobacillus leichmannii*. J. biol. Chem. **241**, 4233 (1966).
35. Abrams, R.: Cytidine 5′-triphosphate as the precursor of deoxycytidylate in *Lactobacillus leichmannii*. J. biol. Chem. **240**, PC3697 (1965).
36. Blakley, R. L., R. K. Ghambeer, P. F. Nixon, and E. Vitols: The cobamide-dependent ribonucleoside triphosphate reductase of lactobacilli. Biochem. biophys. Res. Commun. **20**, 439 (1965).
37. Yphantis, D. A.: Equilibrium ultracentrifugation of dilute solutions. Biochemistry **3**, 297 (1964).
38. Reichard, P.: The synthesis of deoxyribose by the chick embryo. Biochim. biophys. Acta (Amst.) **27**, 434 (1958).
39. Vitols, E., and R. L. Blakley: Hydrogen-donor specificity of ribonucleoside triphosphate reductase from *Lactobacillus leichmannii*. Biochem. biophys. Res. Commun. **21**, 466 (1965).
40. Beck, W. S., M. Goulian, A. Larsson, and P. Reichard: Hydrogen donor specificity of cobamide-dependent ribonucleotide reductase and allosteric regulation of substrate specificity. J. biol. Chem. **241**, 2177 (1966).
41. Orr, M. D., and E. Vitols: Thioredoxin from *Lactobacillus leichmannii* and its role as hydrogen donor for ribonucleoside triphosphate reductase. Biochem. biophys. Res. Commun. **25**, 109 (1966).
42. Monod, J., J. Wyman, and J.-P. Changeux: On the nature of allosteric transitions: a plausible model. J. molec. Biol. **12**, 88 (1965).
43. Beck, W. S.: Regulation of cobamide-dependent ribonucleotide reductase by allosteric effectors and divalent cations. J. biol. Chem. **242**, 3148 (1967).
44. Larsson, A., and P. Reichard: Allosteric effects and substrate specificity of the ribonucleoside diphosphate reductase system from *Escherichia coli* B. Biochim. biophys. Acta (Amst.) **113**, 407 (1966).
45. — — Enzymatic synthesis of deoxyribonucleotides. IX. Allosteric effects in the reduction of pyrimidine ribonucleotides by the ribonucleoside diphosphate reductase system of *Escherichia coli*. J. biol. Chem. **241**, 2533 (1966).
46. — — Enzymatic synthesis of deoxyribonucleotides. X. Reduction of purine ribonucleotides; allosteric behavior and substrate specificity of the enzyme system from *Escherichia coli*. J. biol. Chem. **241**, 2540 (1966).
47. Gottesman, M. M., and W. S. Beck: Transfer of hydrogen in the cobamide-dependent ribonucleotide reductase reaction. Biochem. biophys. Res. Commun. **24**, 353 (1966).

48. Blakley, R. L., R. K. Ghambeer, J. J. Batterham, and C. Brownson: Studies with hydrogen isotopes on the mechanism of action of cobamide-dependent ribonucleotide reductase. Biochem. biophys. Res. Commun. **24**, 418 (1966).
49. Frey, P. A., and R. H. Abeles: The role of the B_{12} coenzyme in the conversion of 1,2-propanediol to propionaldehyde. J. biol. Chem. **241**, 2732 (1966).
50. Beck, W. S., R. H. Abeles, and W. G. Robinson: Transfer of hydrogen from cobamide coenzyme to water during enzymatic ribonucleotide reduction. Biochem. biophys. Res. Commun. **25**, 421 (1966).
51. Abeles, R. H., and W. S. Beck: The mechanism of action of cobamide coenzyme in the ribonucleotide reductase reaction. J. biol. Chem. **242**, 3589 (1967).
52. Vitols, E., C. Brownson, W. Gardiner, and R. Blakley: Cobamides and ribonucleotide reduction. V. A kinetic study of the ribonucleoside triphosphate reductase of *Lactobacillus leichmannii*. J. biol. Chem. **242**, 3035 (1967).
53. Sable, H. Z.: Biosynthesis of ribose and deoxyribose. Advanc. Enzymol. **28**, 391 (1966).
54. Larsson, A.: Enzymatic synthesis of deoxyribonucleotides. VII. Studies on the hydrogen transfer with tritiated water. Biochemistry **4**, 1984 (1965).
55. Moore, E. C., and P. Reichard: Enzymatic synthesis of deoxyribonucleotides. VI. The cytidine diphosphate reductase system from Novikoff hepatoma. J. biol. Chem. **239**, 3453 (1964).
56. —, and R. B. Hurlbert: Regulation of mammalian deoxyribonucleotide biosynthesis by nucleotides as activators and inhibitors. J. biol. Chem. **241**, 4802 (1966).

Discussion

Dr. Horecker: Dr. Beck's interesting paper is now open for discussion.

Dr. Takagi: Have you tested the effect of hydroxyurea on your system?

Dr. Beck: Yes, we have. Hydroxyurea, of course, is thought to inhibit ribonucleotide reductase in animal cells and is known to do so in *E. coli*. We found in preliminary experiments that it does not inhibit *L. leichmannii* reductase. Perhaps this is one more bit of evidence that the enzyme in animal cells is more *E. coli*-like than *L. leichmannii*-like.

Dr. Horecker: It is correct that your enzyme binds a total of eight ligands — four substrates and four effectors? Do you know whether these occupy the same two sets of binding sites?

Dr. Beck: As I mentioned, we think there is only one catalytic site, but it is difficult to establish this by kinetic studies. In an experiment designed to test whether two simultaneously added substrates compete for a binding site, the reduction product of one is usually a positive or negative effector for the reduction of the other. Hence, the kinetics are only suggestive. Nevertheless they are suggestive of a single catalytic site. If I may again anticipate Dr. Larsson's paper, I might add that it was much easier for him and Reichard to show competition between two substrates with the *E. coli* reductase, because the reaction products of that enzyme are diphosphates and the allosteric effectors triphosphates. Hence, the kinetics were unambiguous and they indicated a single catalytic site. We tried to get around this problem by rigging the system in various ways and, as I said, we believe there is one catalytic site. We are doing binding studies to investigate the regulatory site (or sites), but results are not yet available.

Dr Larsson: When the level of vitamin B_{12} in the medium was decreased in cultures of *L. leichmannii*, you observed a parallel decrease of the intracellular level of acid-soluble deoxyribosyl compounds. Have you noticed an increase in the level of acid-soluble ribosyl compounds under these conditions?

Dr. Beck: As in most organisms, the intracellular ribosyl pool is more than a 100 times as large as the deoxyribosyl pool, and it includes ATP and other nucleotides, whose concentrations are significantly affected by other pathways. Nevertheless, we have some evidence that in vitamin B_{12} deficiency there is some increase in the pyrimidine ribosyl pool. I don't know the significance of that observation.

Dr. Horecker: In one of your mechanisms you suggest that the 5′ carbon atom of the adenyl portion of the DBC coenzyme is oxidized to a hydroxyl and then reduced. In this case the corresponding adenosine derivative should be active. Has this been synthesized and tested?

Dr. Beck: That's an interesting thought.

Enzymatic Reduction of Ribonucleotides in Escherichia coli

Agne Larsson

Department of Chemistry II, Karolinska Institutet, Stockholm, Sweden

The fundamental role of deoxyribonucleotides as precursors in the cellular synthesis of DNA has prompted an extensive investigation of the biochemical processes involved in their formation. Tracer experiments performed *in vivo* have established that in *Escherichia coli* — as in several bacterial and animal cells — deoxyribonucleotides are synthesized exclusively by reduction of the corresponding ribonucleotides (*cf.* [1]). *In vivo* experiments have also demonstrated that in the cell the conversion of a ribosyl to a deoxyribosyl compound is an irreversible process.

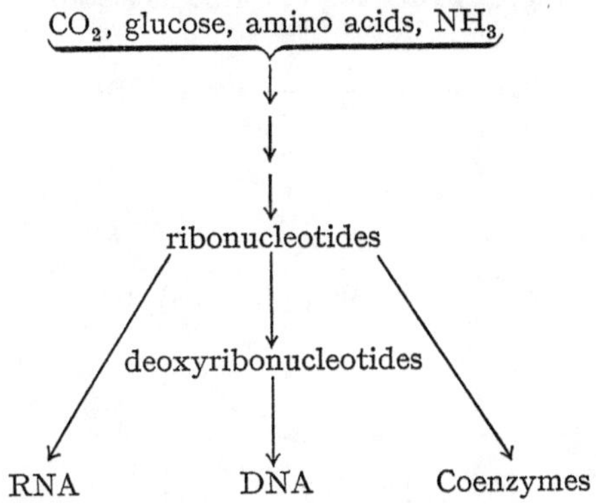

Fig. 1. General scheme of nucleotide metabolism

The general metabolic implications of ribonucleotide reduction are outlined in Fig. 1. Ribonucleotides are formed from compounds commonly occurring in the metabolism of the cell, such as carbon dioxide, glucose, amino acids, and ammonia. They are then utilized for essentially three different purposes: first, they serve as precursors for RNA; second, they are transformed into coenzymes; and finally, they function as precursors for DNA. The initial step in this latter sequence is the reduction of ribonucleotides to deoxyribonucleotides.

From a chemical point of view the conversion of a ribosyl compound to a deoxyribosyl seems to be a very simple reaction. In principle it involves the replacement of the 2'-hydroxyl group in a ribonucleotide by hydrogen (Fig. 2).

Reduction of ribonucleotides has been studied with purified enzymes from three different sources (*cf.* [2]) — from *E. coli* in our laboratory, from *Lactobacillus leichmannii* in the laboratories of Blakley, Beck, and Abrams, and from Novikoff hepatoma in the rat by Moore and coworkers.

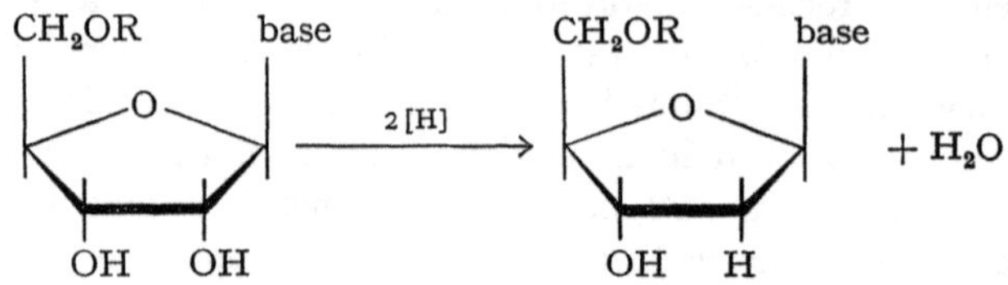

Fig. 2. Reduction of a ribonucleotide

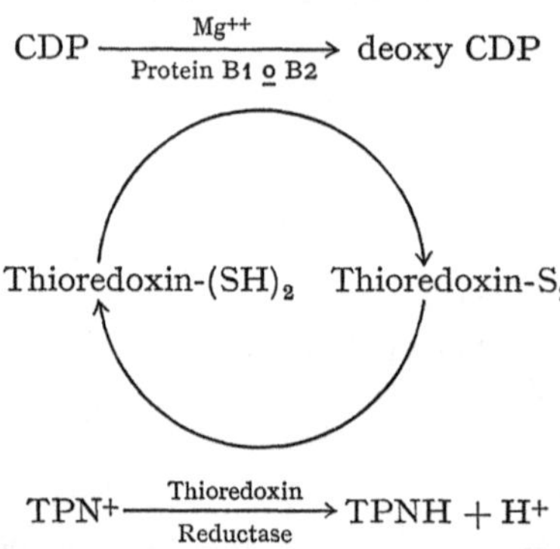

Fig. 3. The ribonucleoside diphosphate reductase system of *Escherichia coli*

Our investigations was started about 1960 with an attempt to demonstrate the reduction of ribonucleotides in cell-free extracts of *E. coli*. For technical reasons we used radioactively labeled cytidine nucleotides, and measured the formation of deoxycytidine nucleotides. After some initial purification it became apparent that the conversion occurred at the diphosphate level, *i.e.* CDP was reduced to dCDP [3]. Although the basic reaction, shown in Fig. 2, seems to be a simple one, it turned out to be a rather complex process in the cell, and, in fact, required the participation of four different protein components — thioredoxin, thioredoxin reductase, proteins B1 and B2 — in addition to three cofactors — ATP, Mg-ions, and TPNH. The way these components interact during CDP reduction is outlined in Fig. 3.

The ultimate hydrogen donor in the reaction is TPNH. However, a hydrogen transferring agent, called *thioredoxin*, is interposed between TPNH and the substrate [4]. Thioredoxin has been characterized as a small protein with a molecular weight of approximately 12,000. The functional groups of this protein consist of two sulfhydryls. When one molecule of CDP is converted to dCDP, these sulfhydryl groups are oxidized to form a disulfide bridge.

The oxidized form of thioredoxin is reactivated by a specific enzyme, called *thioredoxin reductase* [5], which has now been isolated in a pure state [6]. This enzyme was found to be a flavoprotein with a molecular weight of 66,000, containing two molecules of FAD per molecule of enzyme.

TPNH, thioredoxin, and thioredoxin reductase constitute the "thioredoxin system", which we believe to be a new type of hydrogen carrier system. Hydrogen is transferred from TPNH to the FAD group in the catalytic site of thioredoxin reductase, further to the functional sulfhydryl groups of thioredoxin, and finally to the product.

Analogous hydrogen carrier systems, participating during ribonucleotide reduction, have subsequently been isolated from *L. leichmannii* by Orr and Vitols [7] and from Novikoff hepatoma by Moore [8]. It is reasonable to believe that the thioredoxin system might be involved in metabolically important redox reactions other than ribonucleotide reduction.

In addition to the requirement for the thioredoxin system, the reduction of CDP shows an absolute requirement for Mg-ions, and a partial requirement for ATP [3].

The ribonucleotide reductase was purified from *E. coli* using the conversion of CDP to dCDP as an assay. During this purification the enzymatic activity was separated into two protein fractions, called B1 and B2. These fractions were inactive when tested separately but fully active upon recombination. Thus B1 activity could be assayed in the presence of an excess of B2 and *vice versa*. With this assay further purification of B1 and B2 has been possible [9]. At present both protein preparations are approximately 75% pure.

How are the two proteins, B1 and B2 — in addition to thioredoxin and thioredoxin reductase — involved in the reaction? For some time we thought of proteins B1 and B2 as two discrete enzymes, catalyzing partial reactions. For instance, protein B1 might catalyze the formation of an intermediate, which was subsequently converted to the product by protein B2. We were unable, however, to obtain experimental support for a sequential action of the two proteins. Therefore we considered as an alternative hypothesis that proteins B1 and B2 might be nonidentical subunits of the ribonucleotide reductase, *i.e.* a complex of proteins B1 and B2 might constitute the active enzyme. We have recently obtained some evidence for this concept by experiments employing sucrose gradient centrifugations [10].

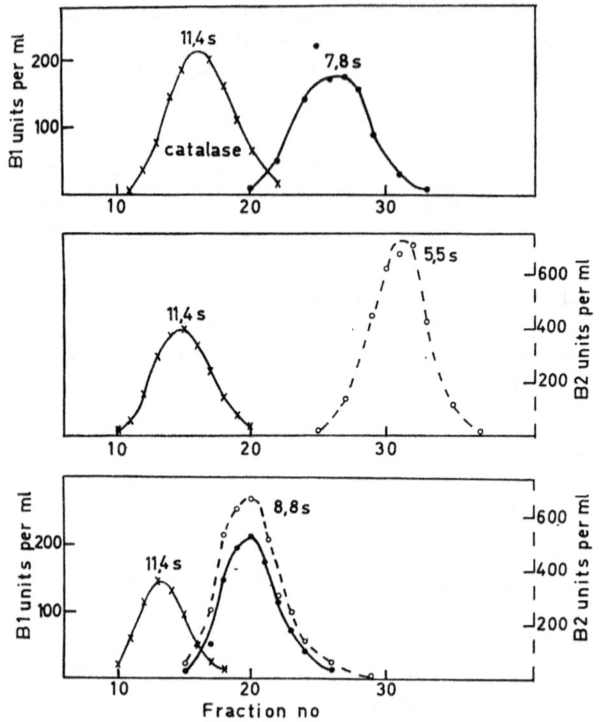

Fig. 4. Sucrose gradient centrifugations of protein B1 (upper graph), protein B2 (middle graph), and a mixture of proteins B1 and B2 (lower graph). Catalase was used as a marker

The result of a typical centrifugation experiment is shown in Fig. 4. A small amount of catalase with an $s_{20,w}$ value of 11.4 S was included as a marker, and proteins B1 and B2 were localized by their enzymatic activities. When each protein was centrifuged alone it sedimented as a single peak, protein B1 was found to have an $s_{20,w}$ value of 7.8 S and B2 a value of 5.5 S. When proteins B1 and B2 were centrifuged together, they did not sediment as separate peaks. Instead, they sedimented together with an $s_{20,w}$ value of 8.8 S. The formation of this complex required the presence of Mg-ions in the gradient. In the absence of Mg-ions proteins B1 and B2 sedimented separately with $s_{20,w}$ values of 7.8 S and 5.5 S, respectively. Mg-ions had no effect on the individual sedimentation patterns of proteins B1 and B2.

On the basis of the experiment shown in Fig. 4 — and similar sucrose gradient centrifugations — it seems likely that proteins B1 and B2 are non-identical subunits of the ribonucleotide reductase, and that the formation of the active complex is dependent on the presence of Mg-ions. This might actually explain the absolute requirement for Mg-ions in the reduction of CDP.

Blakley and Barker demonstrated the involvement of a cobamide coenzyme in the ribonucleotide reductase system of *L. leichmannii* [11]. In *E. coli*, on the other hand, ribonucleotide reduction appears to proceed without participation of any cobamide derivative [9]. The exclusion of a cobamide coenzyme as a cofactor in the enzyme system from *E. coli* is based on two observations. First, addition of cobamide coenzymes

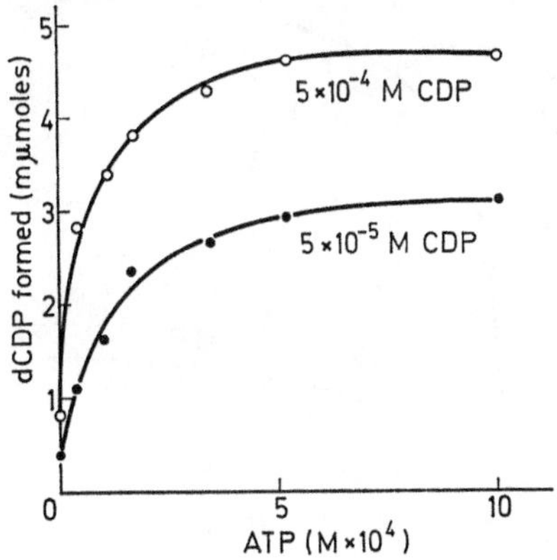

Fig. 5. Effect of ATP on the reduction of CDP

to the incubation mixtures does not stimulate the reaction, and, second, only insignificant amounts of vitamin B_{12} derivatives are present in the highly purified proteins B1 and B2.

It was noted even in early studies with crude extracts from *E. coli* that the reduction of CDP requires the addition of ATP [3], and this requirement was present also with the purified proteins B1 and B2 [9] (Fig. 5). The requirement for ATP, however, was not an absolute one. It seems more appropriate to talk about a stimulation. In general optimal stimulation by ATP was observed at a concentration of approximately 10^{-3} M.

Originally, ATP was thought to participate in the formation of an intermediate — for instance, a phosphorylated, pyrophosphorylated, or adenylated derivative of CDP — which was then reduced [12]. When contaminating phosphatases and kinases had been removed from the enzyme preparation it became evident that this hypothesis was incorrect, since the reduction of CDP was not associated with any consumption of ATP [9].

The function of ATP in the reduction of CDP is further illustrated in Fig. 6 which shows CDP concentration curves in the presence and in the absence of ATP. This nucleotide influences the kinetic parameters of the reaction, increasing both the maximal velocity of the system and the affinity of the reductase for CDP — the K_m value for CDP in the presence of ATP was 3×10^{-5} M compared with 3×10^{-4} M in the absence of ATP.

The effects of ATP on the kinetics of the reaction, in addition to its catalytic participation, indicated that ATP might be acting as a

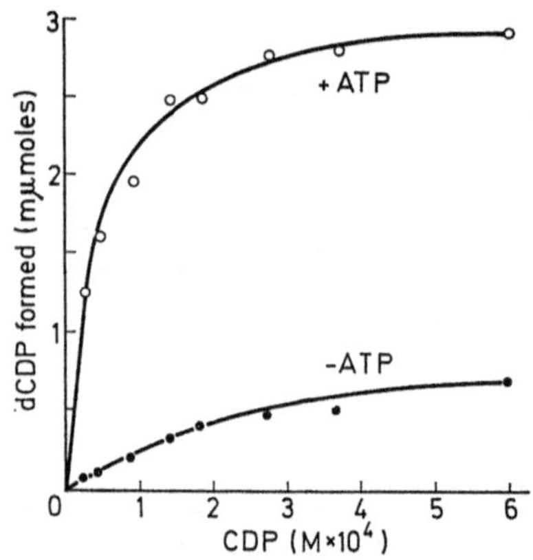

Fig. 6. CDP concentration curves in the absence and in the presence of ATP

positive effector, presumably *via* an *allosteric* mechanism as defined by Monod *et al.* [13, 14].

The reduction of CDP was influenced not only by ATP but also by other nucleotides [15] (Fig. 7). Pronounced stimulation of the reaction was caused by dTTP. In fact, this nucleotide had the same effect as ATP on the kinetics of the reaction, *i.e.* both effectors increased the maximal velocity and the affinity of the enzyme for CDP to the same extent. For maximal effect, however, a concentration of only 10^{-5} M dTTP was required in contrast to 10^{-3} M ATP.

The specificity of ATP as a positive effector is further illustrated in Fig. 7; dATP, in spite of its close structural relationship to ATP, inhibited the reaction. dCTP and dGTP had very little effect. We also observed that nucleoside *tri*phosphates were consistently more active than the corresponding *mono-* and *di*phosphates; for instance, dATP was a more potent inhibitor of CDP reduction than dADP (Fig. 7).

Although proteins B1 and B2 were extensively purified with respect to the reduction of CDP, they were, however, found to catalyze the reductions of adenosine, guanosine, and uridine nucleotides as well [15,

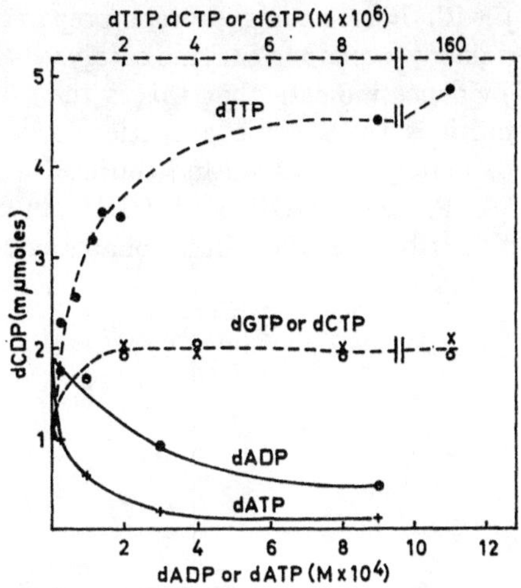

Fig. 7. Effect of deoxyribonucleotides on the reduction of CDP

Reaction	Effector			
	ATP	dTTP	dGTP	dATP
CDP → dCDP	+	+	0	—
UDP → dUDP	+	+	0	—
GDP → dGDP	0	+	+	—
ADP → dADP	0	+	+	—

Fig. 8. Summary of allosteric effects in the ribonucleoside diphosphate reductase system of *Escherichia coli*. + = stimulation; 0 = no or little effect; — = inhibition

16]. The basic requirement for these reactions were the same as those found previously for the reduction of cytidine nucleotides, *i.e.* the reactions occurred at the diphosphate level, and the simultaneous presence of proteins B1 and B2, the thioredoxin system and Mg-ions was required.

Similar to the conversion of CDP to dCDP, the reductions of ADP, GDP, and UDP were influenced markedly by the presence of certain nucleotide effectors [15, 16]. Again, different nucleoside *tri*phosphates modified the kinetic parameters of the reactions. Fig. 8 depicts a summary of these observations — ATP acted as a positive effector in the reduction of pyrimidine ribonucleotides but had no effect on the reduction of purine ribonucleotides; dTTP stimulated all four reactions; dGTP functioned as a positive effector in the reduction of purine ribonucleotides; and dATP, finally, acted as a general negative effector.

Proteins B1 and B2, at a purity of approximately 75%, catalyzed the reductions of the four different ribonucleotide substrates. Since we were not working with homogenous protein preparations, it became important to determine whether one enzyme catalyzed all four reactions. Different lines of evidence indicate that this is the case. For instance, under optimal conditions the specific activities toward the four substrates were approximately the same, and during the purification the activities toward ADP, CDP, GDP, and UDP ran in parallel [16]. We also found that the ribonucleoside diphosphates acted as competitive

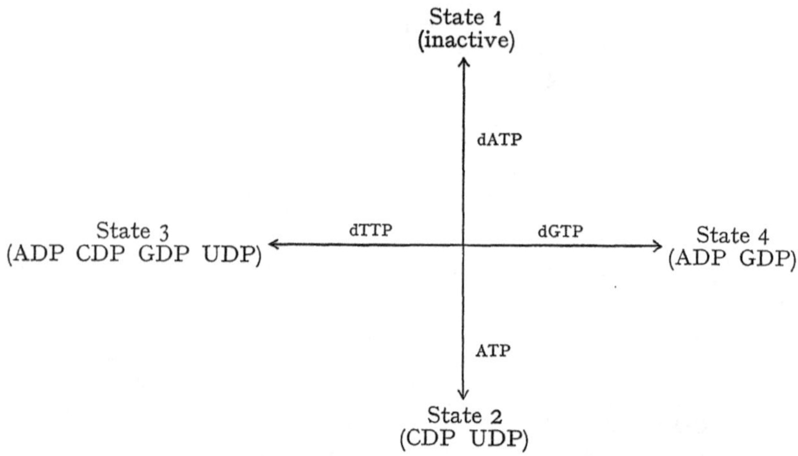

Fig. 9. Schematic representation of allosteric transitions of the ribonucleoside diphosphate reductase from *Escherichia coli*

inhibitors in the reductions of one another, provided that the proper allosteric effector was present [15]. The K_i value for each ribonucleoside diphosphate acting as inhibitor was in good agreement with the apparent K_m value for the same nucleotide acting as substrate. It therefore seemed reasonable to believe that ADP, CDP, GDP, and UDP were reduced at the same catalytic site. A proper name for this enzyme would then be the *ribonucleoside diphosphate reductase*. We believe that the substrate specificity of this enzyme is regulated by the different effectors. As shown in Fig. 9 it is proposed that the enzyme can assume several conformational states, and that these states are members of an equilibrium system. By addition of an effector, the equilibrium is shifted such that one conformational state becomes predominant, and this governs the subsequent activity of the enzyme. For example, ATP stabilizes a pyrimidine-specific state, whereas dGTP stabilizes a purine-specific conformation.

The observation that the ribonucleoside diphosphate reductase is composed of two different types of subunits, called proteins B1 and B2, has prompted an attempt to determine the actual contribution of each subunit to the function of the catalytically active enzyme complex. This investigation has been guided to a large extent by the work of Gerhart and Schachman on the enzyme aspartate carbamoyl transferase [17]. This enzyme also contains two types of subunits and is subject to allosteric regulation by low molecular weight effectors. One subunit, called the regulatory subunit, carries the effector-binding sites, whereas the other, called the catalytic subunit, is responsible for the catalytic activity. Neither of the subunits B1 and B2 of the ribonucleoside

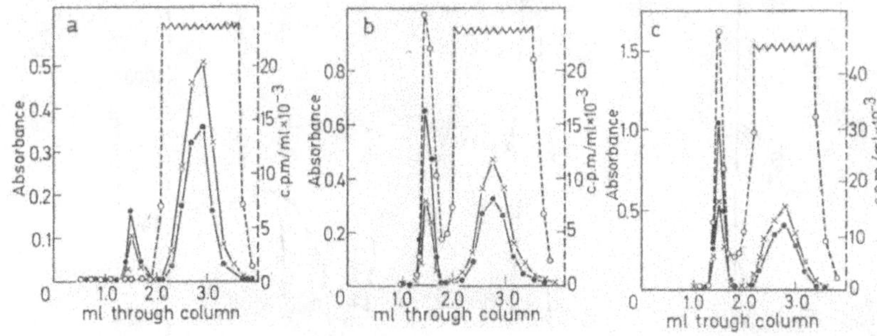

Fig. 10a—c. Chromatography on Sephadex G-25 columns of (a) protein B2 + ^{3}H-dTTP; (b) protein B1 + ^{3}H-dTTP; and (c) protein B1 and B2 + ^{3}H-dTTP. Dashed lines represent radioactivity and continuous lines absorbance at 260 and 280 mμ, respectively

diphosphate reductase, however, exhibits any catalytic activity in the absence of the other and therefore does not appear to be equivalent to the catalytic subunit of the aspartate carbamoyl transferase. Thus far we have instead investigated the possibility that proteins B1 or B2 might correspond to the regulatory subunit by studies of the capacity of the two proteins to bind different nucleotide effectors [10]. Therefore proteins B1 and B2, either separately or in combination, were preincubated in the presence of a radioactive effector and then chromatographed on columns of Sephadex G-25 to separate the proteins from free effectors [18]. Fig. 10 shows such an experiment with ^{3}H-dTTP as effector. In the case of protein B2, dTTP was completely separated from the protein (Fig. 10A). With protein B1, however, either alone or together with protein B2, a small but significant amount of dTTP was eluted with the protein (Fig. 10B and C). Similar observations were also made with dATP and dGTP. On the basis of these preliminary studies it appeared that protein B1 possessed the specific binding sites for the regulatory effectors.

To study quantitative aspects of the binding of effectors to protein B1 we used the technique described by Hummel and Dreyer [18]. This involves filtration of the protein through Sephadex columns equilibrated with different concentrations of the low molecular weight ligand. Fig. 11 gives the result of a typical experiment in which protein B1 was filtered through a column of Sephadex equilibrated with ^{3}H-dTTP. The binding of dTTP to the protein is indicated by the peak of radioactivity which emerges with the protein peak. By calculation of the amount of dTTP bound per absorbance unit of protein at a variety of effector concentrations it was possible to determine both the dissociation constants

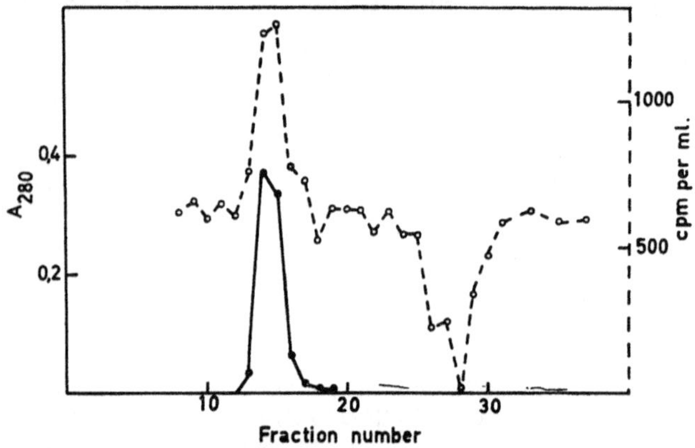

Fig. 11. Binding of ^{3}H-dTTP to protein B1 measured by the technique of Hummel and Dreyer [18]. Dashed line represents radioactivity and continuous line absorbance at 280 mμ

for the effector-protein complexes and the maximal amounts of effector bound per absorbance unit of protein B1. Fig. 12 summarizes the results obtained with the effectors dATP, dGTP, and dTTP. Dissociation constants ranging from 1—3 $\times 10^{-6}$ M were found. These values are in good agreement with the corresponding apparent K_m values calculated in kinetic experiments [15]. Approximately 15 mμmoles of each effector was bound per absorbance unit of protein B1, which corresponds to approximately 2 molecules of effector per molecule of protein. The molecular weight of protein B1, determined by sedimentation equilibrium centrifugation, was found to be 170,000 on the assumption of a partial specific volume of 0.75.

The scheme in Fig. 13 represents a tentative picture of the ribonucleoside diphosphate reductase from *E. coli*. The B1 subunit seems to be larger than the B2 subunit as determined by sedimentation in sucrose gradients [10]. The catalytic site has been drawn on the B2 subunit, although at the present time we have only tentative evidence to support this assumption. Furthermore, we do not yet know the exact

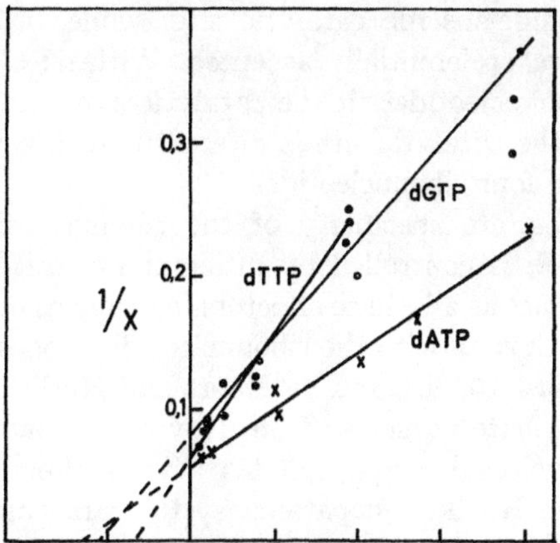

Fig. 12. Binding of dATP, dGTP, and dTTP to protein B1. The data are compiled from different experiments with the technique of Hummel and Dreyer [18]. "s" symbolizes effector concentration, and "K_x" the dissociation constant for effector-protein B1 complex

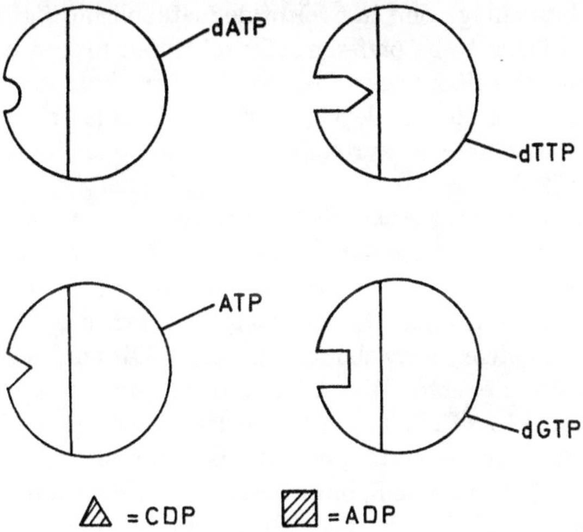

Fig. 13. Schematic representation of the ribonucleoside diphosphate reductase from *Escherichia coli*

stoichiometric relationship between proteins B1 and B2 in the active enzyme complex. Thus there are many hypothetical features in this picture. A number of aspects seem clear, however. First, it is clear that the allosteric effectors bind to protein B1. Furthermore, it also appears certain that the conformation of the catalytic site is determined by the allosteric effectors bound to protein B1. For instance, binding of ATP creates a pyrimidine-specific catalytic site, while with dGTP purine ribonucleotides are preferentially accepted. With dTTP purine as well as pyrimidine ribonucleotides fit the catalytic site, and with dATP the conformation of the catalytic site is altered in such a way as to lower the affinity for all four ribonucleotides.

Thus, the substrate specificity of the ribonucleoside diphosphate reductase of *E. coli* is controlled by different nucleoside triphosphates which apparently act as allosteric effectors. Analogous observations have been made by Beck *et al.* with the ribonucleoside triphosphate reductase from *L. leichmannii* [19, 20], and by Moore and Hurlbert with the ribonucleoside diphosphate reductase from Novikoff hepatoma [21]. In all three systems nucleoside *tri*phosphates act as effectors. The results obtained with the Novikoff hepatoma system are particularly similar to those obtained with *E. coli*.

It is reasonable to assume that the allosteric regulation of the ribonucleotide reductase has some biological function. Accordingly, it might be a mechanism regulating the balanced supply of deoxyribonucleotides which is necessary for DNA synthesis. The ribonucleotide reduction represents a branching point in nucleotide metabolism (Fig. 1), and such reactions are known to be preferentially subjected to feed-back control. A regulation of the ribonucleotide reduction does indeed seem to exist, since in comparison to the levels of ribonucleotides only very small amounts of acid soluble deoxyribonucleotides are detectable in different types of cells [22].

On the basis of our observations with the enzyme system from *E. coli* we have attempted to formulate a scheme which might account for a balanced supply of all four deoxyribonucleotides (Fig. 14). This rather complicated mechanism has the following main features: The formation of the two pyrimidine deoxyribonucleotides, dCDP and dUDP, requires ATP as a positive effector. *Via* a series of enzymatic reactions dUDP is then converted to dTTP. The latter nucleotide occupies a key position in the regulation; under certain conditions it can stimulate further the formation of dCDP and dUDP, but, what is more important, it stimulates the reductions of GDP and ADP. Thus, dTTP switches the substrate specificity of the ribonucleoside diphosphate reductase. dGDP, after phosphorylation to dGTP, can then stimulate further the reduction of purine ribonucleotides. This, however, leads to the accumulation of dATP, which is the second key compound in the scheme. This nucleotide functions as a feed-back inhibitor to all four reactions. We consider the effectors as physiological signals which can coordinate the four different reactions and, finally, when the requirement of DNA synthesis for the four different deoxyribonucleotides is fulfilled, dATP can block the whole system. It should be emphasized that the scheme in Fig. 14

is only an attempt to interprete the observations we have made *in vitro*. Nevertheless it represents an intriguing mechanism for the regulation of DNA synthesis in the cell. An almost identical scheme has been postulated independently by Moore and Hurlbert on the basis of their studies with the ribonucleotide reductase system from Novikoff hepatoma [21].

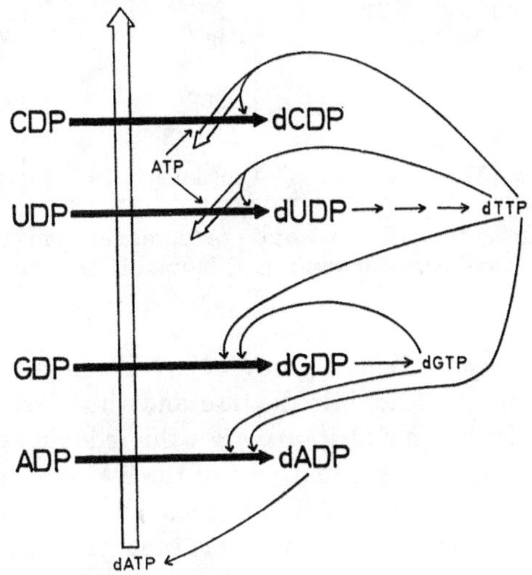

Fig. 14. Hypothetical scheme for the physiological regulation of deoxyribonucleotide biosynthesis

The information available about the regulation of deoxyribonucleotide synthesis *in vivo* in *E. coli* is meager. Results obtained with other types of cells can, however, be explained by the mechanism shown in Fig. 14. For instance, the observations made by Klenow with intact Ehrlich ascites cells support the concept of dATP as a general inhibitor of ribonucleotide reduction [23].

Fig. 15, finally, is a short summary of the present knowledge about the three different purified ribonucleotide reductase systems, namely that from *E. coli*, the one from *L. leichmannii* — reviewed by Dr. W. S. Beck on p. 26—62 in this publication — and the mammalian tumor system studied by Moore and collaborators [8, 21, 24, 25]. It should be pointed out that the work with the two microbiological systems has been carried out with highly purified enzymes, whereas the results with the Novikoff hepatoma were obtained with only a partially purified system. For simplicity the reduction of cytidine nucleotides catalyzed by the different enzymes is compared.

Substrate. In *E. coli* and Novikoff hepatoma the reduction occurs at the *di*phosphate level, whereas in *L. leichmannii tri*phosphates are reduced.

Parameter	Ribonucleotide reductase system from		
	E. coli	*L. leichmannii*	Novikoff hepatoma
Substrate	CDP	CTP	CDP
Hydrogen transport system	TPNH → FADH → → $(SH)_2$	TPNH → FADH → $(SH)_2$ → DBC·H	TPNH → S—B → → S—C
Metal requirement	Mg^{2+}	(Mg^{2+})	Mg^{2+}, Fe^{3+}
Activator	ATP	(ATP)	ATP
Inhibitor	dATP	—	dATP
Subunits	B1 + B2	not known	(P1 + P2)

Fig. 15. Comparison of the ribonucleotide reductase systems from *Escherichia coli, Lactobacillus leichmannii,* and Novikoff hepatoma

Hydrogen Transport System. In *E. coli* hydrogen is transferred from TPNH to FAD in thioredoxin reductase and then further to the SH-groups of thioredoxin. In *L. leichmannii* a thioredoxin system has been isolated [7], and therefore the first part of the hydrogen transfer system is analogous to the one in *E. coli.* The lactobacillus thioredoxin, however, functions as a reductant for the cobamide coenzyme, whereas in *E. coli* no vitamin B_{12} derivative is involved. In the tumor system two protein fractions, called S-B and S-C, have recently been purified by Moore [8]. Fraction S-B is apparently equivalent to thioredoxin reductase and S-C to thioredoxin. No involvement of cobamide coenzymes in the ribonucleotide reduction in Novikoff hepatoma has yet been revealed, despite serious attempts to discover their participation.

Metal Ion Requirement. In the *E. coli* system there is an absolute requirement for Mg-ions. In the *L. leichmannii* system the situation with respect to metal requirement is somewhat more complicated, although under certain experimental conditions Mg-ions can stimulate the reduction of CTP. There is a dual metal requirement in the tumor system; both Mg-ions and Fe-ions are necessary for optimal activity [24].

Activator. In all three systems the reduction of cytidine nucleotides is stimulated by ATP. With the enzymes from *E. coli* and Novikoff hepatoma the reduction of CDP is also stimulated by dTTP. The "prime effector" for CTP reduction catalyzed by the *L. leichmannii* enzyme is dATP [20].

Inhibitor. dATP acts as a negative effector in the reduction of CDP in *E. coli* and Novikoff hepatoma. In *L. leichmannii,* however, dATP is a positive effector, and no specific inhibitor of CTP reduction has been identified.

Subunits. The ribonucleoside diphosphate reductase of *E. coli* apparently consists of non-identical subunits — proteins B1 and B2. This may also be the case with the tumor enzyme. Moore has namely recently been able to separate the enzymatic activity into two protein fractions, called P1 and P2, which might be subunits of the enzyme [25]. Further work is, however, required to establish this. The occurrence of subunits in the *L. leichmannii* enzyme is still an open question.

The above comparison leads to the conclusion that the ribonucleotide reductase system from Novikoff hepatoma is more similar to the system from *E. coli* than that from *L. leichmannii*. Time should show if mammalian reductases in general are patterned after the cobamide-independent ribonucleoside diphosphate reductase system of *E. coli* or after the cobamide-dependent ribonucleoside triphosphate reductase system of *L. leichmannii*.

References

1. Sable, H. Z.: Advanc. Enzymol. **28**, 391 (1966).
2. Larsson, A., and P. Reichard in,: J. N. Davidson and W. E. Cohn (editors), Progress in nucleic acid research and molecular biology, vol. 7, p. 303. New York and London: Academic Press 1967.
3. Reichard, P., A. Baldesten, and L. Rutberg: J. biol. Chem. **236**, 1150 (1961).
4. Laurent, T. C., E. C. Moore, and P. Reichard: J. biol. Chem. **239**, 3436 (1964).
5. Moore, E. C., P. Reichard, and L. Thelander: J. biol. Chem. **239**, 3445 (1964).
6. Thelander, L.: J. biol. Chem. **242**, 852 (1967).
7. Orr, M. D., and E. Vitols: Biochem. biophys. Res. Commun. **25**, 109 (1966).
8. Moore, E. C.: Biochem. biophys. Res. Commun. **29**, 264 (1967).
9. Holmgren, A., P. Reichard, and L. Thelander: Proc. nat. Acad. Sci. (Wash.) **54**, 830 (1965).
10. Brown, N. C., A. Larsson, and P. Reichard: J. biol. Chem. **242**, 4272 (1967).
11. Blakley, R. L., and H. A. Barker: Biochem. biophys. Res. Commun. **16**, 391 (1964).
12. Reichard, P.: J. biol. Chem. **237**, 3513 (1962).
13. Monod, J., J.-P. Changeux, and F. Jacob: J. molec. Biol. **6**, 306 (1963).
14. — J. Wyman, and J.-P. Changeux: J. molec. Biol. **12**, 88 (1965).
15. Larsson, A., and P. Reichard: J. biol. Chem. **241**, 2533 (1966).
16. — — J. biol. Chem. **241**, 2540 (1966).
17. Gerhart, J. C., and H. K. Schachman: Biochemistry **4**, 1054 (1965).
18. Hummel, J. P., and W. J. Dreyer: Biochim. biophys. Acta (Amst.) **63**, 530 (1962).
19. Beck, W. S., M. Goulian, A. Larsson, and P. Reichard: J. biol. Chem. **241**, 2177 (1966).
20. — J. biol. Chem. **242**, 3148 (1967).
21. Moore, E. C., and R. B. Hurlbert: J. biol. Chem. **241**, 4802 (1966).
22. Lark, K. G., in: J. H. Taylor (editor), Molecular genetics, part I, p. 153. New York and London: Academic Press 1963.
23. Klenow, H.: Biochim. biophys. Acta (Amst.) **61**, 885 (1962).
24. Moore, E. C., and P. Reichard: J. biol. Chem. **239**, 3453 (1964).
25. — Personal communication.

Discussion

Dr. Takagi: Have you isolated a mutant of *E. coli* which requires deoxyribonucleotides for the growth? If so, what kind of alteration in the reductase system have you found?

Dr. Larsson: The answer to your first question is: No. For approximately three years Dr. O. Karlström working in our laboratory has tried to isolate a mutant of *E. coli* requiring deoxyribosyl compounds for growth. Unfortunately his efforts have been unsuccessful so far. We think that this is due to the fact that such mutants do not survive the selection process. This, in turn, might be the result of a phenomenon called "thymineless death". Cells that are starved for thymine during a long enough time do not survive. This might be true also when the cells are deprived of other DNA precursors. If this is so, one has to design a process for isolation of ribonucleotide reductase-minus mutants using very short selection times.

Dr. Pogell: I was interested in the fact that although you found nonsigmoidal behavior for your various effectors, you find two molecules of effector bound per molecule of protein. Do you have any comment on this point? Have you tested various mixtures of negative and positive effectors?

Dr. Larsson: We have tried to demonstrate sigmoidal ligand saturation curves for effectors as well as substrates, since this type of behavior is generally accepted to be characteristic of an allosteric enzyme. According to the terminology of Monod *et al.* [14] this is called a *homotropic* interaction. So far we have been able to show *heterotropic* interaction in the ribonucleoside diphosphate reductase system of *E. coli*, *i.e.* the binding of one ligand (effector) influences the binding of the other (substrate). Therefore we believe that it is justified to call our enzyme system allosteric.

Concerning the mixing of positive and negative effectors, we have performed a series of such experiments, combining for instance dATP and ATP, dATP and dTTP, ATP and dTTP, etc. [15, 16]. These results are not easy to summarize. For instance, dATP inhibition of the reduction of purine as well as pyrimidine ribonucleotides can be reversed by ATP, in spite of the fact that ATP *per se* is a positive effector only in the reduction of pyrimidine ribonucleotides. The only clearcut sigmoidal ligand saturation curve that we have observed with the ribonucleotide reductase from *E. coli* was when the inhibition of CDP reduction caused by dATP was reversed by ATP [9].

The Uronic Acid Pathway and its Defect in Essential Pentosuria*

OSCAR TOUSTER

Department of Molecular Biology
Vanderbilt University, Nashville, Tennessee, U.S.A.

I shall briefly review the experimental basis for the glucuronate-xylulose pathway and then discuss some conceptual problems in part related to current studies in our laboratory.

Elucidation of the Glucuronate-Xylulose Pathway

The elaboration of the pathway followed two lines of investigation: the biosynthesis of L-ascorbic acid and the study of L-xylulose metabolism and essential pentosuria. The work of King, Burns, Conney and others led to the discovery that D-glucuronate and L-gulonate are precursors of L-ascorbic acid in rats, the evidence initially being derived from tracer studies, which were later supported by enzymatic experiments (Fig. 1) [1]. While the work on ascorbic acid formation was in

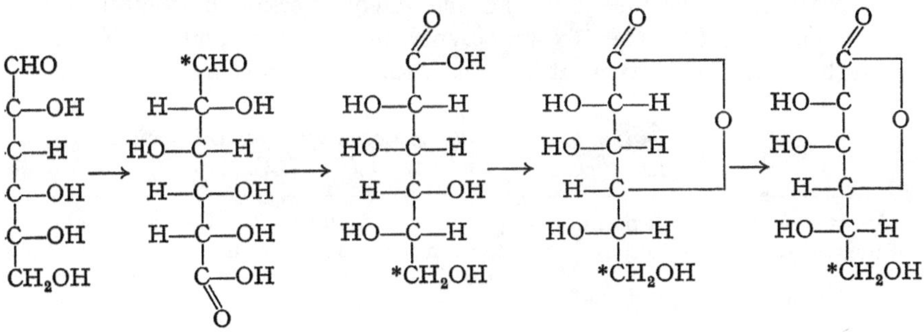

D-Glucose D-Glucuronic Acid L-Gulonic Acid L-Gulonolactone L-Ascorbic Acid

Fig. 1. The biosynthesis of L-ascorbic acid from D-glucose

progress in other laboratories, we were engaged in a study of L-xylulose metabolism. Our studies included the demonstration that L-xylulose, which is excreted daily in gram quantities by pentosuric individuals, is excreted in milligram amounts by normal animals, including non-

* Unpublished work described in this paper was supported in part by Grant No. GB-3541 from the National Science Foundation.

pentosuric humans, and that D-glucuronolactone feeding increases the urinary xylulose level [2, 3]. Enklewitz and Lasker [4] had already observed in 1935 that glucuronolactone enhances pentose excretion in pentosuric subjects. That the conversion was a direct one, and resembled the carbon chain inversion observed in the biosynthesis of ascorbic acid, was indicated by the demonstration that 1-^{13}C-D-glucuronolactone is converted to 5-^{13}C-L-xylulose in a subject with essential pentosuria [5]. In addition to this work on the formation of the ketopentose, we showed that the sugar is rapidly utilized by guinea pig liver, that is, as well as glucose or fructose [2]. We were then able to elucidate the enzymatic basis of the transformation. With the help of Vernon Reynolds, then a medical student, the enzyme system responsible for L-xylulose utilization was localized in the mitochondrial fraction, indeed, in an insoluble fraction obtained after subjecting the particles to osmotic shock [6]. After some delay in learning the nature of the reaction, the reaction product was identified as the polyol xylitol [6], rather than the alternative possible product, L-arabitol. Xylitol, but not L-arabitol, is well utilized in animals in a manner consistent with its assigned position in the cycle shown in Fig. 2 [7, 8] (Table I).

When Dr. Hollmann came to our laboratory for a year, we were able to advance in our progress through the pathway by showing that two mitochondrial xylitol dehydrogenases existed, an NADP-linked enzyme catalyzing the interconversion of L-xylulose and xylitol, and an NAD-linked enzyme effecting the interconversion of xylitol and D-xylulose [9, 10]. The L-xylulose enzyme has very great specificity, but the D-xylulose enzyme utilizes many substrates and closely resembles

Table I. *Distribution of ^{14}C in liver glycogen after administration of D-ribose-1-^{14}C and xylitol-1-^{14}C* [12]

Substance injected	Experimental animal	% administered ^{14}C in glycogen	Relative % ^{14}C of glucose carbon atom No.				
			1	2	3	4+5	6
D-Ribose-1-^{14}C (4.30 μc, 10.4 mg)	Rat ♀ (3.56 gm)	7.1	69.5	2.4	28.5	2.9	1.4
Xylitol-1-^{14}C (6.70 μc, 10 mg)	Guinea pig ♀ (765 gm)	23.8	67.1	0.4	25.6	2.0	2.1
Xylitol-1-^{14}C (6.70 μc, 10 mg)	Rat ♂ (474 gm)	11.8	67.1	0.6	25.1	2.2	3.3

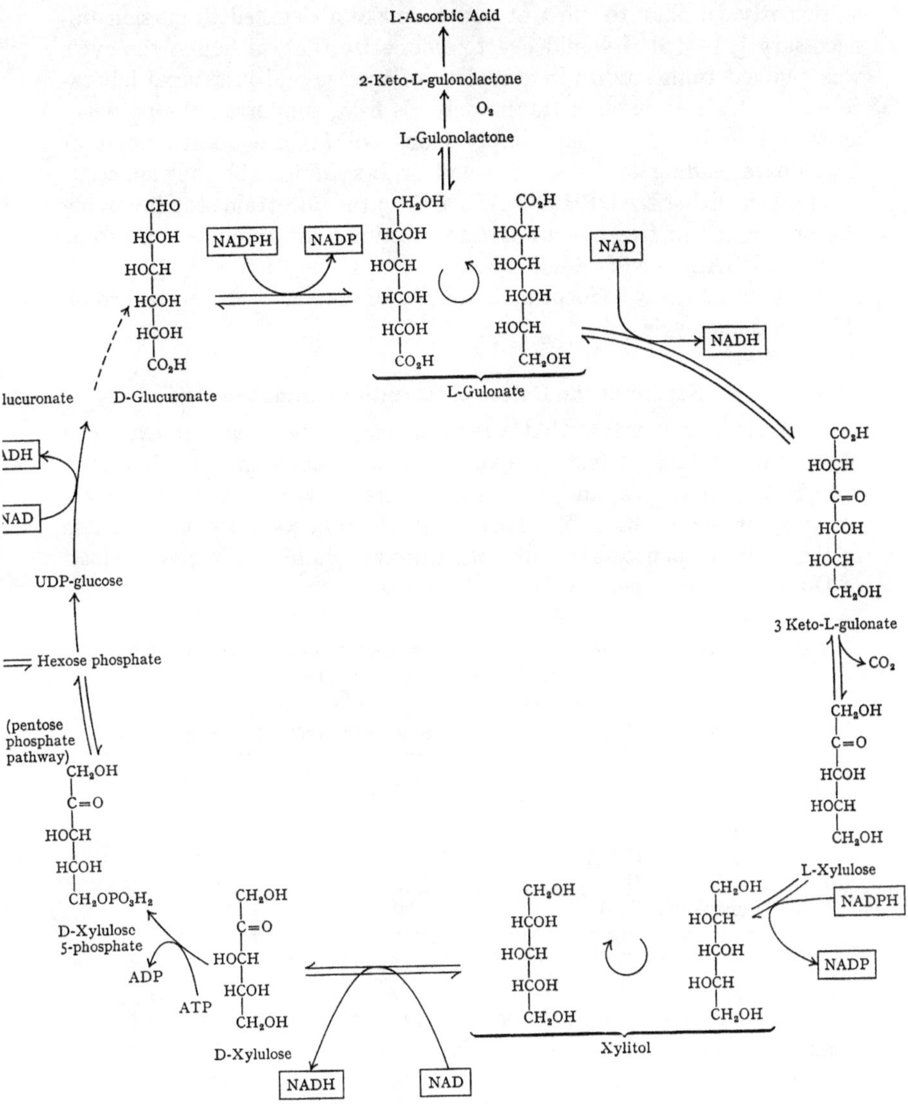

Fig. 2. The glucuronate-xylulose cycle

a long known soluble polyol dehydrogenase of the cytoplasm [10, 11]. With the report of the presence of D-xylulokinase in calf liver by Hickman and Ashwell [12], a meeting of the new pathway with the pentose phosphate pathway was effected. The new pathway, including some steps subsequently elucidated, is shown in Fig. 2. The reactions are sufficiently familiar to most of you to make a detailed discussion unnecessary [13—16]. I would like to emphasize that, although the cycle was evolved from studies in a variety of species and in several laboratories, all tests employing tracer methods have supported the pathway as shown in Fig. 2. I shall simply point out (a) the branch point at L-gulonate leading to L-ascorbic acid or L-xylulose, (b) the numerous reactions utilizing NADPH or NAD, and (c) the uncertain step involving the production of free glucuronate from uridine diphosphate glucuronic acid (UDPGA).

I should like now to focus on a number of specific problems regarding this metabolic cycle.

Nature of the Defect in Essential Pentosuria

I should first mention that a renal defect, rather than the existence of a true metabolic defect, was ruled out by studies in our laboratory with R. C. Bozian [12] and by studies reported by Freedberg *et al.* [18]. Our results, shown in Table II, demonstrate that xylulose accumulates in the blood of pentosuric, but not of normal, individuals given a load of D-glucuronolactone.

Table II. *Serum xylulose levels in normal and pentosuric subjects before and after 5 gm glucuronolactone taken orally* [12]

Subject	Serum xylulose (mg-%) [a]		
	fasting	1 hr	2 hr
Normal (M.K.P.)	<0.3	<0.3	<0.3
Normal (R.C.P.)	<0.3	<0.3	<0.3
Pentosuric (I. B.)	<0.3	9	8
Pentosuric (I.B.)	2	7	11

[a] Analysis by paper chromatography after deionization.

The fact that L-xylulose is excreted in gram quantities by pentosuric humans points to a deficiency of the NADP-linked xylitol dehydrogenase. However, it occurred to us that since reducing tests had always been used for examining pentosuric urine samples, and since the L-xylulose-xylitol reaction is reversible, it was conceivable that the enzymatic defect resided in the deficient utilization of the non-reducing substance xylitol, and that xylitol might be the principal unused metabolite and L-xylulose a secondary product. We therefore examined pentosuric urine for the presence of pentitols, utilizing chromatographic methods and periodate-reactivity as the basis for analysis. A considerable amount of a pentitol was isolated, but it was shown to be L-arabitol rather than xylitol [19] (Fig. 3). The finding was surprising not only because it was the wrong pentitol, but also because L-arabitol had not previously been found in nature. Since the relatively unspecific NAD-linked polyol dehydrogenases appeared to have a slow but definite action on L-xylulose, perhaps forming L-arabitol, as well as on L-arabitol itself, we considered that this pentitol was merely a detoxication product of the

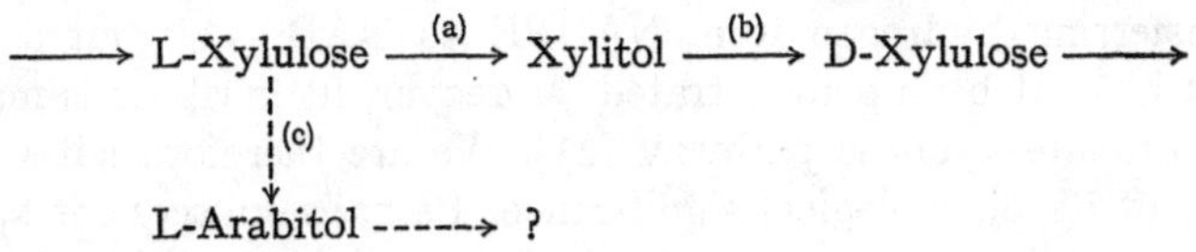

Fig. 3. The formation of L-arabitol as well as xylitol from L-xylulose

accumulated L-xylulose. The urinary L-arabitol is in fact labeled when a pentosuric is given 1-^{13}C-glucuronolactone [19].

Hiatt [20] has presented isotopic evidence consistent with the assumption that a block exists at the L-xylulose step in pentosuric individuals (Table III). It will be noted that the label in the adminis-

Table III. *Urinary L-xylulose and ribose in normal and pentosuric patients given imidazoleacetic acid (IAA) and D-glucuronolactone uniformly labeled with ^{14}C* [20]

	Normal	Pentosuric
Urinary L-Xu (0—10 hr) (μmoles)	0.5	3,950
% administered ^{14}C	0.002[a]	4.3
Ribose from urinary IAA riboside (0—10 hr) (μmoles)	275	208
% administered ^{14}C	0.058	0.002[a]

[a] Significance doubtful because of small quantity of radioactivity.

tered glucuronolactone did not find its way to ribose in the pentosuric, presumably because its progress through the pathway was blocked at L-xylulose. Although there have been no reports of the demonstration of the absence of the L-xylulose reducing enzyme in pentosuric liver, there seems to be little doubt that this is the site of the metabolic defect. You will recall that unlike other classical inborn errors of metabolism, essential pentosuria seems to be a harmless defect [15], and afflicted individuals are seldom in a situation when a sizeable piece of liver can be obtained for analysis. It is possible, from the work of Asakura *et al.* [21], that analysis of erythrocytes of normal and pentosuric humans will be useful.

Role of the Cycle

Since the cycle appears to be widespread in mammals, and may even occur in the crustacean [22], it may be surmised that it is fulfilling a useful function in addition to its role in ascorbic acid biosynthesis. However, since pentosuria is not a disease, it would seem as if the useful function is not crucial to the well-being of the pentosuric individual. We have suggested that the cycle might provide an effective means for transferring hydrogen from NADPH to NAD [13], but a role of this sort has not been demonstrated. Moreover, little ribose is made via the glucuronate-xylulose pathway [23]. We are therefore left with the question of its physiological significance. Perhaps subsequent speakers at this symposium will be able to shed some light on this point.

There are other basic questions about the accepted pathway that I now wish to bring to your attention.

How is Free Glucuronate Formed?

The existence of nucleotide pyrophosphatase activity in liver and kidney, which can effect the cleavage of UDPGA to AMP and glucuronic acid 1-phosphate, together with the assumed presence of phosphatases capable of splitting this phosphate to glucuronic acid and inorganic phosphate, initially led to the assumption [24] that glucuronic acid 1-phosphate is an intermediate between UDPGA and glucuronic acid (Fig. 4). However, Conney and Burns [25] reported the absence of glucuronic acid 1-phosphatase activity in liver homogenates, and Hollmann and I [26] interpreted our studies on enzyme changes induced by drugs that stimulate ascorbic acid biosynthesis as possibly suggestive of a pathway to glucuronic acid through the successive actions of glucuronyl transferase and β-glucuronidase (lower route in Fig. 4). The inducing agents fell into two classes, those elevating UDPG dehydrogenase activity and those elevating glucuronyl transferase. UDPGA pyrophosphatase did not increase when ascorbic acid excretion was enhanced. Nonetheless, it may be said that neither route from UDPGA to glucuronate has very convincing support. β-Glucuronidase is not much affected by the enhancing agents, and Gunn rats, which are unable to raise their genetically depressed glucuronyl transferase levels, respond to the agents with increased ascorbic acid excretion (unpublished work in this laboratory). I should like to mention new evidence that is inconsistent with the glucuronic acid 1-phosphate route.

Schliselfeld [27] in our laboratory obtained preliminary evidence that the weakly alkaline conditions used by Conney and Burns [18] precluded their finding the glucuronic acid 1-phosphatase activity that does indeed exist in liver. At pH 7.01, some disappearance of glucuronic

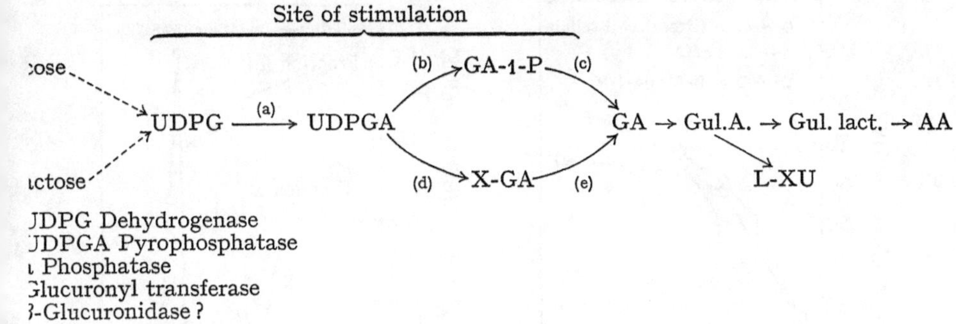

Fig. 4. Postulated pathways for the biosynthesis of L-ascorbic acid in mammals

Table IV. *The hydrolysis of glucuronic acid 1-phosphate by rat liver homogenate at pH 7.01* [27]

Reaction time (minutes)	Alkali-stable uronate in reaction (μmoles)	Total uronate in reaction (μmoles)	Alkali-stable uronate in absence of homogenate (μmoles)	Inorganic phosphate formed in reaction (μmoles)
1.00	11.9	—	12.7	−0.62
60.00	9.3	—	12.9	−0.51
239.00	8.7	—	13.2	−3.62
428.00	7.0	8.5	13.0	0.30

Table V. *The hydrolysis of glucuronic acid 1-phosphate by rat liver homogenate at pH 6.30* [27]

Reaction time (minutes)	Alkali-stable uronate in reaction (μmoles)	Total uronate in reaction (μmoles)	Alkali-stable uronate in absence of homogenate (μmoles)	Inorganic phosphate formed in reaction (μmoles)
1.00	11.6	—	13.2	0.22
60.00	10.1	—	13.2	0.00
239.00	7.0	—	13.2	4.2
428.00	5.5	11.4	13.8	5.4

1-phosphate occurred in a rat liver homogenate (Table IV), and a greater loss occurred at pH 6.3 (Table V). In this work it was necessary to measure glucuronic acid 1-phosphate disappearance, which we did indirectly, by estimation of carbazole-reacting material remaining after heating aliquots with alkali [27], since inorganic phosphate formation was not a reliable index of the extent of reaction when whole homogenates were used. These results were subsequently confirmed in our

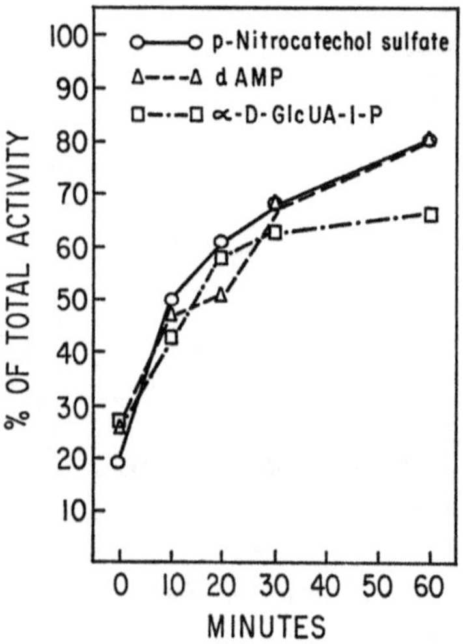

Fig. 5

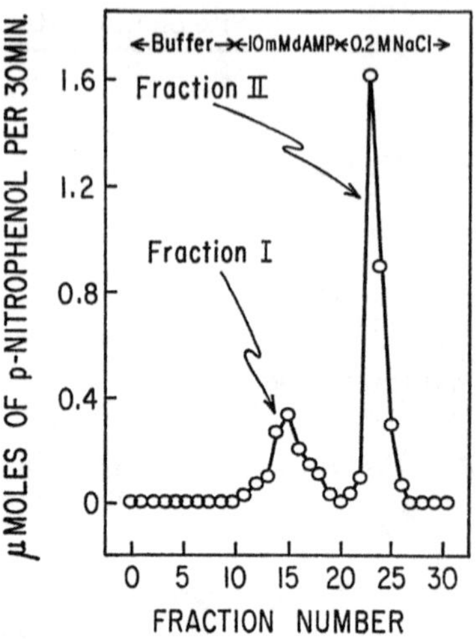

Fig. 6

Fig. 5. Release of phosphatase and aryl sulfatase activities from rat liver lysosomes, in 0.25 M sucrose containing 0.1 M sodium malonate buffer, pH 7.0, as a function of repliminary incubation time at 37° [30]

Fig. 6. Chromatography of the 50—60% ammonium sulfate fraction on DEAE-cellulose equilibrated with 10 mM Tris HCl buffer, pH 7.4 [30]

laboratory by Dr. Hollmann, using an enzymatic method [28] for estimating the free glucuronate formed[1]. We then found that the optimum pH for hydrolysis occurred between 4 and 5 and later demonstrated that the enzymatic system responsible for the hydrolysis resided in the lysosomal fraction [29]. Of all the sugar phosphates tested, glucuronic acid 1-phosphate was cleaved most rapidly. In view of our interest in glucuronic acid formation, and in view of the fact that lysosomal phosphatase had not been studied directly, we decided to purify the phosphatase responsible for the hydrolysis of the uronic acid derivative [30].

Fig. 5 shows that the glucuronic acid 1-phosphatase is activated at the same rate as d-AMPase and aryl sulfatase when rat liver lysosomes are incubated at 37° C. These results are consistent with the lysosomal occurrence of the enzymatic activities being assayed. By ammonium sulfate precipitation followed by DEAE-cellulose chromatography, Dr. Charalampos Arsenis was able to resolve rat liver lysosomal phosphatases into a nucleotidase and a sugar phosphate phosphohydrolase of broad specificity [30]. It was necessary to use 10 mM d-AMP as a very specific agent to selectively elute the nucleotidase. Fig. 6 shows

1. We thank Dr. M. P. Starr for a culture of *Pseudomonas syringae*, from which we isolated uronate dehydrogenase, after the organism was grown on glucuronate.

Table VI. *Fractionation of phosphatase activity from rat liver lysosomes by ammonium sulfate and column chromatography on DEAE-cellulose* [30]

Fraction	Volume (ml)	Total protein (mg)	Total units (units)	Specific activity (units/mg)
Lysosomes treated with Triton X-100 (0.1% final concentration)	60	93.6	2,556	27.4
Ammonium sulfate precipitate (50—60%)	7	5.11	478.8	93.7
DEAE-cellulose				
Fractions 14 and 15	20	0.040	48.7	1,220
Fractions 23 and 24	20	0.200	201.6	1,001

Table VII. *Relative rate of hydrolysis of phosphate esters by phosphatase Fractions I and II from rat liver lysosomes* [30]

Substrate	Fraction I	Fraction II
5′-AMP	67	85
3′-AMP	85	83
2′-AMP	55	70
5′-GMP	40	50
5′-CMP	72	55
5′-UMP	45	50
5′-dAMP	100	100
5′-dGMP	74	76
5′-dCMP	74	60
5′-TMP	64	78
α-D-GlcUA-1-P	5	115
α-D-Gal-1-P	4	95
α-D-Glc-1-P	3	90
α-D-Man-1-P	3	74
α-D-Xyl-1-P	4	68
D-Fru-1-P	5	61
D-Glc-6-P	2	80
D-Fru-6-P	5	30
D-Man-6-P	1	75
D-Gal-6-P	1	53
D-Rib-5-P	3	78
β-Glycerol-P	18	73
p-Nitrophenyl phosphate	72	200

the pattern of elution, with *p*-nitrophenyl phosphate as substrate in assaying the phosphatase activity in each tube. Fraction I is a nucleotidase having a pH optimum of 5.0, while Fraction II is a general sugar phosphate phosphohydrolase with a pH optimum of 4. A table of purification is shown in Table VI. The substrate specificity of each fraction is shown in Table VII. It is very interesting that glucuronic acid

1-phosphate is the best sugar phosphate substrate for the second enzyme. However, it seems likely that this is pure coincidence without metabolic significance. Although these studies have contributed to the area of lysosome enzymology, they do not yield much support to the idea that glucuronic acid 1-phosphate is a metabolic intermediate. We have not as yet found a moderately specific liver enzyme for this substance. Moreover, lysosomes are generally considered to have mainly degradative functions. It should also be mentioned that Takanami *et al.* [31] have shown that, at high pH, the alkaline phosphatase of liver cleaves glucuronic acid 1-phosphate.

1. This enzyme has been reported to occur in
   rat liver nuclei,
   rat liver microsomes.
2. Similarity of nuclear and microsomal enzyme reported [33].
3. Microsomal enzyme attributed to plasma membranes in microsomal preparations [34].
4. Preparation of nuclei in 2.1 M sucrose showed that nuclei contain <10% of nucleotide pyrophosphatase activity in whole homogenate. The same holds for phosphodiesterase I, which had also been reported to occur in the nuclear and microsomal fractions prepared by the Schneider-Hogeboom procedure[2].
5. 60—80% of liver nucleotide pyrophosphatase activity was found in a fraction, prepared by discontinuous sucrose gradient centrifugation, with the properties of plasma membranes[2].

Fig. 7. UDPGA pyrophosphatase (nucleotide pyrophosphatase) of rat liver

Another line of research leads to doubt regarding a role of UDPGA pyrophosphatase. This enzyme was originally reported to occur in rat kidney microsomes by Ginsberg [32]. Nucleotide pyrophosphatases have been found in various tissues, and in some studies the enzyme was localized in either the nuclear or microsomal fractions or both (Fig. 7). In our own report [33], we pointed out the similarity between the nuclear and microsomal nucleotide pyrophosphatases of rat liver. Emmelot and his coworkers [34] later showed that the microsomal fraction possessed this activity primarily because of contamination with plasma membranes. At about the same time, my student, Mr. Herman Hendrickson, showed that properly prepared nuclei have little activity, and he traced almost all of the enzyme to a fraction that, on the basis of results from assay

2. Experiments of H. Hendrickson in the author's laboratory.

of specific marker enzymes, appears to be derived from the plasma membrane of the liver cells. He has developed a highly efficient procedure for preparing this membrane fraction. By the procedure given in Fig. 8, he was able to demonstrate that nucleotide pyrophosphatase, as well as the accepted plasma membrane enzyme 5′-nucleotidase [34], accompany each other during fractionation steps. It must be mentioned that the only basis for assuming that our fraction does not consist of the Golgi apparatus is that thiamine pyrophosphatase [35] does not show a discrete peak in the sucrose density centrifugation.

---

1. Nuclei sedimented in 2.1 M sucrose (pH 8.0 with 0.005 M Tris-HCl). (Other particles float on sucrose.)
2. Nuclei-free preparation centrifuged in discontinuous sucrose gradient (ϱ 1.03 to 1.18) buffered with 0.005 M Tris-HCl at pH 8.0.
3. Fractions analyzed for marker enzymes for various cellular organelles. Nucleotide pyrophosphatase occurs at ϱ 1.135 with most of the 5′-nucleotidase and phosphodiesterase I, but little (<10%) of the glucose 6-phosphatase of the original homogenate. Mitochondrial, lysosomal, and microsomal marker enzymes are concentrated at higher sucrose densities. Thiamine pyrophosphatase, used as Golgi apparatus marker, was spread through gradient. Sialic acid, lipid analyses, and electron micrographs consistent with plasma membrane origin of 1.135 fraction.
4. Solubilization of nucleotide pyrophosphatase with deoxycholate followed by filtration through Sephadex G-75 and then Sepharose 4B, as well as substrate competition studies and inactivation rate studies, suggest that nucleotide pyrophosphatase and phosphodiesterase I may be identical.

Fig. 8. Preparation of rat liver membrane fraction containing nucleotide pyrophosphatase[3]

I have been using the term nucleotide pyrophosphatase in place of UDPGA pyrophosphatase because there has never been any evidence for the existence of a specific enzyme for the cleavage of UDPGA into UMP and glucuronic acid 1-phosphate in rat liver. Moreover, we now have direct evidence that phosphodiesterase I, which was reported by deLamirande *et al.* [36] to be concentrated in the nuclear and microsomal fractions of rat liver prepared by the Hogeboom and Schneider method (Fig. 9), is also almost exclusively a plasma membrane enzyme. In addition, we have suggestive preliminary evidence that phosphodiesterase I and nucleotide pyrophosphatase may be identical. Further work remains to be done on this question.

3. H. Hendrickson and O. Touster.

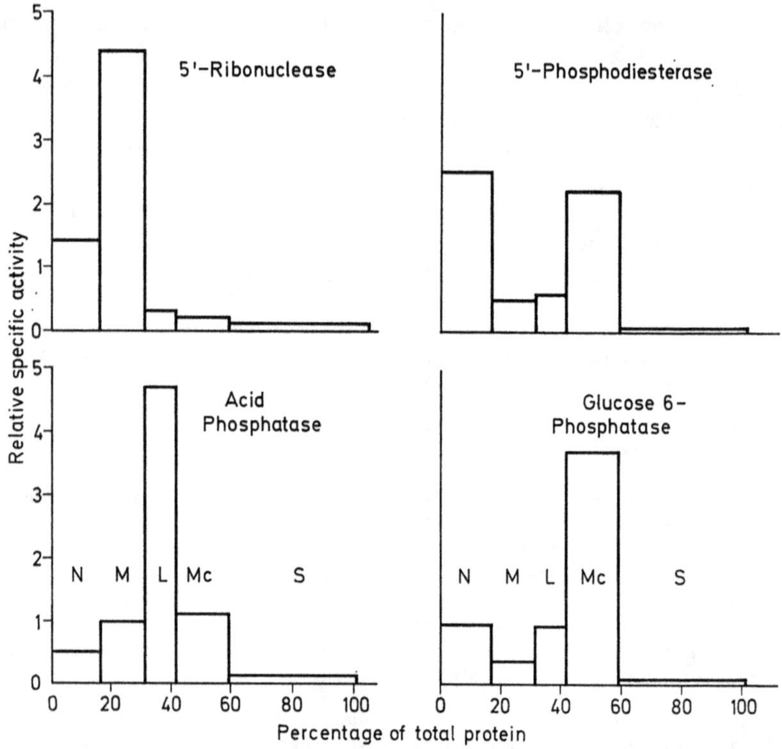

Fig. 9. Intracellular distribution of phosphodiesterase I and three other enzymes as reported by deLamirande *et al.* [36]

Intracellular Localization of the Xylitol Dehydrogenases in Guinea Pig Liver

The glucuronate-xylulose cycle includes two successive reactions catalyzing the conversion of L-xylulose to D-xylulose via xylitol. Although a soluble NAD-linked polyol dehydrogenase is present in liver, it has seemed likely that the interconversion is effected by the mitochondrial NAD-linked dehydrogenase together with the mitochondrial NADP-linked enzyme. The particulate nature of these two enzymes has recently been substantiated in a histochemical study reported by Stiller and Gorski [37]. For the enzymes to participate in a metabolic pathway in which both prior and subsequent steps are carried out in non-mitochondrial compartments of the cell, it appeared to us that the enzymes might need to be localized in or near the outer mitochondrial membrane.

To investigate this point, as well as to disclose a possible role in hydrogen transfer into mitochondria, the procedure of Sottocasa *et al.*

[38] has been employed by Mr. Thomas Maniatis to prepare sub-mitochondrial fractions. Both xylitol dehydrogenases were solubilized by the swelling-shrinking-sonication procedure, a result suggesting a very loose attachment to membranes or unusually facile leakage from a soluble compartment or from the matrix. In Table VIII it is seen that the swelling-shrinking-sonication procedure liberates in soluble form a large portion of malic dehydrogenase and all of the xylitol dehydrogenase activity, while monamine oxidase can still be largely localized in the outer membrane fraction. From subsequent experiments it appears that NAD-xylitol dehydrogenase and adenylate kinase are almost completely released in soluble form by swelling and shrinking alone, whereas the NADP-xylitol dehydrogenase requires sonication for full release.

This approach should yield information which may permit an appraisal of the function of the xylitol dehydrogenases in the glucuronate-xylulose pathway and in mitochondrial metabolism.

Table VIII. *Release of mitochondrial enzymes by swelling, shrinking, and sonication by method of Sottocasa et al.* [38][a]

	Soluble (supernatant)		Outer membrane (interface)		Core (pellet)	
	expt. 1	expt. 2	expt. 1	expt. 2	expt. 1	expt. 2
MDH	280,000	61,800	53,800	19,200	22,000	70,000
MAO	351	254	300	430	1	544
NAD-XDH	3,276	3,670	0	0	0	0
NADP-XDH	1,872	1,114	0	0	0	0

[a] Experiments of T. Maniatis in the author's laboratory.

Abbreviations: MDH, malic dehydrogenase; MAO, monoamine oxidase; NAD-XDH, NAD-linked xylitol dehydrogenase; NADP-XDH, NADP-linked xylitol dehydrogenase.

Conclusion

I can best summarize my talk with the diagram in Fig. 10. As drawn, the glucuronate-xylulose pathway includes enzymes which occur in a variety of intracellular membranes and organelles, and some of the enzymes have very broad specificity. This situation leads me to wonder whether the cycle is correct as currently accepted, whether some reactions are undiscovered, or whether we have sufficient understanding of the metabolic interrelationships among the various compartments within cells.

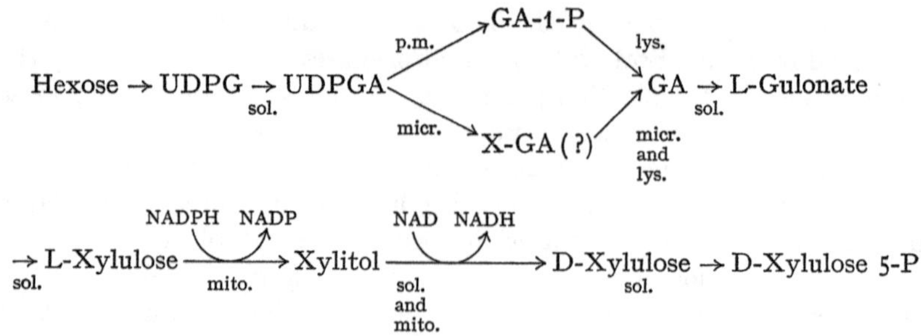

(sol. = soluble; mito. = mitochondria; lys. = lysosomes; micr. = microsomes; p.m. = plasma membrane).

Fig. 10. Intracellular localization of reactions presumably involved in the glucuronate-xylulose pathway

One problem regarding the intracellular distribution of enzymes of the glucuronate-xylulose cycle has been eliminated by the recent demonstration of a high level of NADP-linked xylitol (L-xylulose) dehydrogenase activity in guinea pig liver cytosol [Arsenis, C., and Touster, O., J. Biol. Chem. **244**, 3895 (1969)].

References

1. Burns, J. J.: Biosynthesis of L-ascorbic acid; basic defect in scurvy. Amer. J. Med. **26**, 740 (1959).
2. Touster, O., R. M. Hutcheson, and V. H. Reynolds: The formation of L-xylulose in mammals and its utilization by liver preparations. J. Amer. chem. Soc. **76**, 5005 (1954).
3. — —, and L. Rice: The influence of D-glucuronolactone on the excretion of L-xylulose by humans and guinea pigs. J. biol. Chem. **215**, 677 (1955).
4. Enklewitz, M., and M. Lasker: The origin of L-xyloketose (urine pentose). J. biol. Chem. **110**, 443 (1935).
5. Touster, O., R. H. Mayberry, and D. B. McCormick: The conversion of 1-^{13}C-D-glucuronolactone to 5-^{13}C-L-xylulose in a pentosuric human. Biochim. biophys. Acta (Amst.) **25**, 196 (1957).
6. — V. H. Reynolds, and R. M. Hutcheson: The reduction of L-xylulose to xylitol by guinea pig liver mitochondria. J. biol. Chem. **221**, 697 (1956).
7. McCormick, D. B., and O. Touster: The conversion *in vivo* of xylitol to glycogen via the pentose phosphate pathway. J. biol. Chem. **229**, 451 (1957).
8. — — The conversion of D-arabitol-1-^{14}C, L-arabitol-1-^{14}C, and D-ribitol-1-^{14}C to liver glycogen in the rat and guinea pig. Biochim. biophys. Acta (Amst.) **54**, 598 (1961).
9. Hollmann, S., and O. Touster: An enzymatic pathway from L-xylulose to D-xylulose. J. Amer. chem. Soc. **78**, 3544 (1956).
10. — — The L-xylulose-xylitol enzyme and other polyol dehydrogenases of guinea pig liver mitochondria. J. biol. Chem. **225**, 87 (1957).

11. — Trennung, Reinigung und Eigenschaften der mitochondrialen Xylit-Dehydrogenasen der Meerschweinchenleber. Hoppe-Seylers Z. physiol. Chem. **317**, 193 (1959).

12. Hickman, J., and G. Ashwell: Purification and properties of D-xylulokinase in liver. J. biol. Chem. **232**, 737 (1958).

13. Touster, O.: Essential pentosuria and the glucuronate-xylulose pathway. Fed. Proc. **19**, 977 (1960).

14. Burns, J. J., A. H. Conney, P. G. Dayton, C. Evans, G. R. Martin, and D. Taller: Observations on the drug-induced synthesis of D-glucuronic, L-gulonic and L-ascorbic acids in rats. J. Pharmacol. exp. Ther. **129**, 132 (1960).

15. Touster, O.: Pentose metabolism and pentosuria. Amer. J. Med. **26**, 724 (1959).

16. Burns, J. J., and A. H. Conney: Metabolism of glucuronic acid and its lactone. In: Glucuronic acid (G. J. Dutton, ed.), p. 365. New York: Academic Press 1966.

17. Bozian, R. C., and O. Touster: Essential pentosuria: renal or enzymic disorder. Nature (Lond.) **184**, 463 (1959).

18. Freedberg, I. M., D. S. Feingold, and H. H. Hiatt: Serum and urine L-xylulose in pentosuric and normal subjects and in individuals with pentosuria trait. Biochem. biophys. Res. Commun. **1**, 328 (1959).

19. Touster, O., and S. O. Harwell: The isolation of L-arabitol from pentosuric urine. J. biol. Chem. **230**, 1031 (1958).

20. Hiatt, H. H.: Studies of ribose metabolism. IV. The metabolism of D-glucuronolactone in normal and pentosuric subjects. Biochim. biophys. Acta (Amst.) **28**, 645 (1958).

21. Asakura, T., K. Adachi, S. Minakami, and H. Yoshikawa: Non-glycolytic sugar metabolism in human erythrocytes. I. Xylitol metabolism. J. Biochem. (Japan) **62**, 184 (1967).

22. Puryear, R. L.: The glucuronic acid pathway in the blue crab, *Callinectes sapidus* rathbun: The enzymes of the UDPglucose to glucuronic acid portion of the pathway. Comp. Biochem. Physiol. **20**, 499 (1967).

23. Hiatt, H. H., and J. Lareau: Studies of ribose metabolism. VIII. An assessment of ribose biosynthesis from hexose by way of the C-6 oxidation pathway. J. biol. Chem. **233**, 1023 (1958).

24. Evans, C., A. H. Conney, N. Trousof, and J. J. Burns: Metabolism of D-galactose to D-glucuronic acid, L-gulonic acid and L-ascorbic acid in normal and barbital treated rats. Biochim. biophys. Acta (Amst.) **41**, 9 (1960).

25. Conney, A. H., and J. J. Burns: Metabolism of uridine diphosphate glucuronic acid by liver and kidney. Biochim. biophys. Acta (Amst.) **54**, 369 (1961).

26. Hollmann, S., and O. Touster: Alterations in tissue levels of UDPG dehydrogenase, UDPGA pyrophosphatase, and glucuronyl transferase induced by substances influencing the production of ascorbic acid. Biochim. biophys. Acta (Amst.) **62**, 338 (1962).

27. Schliselfeld, L. H.: A nuclear nucleotide pyrophosphatase and its relation to glucuronic acid biosynthesis in rat liver. Ph. D. diss., Vanderbilt University, 1964.

28. Kilgore, W. W., and M. P. Starr: Uronate oxidation by phytopathogenic *Pseudomonas*. Nature (Lond.) **183**, 1412 (1959).

29. Arsenis, C., S. Hollmann, and O. Touster: Acid phosphatases from rat liver lysosomal fraction with high activity towards α-D-glucuronic acid-1-phosphate. Abstracts of the American Chemical Society Meeting, New York, September 1966, C-286.
30. —, and O. Touster: The partial resolution of acid phosphatase of rat liver lysosomes into a nucleotidase and a sugar phosphate phosphohydrolase. J. biol. Chem. **242**, 3399 (1967).
31. Takanashi, S., K. Ihda, and M. Kawada: Studies on enzymatic hydrolysis of 1-phosphoglucuronate. J. Biochem. (Tokyo) **59**, 78 (1966).
32. Ginsburg, V., A. Weissbach, and E. S. Maxwell: Formation of glucuronic acid from uridinediphosphate glucuronic acid. Biochim. biophys. Acta (Amst.) **28**, 649 (1958).
33. Schliselfeld, L. H., J. van Eys, and O. Touster: The purification and properties of a nucleotide pyrophosphatase of rat liver nuclei. J. biol. Chem. **240**, 811 (1965).
34. Emmelot, P., C. J. Bos, E. L. Benedetti, and P. H. Rumke: Studies on plasma membranes. I. Chemical composition and enzyme content of plasma membranes isolated from rat livers. Biochim. biophys. Acta (Amst.) **90**, 126 (1964).
35. Novikoff, A. B., and S. Goldfischer: Nucleoside diphosphatase activity in the Golgi apparatus and its usefulness for cytological studies. Proc. nat. Acad. Sci. (Wash.) **47**, 802 (1961).
36. DeLamirande, G., R. Morris, and M. Blackstein: Intracellular distribution of 5′-ribonuclease and 5′-phosphodiesterase in rat liver. Arch. Biochem. **118**, 347 (1967).
37. Stiller, D., u. J. Gorski: Histochemische Untersuchungen zum Nachweis der Xylitdehydrogenasen. Histochemie **5**, 407 (1965).
38. Sottocasa, G. L., B. Kuylenstierna, L. Ernster, and A. Bergstrand: An electron-transport system associated with the outer membrane of liver mitochondria. J. Cell Biol. **32**, 415 (1967).

Discussion

Dr. Horecker: It seems that you have implicated every part of the cell, except the nucleous, in your pathway. I wonder if the two steps between the particles and mitochondria are really soluble. It would make sense if L-xylulose were formed in the mitochondria, and the second dehydrogenase, involving the release of D-xylulose were soluble.

Dr. Touster: That part of the pathway was taken from the literature, which reports how Ashwell and the Japanese workers prepared their enzymes. They had centrifuged for 8, 10, or 15,000 × g and had used the supernatant. It is possible that they had only 20% or so of the enzyme actually in the liver and that it was leaking out of the mitochondria. So that I think it might be a good idea to go back and repeat some of these earlier experiments.

Dr. Yoshikawa: We know that the main tissues which have the Glucuronic Acid Pathway are liver, kidney and so on. Is there any activity of the Glucuronic Acid Pathway in muscle?

Dr. Touster: I think someone has looked at muscle. I'm not sure whether it was Hiatt or Dr. Hollmann.

Dr. Hollmann: I would like to tell you that the TPNH enzyme was tested in muscle. This tissue has very low sorbitol dehydrogenase activity.

Dr. Yoshikawa: I am now trying the effects of xylitol on the enzymes of muscle. We found some effect of xylitol on a malic enzyme.

Dr. Touster: In think you have two problems. There are many people here interested in giving xylitol, and we may ask what will happen to the compound? The liver is rich in the soluble, non-specific polyol dehydrogenase and other tissues may also have this enzyme. In that case, there may not be a problem. But if a precursor of xylitol is added, one must know whether the TPN-dehydrogenase is present.

Dr. Horecker: I have the impression from your discussion that you do not believe the action of phosphatase activity on glucuronic acid 1-phosphate to be sufficient to account for the over-all physiological metabolism of the Uronic Acid Pathway *in vivo*.

Dr. Touster: I didn'd mean to imply that. Are you referring to the lysosomal activation experiment?

Dr. Horecker: Yes.

Dr. Touster: That's a special type of experiment in sucrose at 37° to demonstrate that the preparation is lysosomal. There is much more alkaline phosphatase than is apparent from this activity.

Dr. Hollmann: We calculate that the activity is enough to account for even increased synthesis of ascorbic acid.

Dr. Horecker: That answers my question.

Dr. Touster: My skepticism has to do with the lysosome concept. There's a paper by Wilgram and Kennedy that has not been frequently quoted [J. biol. Chem. **238**, 2615 (1963)] which reports that phosphatidic acid phosphatase is most highly concentrated in the lysosomes, although there's some in neighboring fractions. If this enzyme is really phosphatidic acid phosphatase, it would mean that a main enzyme of a biosynthetic pathway is in the lysosomes, and would violate the principle of de Duve that these contain only catabolic reactions. But I believe that he would agree that all lysosomes may not be identical. We don't even know that each lysosome has a full complement of lysosomal enzymes.

In addition there are the peroxosomes. Thus the role of the lysosomal enzyme remains an open question; it is still disappointing when one has a pathway and is unable to find a reasonable enzyme, active at a physiological pH, which can do the job.

Dr. HORECKER: What about phosphoglucomutase?

Dr. TOUSTER: I don't know. You can move the phosphate anywhere from the 1-position in glucuronate.

Dr. POGELL: On the same subject, have you or Dr. Hollmann ever used the whole homogenate with UDP glucuronic acid as substrate and measured glucuronate appearance at physiological pH?

Dr. TOUSTER: I have been thinking about this experiment, for obvious reasons. I think we tried this some time ago but I can't remember the details. There is an enzyme which Luis Glaser has found in bacteria which makes free sugar out of sugar nucleotides.

Pentitol-Metabolizing Enzymes of the Uronic Acid Pathway

SIEGFRIED HOLLMANN

Department of Physiological Chemistry,
University of Düsseldorf, Düsseldorf, West Germany

According to the sequence of reactions of the uronic acid pathway shown in the preceding chapter, two pentitol dehydrogenases (Fig. 1) linking L- and D-xylulose via xylitol are operative in this third pathway of carbohydrate metabolism in mammals, namely the NAD- and the

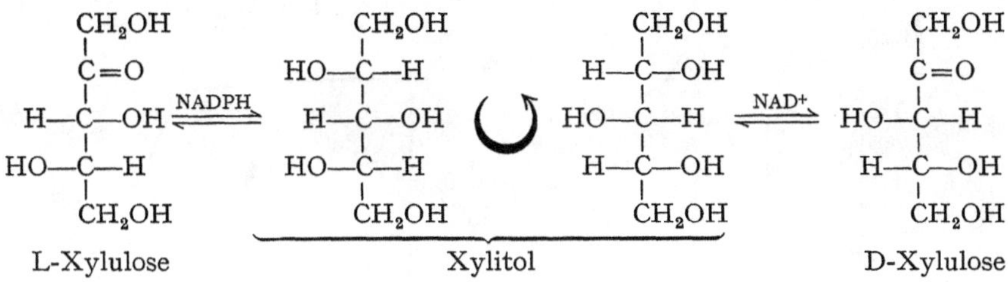

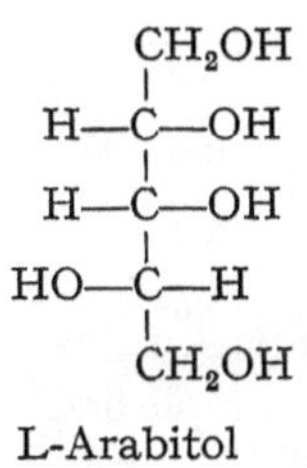

Fig. 1. Pentitol-pentulose interactions in the uronic acid pathway

NADP-specific xylitol dehydrogenase, systematically named xylitol: NAD oxidoreductase (D-xylulose-forming) (EC 1.1.1.9) and xylitol: NADP oxidoreductase (L-xylulose-forming) (EC 1.1.1.10), respectively. Mention has also been made already of the fact that pentosuric individuals, after administration of D-glucuronolactone-1-^{13}C, excrete labeled L-arabitol in their urine [1]. This observation may be taken as suggestive evidence for the existence in mammals of a third pentitol oxidizing enzyme interconverting L-arabitol and L-xylulose. There is no indication whatever for the occurrence in mammals of an enzyme possibly linking D-arabitol to D-xylulose. Moreover, D-arabitol has been shown to be a very poor precursor of glycogen in rats and guinea pigs [2]. Therefore, only the three systems mentioned above have to be discussed in the following sections.

Xylitol Dehydrogenases

Discovery of the two xylitol dehydrogenases was rendered possible by the findings of Touster *et al.* [3] that guinea pig liver mitochondria catalyze the reduction of L-xylulose to xylitol and that this enzymic

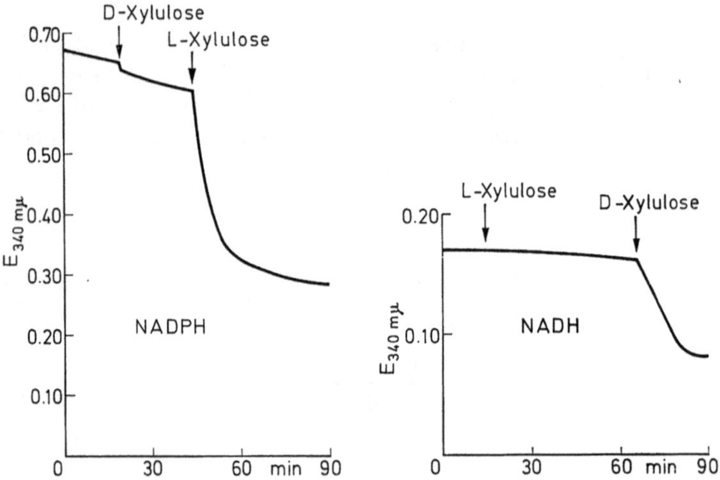

Fig. 2. Reduction of D- and L-xylulose by extracts from guinea pig liver mitochondrial residue in the presence of NADH and NADPH respectively

ability is entirely bound to the particulate fraction of osmotically disrupted mitochondria. Solubilization of the enzyme was first accomplished by treatment of the mitochondrial residue with butanol [4]. Later, liver acetone powder has been used as starting material for preparation of soluble extracts [5]. The latter procedure is preferable for two reasons. The preparation is more stable and the yield is 30 to 50 times higher than that obtained by butanol treatment [6].

The crude extract from mitochondrial residue, when incubated with xylitol, reduces either NAD^+ or $NADP^+$, but the latter more rapidly under identical conditions. When this reaction was studied in the reverse direction, we found [4] that, in the presence of NADPH, L-xylulose is reduced by the extract (Fig. 2), whereas in the presence of NADH L-xylulose is inert, but D-xylulose is reduced. This observation formed the basis for our assumption that liver mitochondria contain two xylitol dehydrogenases, differing in substrate specificity and pyridine nucleotide requirement, as initially depicted in Fig. 1. Subsequently, these two enzymes have been further characterized after a complete separation from each other had been achieved [5, 7].

Xylitol:NADP Oxidoreductase (L-Xylulose-Forming). Among the mammalian polyol dehydrogenases, the NADP-specific xylitol dehydrogenase is one with the highest substrate specificity. Of 19 alditols with 4 to 7 carbon atoms tested, only xylitol [4] and D-threitol [8] are oxidized

Table I. *Activity of xylitol: NADP oxidoreductase (L-xylulose-forming) in guinea pig tissues* [6]

Tissue	mU per g wet tissue	
	Xylitol: NADP oxidoreductase (L-Xylulose-forming) [a]	G-6-P-dehydrogenase
Liver		
male	806.1 ± 125.1	460 ± 30
female	703.3 ± 148.1	1,040 ± 120
Kidney	563.5 ± 88.6	690 ± 70
Lung	176.8 ± 56.8	880 ± 60
Spleen	151.7 ± 28.4	3,050 ± 310
Adrenal gland	198.6 ± 38.6	1,630 ± 250
Testis	39.2 ± 6.9	300 ± 30
Ovary	49.1 ± 18.5	300 (cow)
Brain	41.1 ± 11.4	320 ± 30
Cardiac muscle	11.7 ± 7.2	260 ± 130
Skeletal muscle	no activity detectable	80 ± 10

[a] Mean values from six animals.

Activity of glucose-6-phosphate dehydrogenase in rat tissues, as reported by Glock and McLean, is given in mU for comparison [9].

by the enzyme, forming L-xylulose and D-erythrulose respectively. The equilibrium constant for the xylitol-L-xylulose reaction is 2.97×10^{-11}, K_m for xylitol is reported to be 2.54×10^{-2} M [7], for L-xylulose 2.9×10^{-4} M [5]. 10^{-3} M p-chloromercuribenzoate inhibits the enzyme completely [5].

Since all studies so far reported were carried out on the enzyme derived from guinea pig liver mitochondria, a thorough survey of its tissue distribution seemed desirable. The guinea pig again was chosen for these measurements [6] because it resembles man in having a dietary ascorbic acid requirement. The results, summarized in Table I, show that the activity is highest in liver and kidney and equals that of glucose-6-phosphate dehydrogenase in these tissues, whereas in all other tissues its activity amounts to only 4.5—20% of the latter. The activity is extremely low in testis, ovary, brain, and heart muscle, no activity is detectable in skeletal muscle. Studies on the occurrence in the different orders of vertebrates of the enzyme have therefore been restricted to liver and kidney [6]. From Table II it is evident that the enzyme is present in these organs in all the vertebrates studied, without any discernible relationship between level of activity and phylogenic stage of development. In fishes and amphibia, the activity in liver and kidney is identical, in reptiles and birds, it is significantly higher in kidney,

Table II. *Activity of xylitol: NADP oxidoreductase (L-xylulose-forming) in liver and kidney of vertebrates* [6]

Species	mU per g wet weight[a]	
	liver	kidney
Carp	79.4 ± 14.1	74.0 ± 23.1
Eel	151.7 ± 52.4	155.3 ± 18.4
Pond-Salamander	148.3 ± 48.9	140.1 ± 24.4
Frog	294.2 ± 91.0	not determined
Ring-Snake	36.3 ± 12.7	66.2 ± 24.9
Tortoise	40.6 ± 31.4	104.6 ± 36.3
Cock	208.0 ± 93.4	773.9 ± 316.5
Pigeon	607.7 ± 77.1	1,294.4 ± 53.4
Dog	323.5 ± 101.6	332.6 ± 4.0
Cat	42.2 ± 11.7	98.0 ± 41.4
Rabbit	54.9 ± 20.5	51.1 ± 8.7
Rat	248.9 ± 59.3	360.8 ± 83.9
Guinea pig	806.1 ± 125.1	563.5 ± 88.6
Pig	436.1 ± 192.1	424.7 ± 113.1
Ox	347.3 ± 49.3	198.7 ± 26.4
Man	220.7; 156.5	not determined

[a] Mean values from six animals.

in mammals, no rule can be laid down in this respect. Noteworthy are the high activities in birds, particularly in pigeon kidney, and the low activities found in reptiles, rabbit, and cat. Cats also differ in some other respects in their glucuronic acid metabolism from the other mammalian species [10].

Xylitol: NAD Oxidoreductase (D-Xylulose-Forming). The NAD-specific xylitol dehydrogenase, on the other hand, exhibits a broad spectrum of activity [4, 7], as it follows from the summary in Table III. The list is almost complete, except for some likely heptitol substrates which were not available. Oxidizable polyols can be classified (Fig. 3) as belonging to either the D-xylo- or D-ribo-type. Nevertheless, the constant ratios of activity toward the polyol substrates observed during purification [7]

Table III. *Reactions catalyzed by xylitol:NAD oxidoreductase (D-xylulose-forming)*

L-Threitol	$\rightleftharpoons$	L-Erythrulose
Xylitol	$\rightleftharpoons$	D-Xylulose
Ribitol	$\rightleftharpoons$	D-Ribulose
Sorbitol	$\rightleftharpoons$	D-Fructose
L-Iditol	$\rightleftharpoons$	L-Sorbose
Allitol	$\rightleftharpoons$	D-Allulose
D-Altro-D-glucoheptitol	$\rightleftharpoons$	D-Altroheptulose

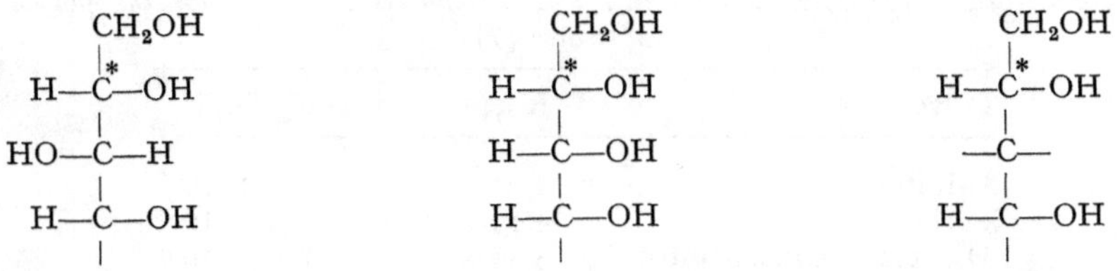

D-Xylo-type D-Ribo-type D-Erythro-1.2.4-configuration

Fig. 3. Structural requirement of substrates of xylitol:NAD oxidoreductase (D-xylulose-forming)

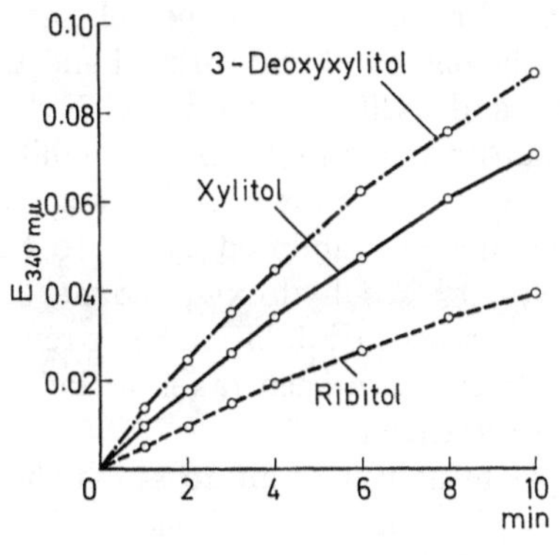

Fig. 4. Influence of C_3-substituent on the pentitol oxidation rate by xylitol:NAD oxidoreductase (D-xylulose-forming) [7]

point to a single enzyme responsible for all the before-mentioned transformations. Since the two types have the D-erythro-1.2.4-configuration in common, we tested whether a hydroxyl on C_3 is at all essential for a polyol to be acted upon by the enzyme [7]. The results show (Fig. 4) that the reaction rate decreases in the order 3-deoxyxylitol, xylitol, ribitol, that means that a C_3-hydroxyl is in fact inhibitory to the enzyme, by steric hindrance probably, a cis- more than a trans-hydroxyl. This observation led to the suggestion that the enzyme be named NAD-D-erythro-1.2.4-polyol dehydrogenase, a nomenclature fitting the experimental results as well as the three-point attachment hypothesis.

The K_m values for the polyol substrates and equilibrium constants for the polyol oxidations are listed in Table IV. K_m values of the D-ribo-type polyols are usually one to two orders of magnitude higher than

Table IV. *Thermodynamic equilibrium constants for the xylitol: NAD oxidoreductase (D-xylulose-forming) and Michaelis constants for the polyol substrates* [7]

Polyol	K	K_m (M)
L-Iditol	4.23×10^{-10}	4.18×10^{-3}
Sorbitol	4.19×10^{-10}	3.38×10^{-3}
D-Altro-D-Glucohepitol	5.54×10^{-11}	1.47×10^{-2}
Xylitol	4.18×10^{-11}	5.96×10^{-4}
Ribitol	1.75×10^{-11}	1.78×10^{-2}
Allitol	1.44×10^{-11}	3.70×10^{-2}
L-Threitol	6.91×10^{-12}	1.24×10^{-2}

those of the corresponding D-xylo-type polyols, as is evident from the values for allitol on the one hand and sorbitol and L-iditol on the other hand, and for ribitol and xylitol respectively. Xylitol, showing th lowest K_m, is the only substrate causing substrate inhibition at higher concentration. Its optimal concentration is in the range of $4-8 \times 10^{-3}$ M [7].

The enzyme is completely inhibited by 2×10^{-4} M p-chloromercuribenzoate and by 2.5×10^{-3} M 8-hydroxyquinoline. 2.5×10^{-3} M cysteine inhibits 34% on the average, the inhibition being reversed by 2.5×10^{-4} M zinc sulfate. These effects have been taken as indication for a possible metal dependence of the enzyme [7].

It seems appropriate at this point to stress that the NAD-specific xylitol dehydrogenase from liver mitochondria closely resembles many of the known non-particulate mammalian polyol dehydrogenases, in substrate specificity and in other properties as well, such as the sorbitol or L-iditol dehydrogenase from liver cytoplasm [11] and the so-called "ketose reductase" derived from male accessory organs [12, 13] and from spermatozoa [14]. The most extensive specificity studies have been performed on the L-iditol dehydrogenase [15] which is so far the only one obtained in crystalline form [16]. Distribution in human tissues of the "sorbitol dehydrogenase" has been studied [17]. In spite of many similarities, the difference in cellular location and differences in the relative rates of oxidation of the polyol substrates justify the assumption that the NAD-specific xylitol dehydrogenase and the soluble L-iditol dehydrogenase are not identical.

Metabolism of L-Arabitol in Mammals

The problem of transformation of L-arabitol in mammalian metabolism is quite a puzzling one and far from being solved. Only 13% of ^{14}C injected intraperitoneally into rats as 1-^{14}C-L-arabitol are accounted for by $^{14}CO_2$ within 24 hrs, 37% are excreted in urine in an unidentified form, and only 1—3% are incorporated into liver glycogen. The distribution of ^{14}C and the asymmetric pattern of glycogen labeling cannot be explained by known enzymic reactions and have been interpreted as indication for extensive chain cleavage of the pentitol after preliminary oxidation to a pentulose [2].

Reports on the oxidation of L-arabitol by mammalian polyol dehydrogenases are scarce and contradictory. Oxidation has constantly been observed in experiments using preparations from liver mitochondria [7], liver cytoplasm [18] or liver acetone powder [5], or even using the crystalline enzyme from sheep liver [16]. But, never has it been possible to separate the activity toward L-arabitol from that toward sorbitol, even after a several hundred fold purification. Since Williams-Ashman *et al.* [13] had shown that the "ketose reductase" from seminal vesicle, after partial purification, was no longer able to oxidize L-arabitol, we hoped to achieve a separation of the two postulated enzymes in extracts of that tissue.

Our purest preparation from sheep seminal vesicle [19] showed a 250-fold increase in its specific activity with sorbitol as substrate, but, as observed for the liver enzyme, no change in the sorbitol-L-arabitol ratio, the ratio remaining constant at 100/28 on an average. First doubts concerning the homogeneity of our preparation emerged, when small but always reproducible differences in thermal stability of the two activities were observed. L-arabitol activity is more susceptible to inactivation on either storage at —15°C or heating at 54°C than sorbitol activity. Looking for additional physicochemical or kinetic peculiarities of the two activities, we found that in competition experiments with the two polyols added together, the rate of NAD^+ reduction is always additive and that the oxidation rates for sorbitol and L-arabitol are maximal at a different pH (Fig. 5). L-arabitol oxidation is much more dependent on pH than sorbitol oxidation, particularly in the weakly basic region. It is almost undetectable at pH 7, where sorbitol oxidation is more than half maximal. It has to be concluded, therefore, that the two substrates are either bound at different sites of the enzyme or by

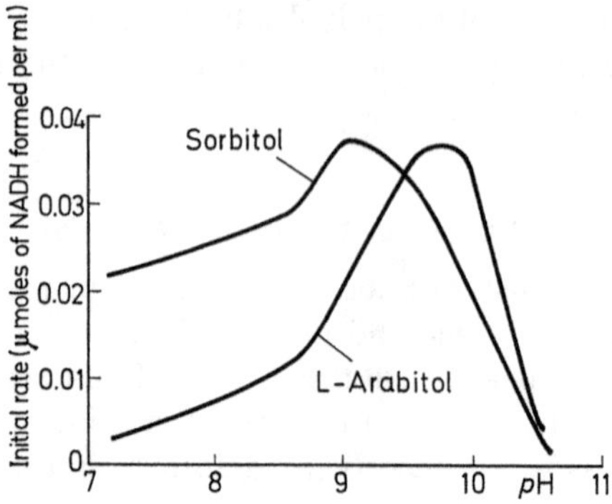

Fig. 5. Effect of pH on initial rate of sorbitol and L-arabitol oxidation by a purified preparation from sheep seminal vesicle [19]

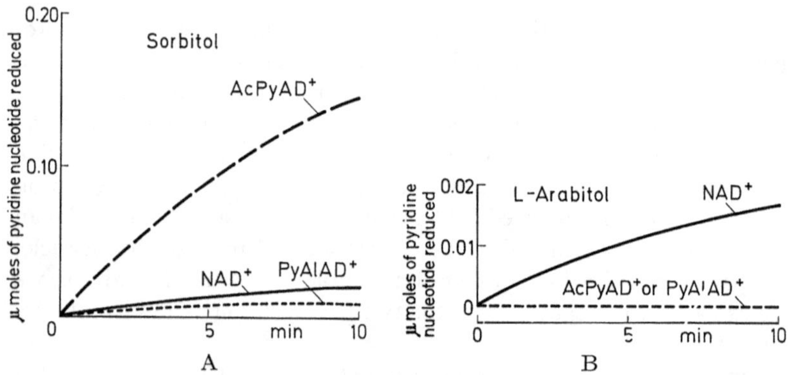

Fig. 6A and B. Reactivity of NAD analogs in the polyol oxidation by purified preparation from sheep seminal vesicle [19]. $AcPyAD^+$ = 3-acetyl-pyridine-AD^+; $PyAlAD^+$ = 3-Pyridinealdehyde-AD^+. A, substrate sorbitol; B, substrate L-arabitol

two distinct protein entities. Finally, differences in the binding of the pyridine nucleotide were indicated in experiments using NAD analogs. As shown in Fig. 6A on sorbitol as example, typical polyol substrates of L-iditol dehydrogenase, corresponding to the before-mentioned D-erythro-1.2.4-configuration, are oxidized 7- to 9-times more rapidly by our preparation when NAD^+ is substituted by 3-acetylpyridine-AD^+ ($AcPyAD^+$), whereas in the presence of 3-pyridinealdehyde-AD^+ ($PyAlAD^+$), their oxidation rate amounts to only 60—80% that in the presence of NAD^+. In contrast, L-arabitol (Fig. 6B) is not oxidized at all on addition of either one of the two analogs. Galactitol behaves identically in this respect.

All these results taken together are presumptive evidence for molecular heterogeneity of our purified dehydrogenase preparation from sheep seminal vesicle. Besides a NAD-D-erythro-1.2.4-polyol dehydrogenase, we postulate a second one oxidizing L-arabitol and galactitol. If, in analogy to the structural requirements of the "sorbitol enzyme", we assume for the "L-arabitol enzyme" also specificity with respect to the configuration on the C-atoms 1, 2 and 4 (Fig. 7), then a specificity

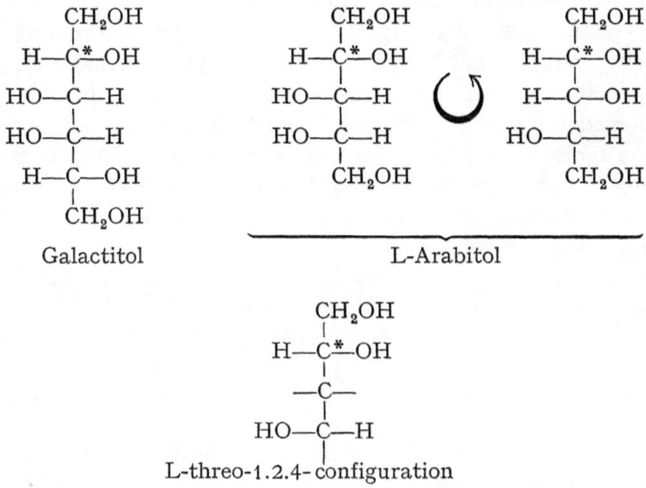

Fig. 7. Structural requirement of substrates of a proposed L-arabitol oxidizing enzyme from sheep seminal vesicle

for the L-threo-1.2.4-configuration could be ascribed to the second enzyme. Considering this rule, D-tagatose should be produced from galactitol and a mixture of L-ribulose and L-xylulose from L-arabitol, with L-ribulose probably prevailing due to a presumed hindring effect of the cis-hydroxyl on C_3 in the second transformation.

At first glance, the results seemed to be in harmony with this hypothesis. The reaction mixture, after incubation of L-arabitol, NAD^+ and the preparation from seminal vesicle in the presence of pyruvate and lactic dehydrogenase, removal of protein, NAD and deionization, was separated on a borate column into two peaks in a ratio of roughly 4:96. But unfortunately, the small first peak is, according to five criteria, certainly not xylulose as expected and as Smith [16] had concluded from his investigations on the sheep liver enzyme. Peak 2 does contain ribulose as the main component, the optical rotation of which has not been determined. But in addition, it contains a second substance of unidentified nature. Since only 70% of the carbazole reacting material placed onto the borate column was recovered in the eluates, probably a fourth component has to be claimed in the reaction mixture. Ribulose, the main and only identified component, cannot be a primary product of L-arabitol oxidation because L-ribulose is not at all reduced by the enzyme, neither in the presence of NADH nor in the presence of AcPyADH.

Evidently, the problem of the nature of L-arabitol oxidation product is even more complicated than the problem of the enzyme itself catalyzing this reaction. One might speculate, on the basis of the multiplicity of reaction products, that the primary reaction product may be a 3-ketopentose, enolized already under the alkaline condition of incubation and finally forced into the cis-configuration of ribulose by borate. So far, this is no more than just a plausible hypothesis which has to await experimental proof, but which could account for the unexpected and unexplained labeling of glycogen after administration of C_1-labeled L-arabitol [2].

References

1. Touster, O., and S. O. Harwell: J. biol. Chem. **230**, 1031 (1958).
2. McCormick, D. B., and O. Touster: Biochim. biophys. Acta (Amst.) **54**, 598 (1961).
3. Touster, O., V. H. Reynolds, and R. M. Hutcheson: J. biol. Chem. **221**, 697 (1956).
4. Hollmann, S., and O. Touster: J. biol. Chem. **225**, 87 (1957).
5. Hickman, J., and G. Ashwell: J. biol. Chem. **234**, 758 (1959).
6. Hollmann, S., u. G. Laumann: Hoppe-Seylers Z. physiol. Chem. **348**, 1073 (1967).
7. — Hoppe-Seylers Z. physiol. Chem. **317**, 193 (1959).
8. Batt, R. D., F. Dickens, and D. H. Williamson: Biochem. J. **77**, 272 (1960).
9. Glock, G. E., and P. McLean: Biochem. J. **56**, 171 (1954).
10. Robinson, D., and R. T. Williams: Biochem. J. **68**, 23P (1958).
11. Blakley, R. L.: Biochem. J. **49**, 257 (1951).
12. Williams-Ashman, H. G., and J. Banks: Arch. Biochem. **50**, 513 (1954).
13. — —, and S. K. Wolfson Jr.: Arch. Biochem. **72**, 485 (1957).
14. King, T. E., and T. Mann: Nature (Lond.) **182**, 868 (1958).
15. McCorkindale, J., and N. L. Edson: Biochem. J. **57**, 518 (1954).
16. Smith, M. G.: Biochem. J. **83**, 135 (1962).
17. Gerlach, U.: Klin. Wschr. **37**, 93 (1959).
18. Touster, O., and G. Montesi: In: S. P. Colowick and N. O. Kaplan, Methods in enzymology, vol. V, p. 317. New York: Academic Press 1962.
19. Hollmann, S., u. D. Lattermann: Hoppe-Seylers Z. physiol. Chem. **350**, 51 (1969).

Discussion

Dr. Horecker: Have you looked at the products formed from sorbitol at high pH?

Dr. Hollmann: The product formed from sorbitol is always fructose, exclusively.

Dr. Horecker: Would it be possible to trap the primary product formed from L-arabitol by carrying out the reaction in the presence of borate?

Dr. Hollmann: This might be possible, in case the enzyme is not inhibited by borate, as some of the polyol dehydrogenases are. Your idea is worth testing.

Dr. Hosoya: You presented the NADP-linked xylitol dehydrogenase only in the mitochondria. In my laboratory, we have measured a small amount of NADP-linked DH in rat liver mitochondria and considerable amount of this enzyme in cytoplasma. Have you investigated the cell particles of rat liver?

Dr. Hollmann: In all my experiments, I found the activity entirely bound to mitochondria in all tissues studied.

Dr. Ninomiya: Did you study the ontogeny of the enzyme?

Dr. Hollmann: Studies on ontogenetic development of the enzymes have not been performed.

Dr. Ninomiya: How many molecules of DPNH are located on one molecule of polyol dehydrogenase?

Does polyol dehydrogenase make some contribution to the respiration of the mitochondria?

Dr. Hollmann: Nothing can be said so far to these questions.

Dr. Minakami: Did you study the difference in the cellular localization of NAD and NADP xylitol dehydrogenase in different animals?

Dr. Hollmann: All localization studies have been done on guinea pig tissues only.

Dr. Macdonald: Since fructose is found in seminal fluid, are the reactions you described sex dependent?

Dr. Hollmann: Dependency of the "ketose reductase" in the prostate gland on testosterone has been shown by Mann.

Dr. TOUSTER: Isn't your difference in results as compared with Smith possibly due to the fact that he used a liver enzyme with L-arabitol whereas you employed the seminal vesicle enzyme?

Dr. HOLLMANN: This is, of course possible. But, the similarities of our results in all other respects are so striking that I have reason to believe that he is probably wrong in his assumption, one of the products formed from L-arabitol by the sheep liver enzyme being L-xylulose.

Metabolism of Xylose by the Lens of the Eye

RUTH VAN HEYNINGEN

Nuffield Laboratory of Ophthalmology, University of Oxford
Oxford, England

Sugar Cataracts

Interest in the metabolism of xylose by the lens stems from the finding, by Darby and Day in 1940, that this sugar when fed to weanling rats, rapidly causes cataract (an opaque lens). It had earlier been shown by Mitchell and Dodge in 1935 that galactose was similarly cataractogenic, and the third type of these so-called "sugar" cataracts is brought about by rendering an animal permanently diabetic. L-arabinose, when fed simultaneously with D-galactose, will hasten the onset of cataract (Patterson, 1955), although this sugar has not itself been shown to be cataractogenic. In 1959 it was shown that all these four sugars (D-xylose, D-galactose, D-glucose and L-arabinose) were reduced by an aldose reductase in the lens to the corresponding polyhydric alcohol, the coenzyme being reduced NADP (van Heyningen, 1959a and b, 1962). All have the same configuration of carbon atoms 1—3.

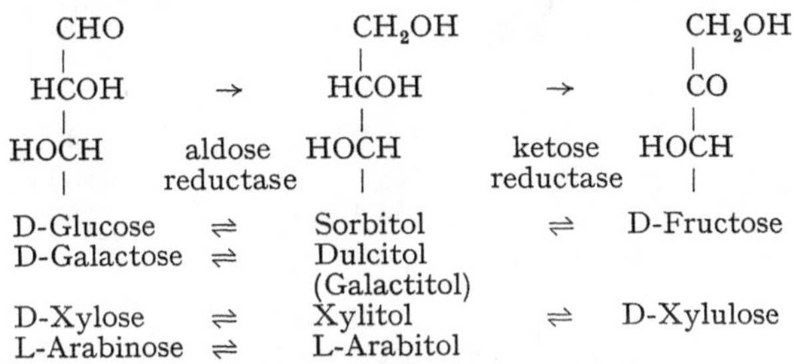

Fig. 1. Sorbitol pathway. Configuration of the first three carbon atoms of the sugar and polyols involved

Fig. 1 shows the formation of the four polyols; sorbitol and xylitol are further metabolised in the lens, by the enzyme ketose reductase (L-iditol dehydrogenase, E.C. 1.1.1.14) to form fructose and xylulose, but dulcitol (galactitol) is not a substrate of this enzyme.

It has usually been assumed that the production of polyols in the lens is in some way involved in the formation of sugar cataract. In the last few years a great deal has been found out about this cataractogenic process, mostly by Kinoshita and his co-workers in Boston (see Kinoshita, 1965). However, this work has been mostly on galactose cataract. This paper will concentrate on what is known about xylose cataract, which differs in some particulars from galactose and from diabetic cataracts, although there is much evidence for the belief that the formation of the polyol is the initiating cause in all cases (Lerman *et al.*, 1961, 1965, have an alternative theory).

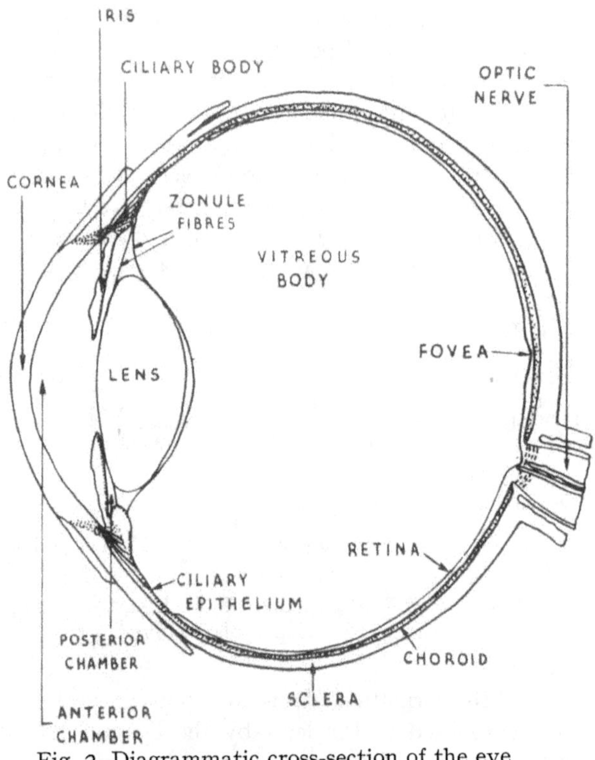

Fig. 2. Diagrammatic cross-section of the eye

The Lens

Fig. 2 shows a cross-section of the eye. The lens has no blood supply and relies primarily on the aqueous humour flowing slowly through the anterior chamber for its nourishment and for the elimination of metabolic waste products. Xylose (and the other monosaccharides) readily penetrate from the blood into the aqueous humour. The lens grows throughout life, cytologically isolated within its capsule. Mitoses occur almost exclusively in the single layer of cells on the anterior surface of the lens; the daughter cells so produced elongate to form the lens fibres, which elongate and curve round the deeper and older fibres. The old cells at the centre (or nucleus) of the lens become more and more compressed and fresh fibres continue to be formed at the periphery throughout life but at a decreasing rate.

Xylose Cataract

To produce xylose cataract weanling Wister-albino rats (21—24 days old) are fed on a diet containing 35% xylose. Most types of experimental

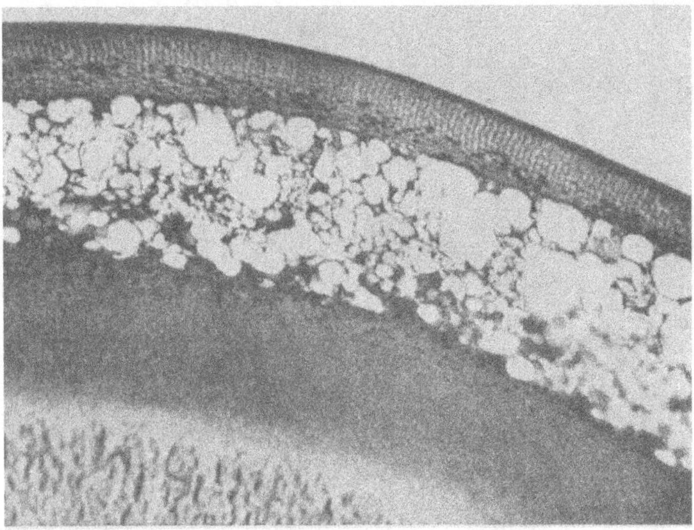

Fig. 3. Photomicrograph of a portion of the lens (anterior cortical zone) from a rat maintained on the xylose diet for 4 days. Note extensive hydropic degeneration of anterior cortex (Lerman *et al.*, 1965)

cataract can be produced more readily in young animals but xylose cataract can only be produced in weanling rats. Diets containing 25 to 50% xylose have been used in other laboratories and different strains of rats seem to respond differently (Lerman and Heggeness, 1960). The eyes are examined daily (by ophthalmoscope). On the 4th day the lens usually shows a trace of opacity around the periphery. This opacity formation progresses daily until the 8th or 9th day, when the lens is usually opaque and shining-white, visible to the naked eye. Thereafter the opacity regresses until, by the 3rd week, only minimal cortical opacity can be seen.

Fig. 3 shows a photomicrograph of the water-clefts and swelling which occurs under the capsule of the xylose-fed rat. These vacuoles become much smaller as the lens recovers but even after 60 days, when the lens is apparently transparent, they are still detectable though more centrally located. Kinoshita has shown that in galactose cataract the lens acts as an osmometer; the fact that the lens is relatively impermeable to polyols means that once formed, they do not readily leak out. Instead, water is drawn into the lens to maintain osmotic equilibrium with the external medium and the increased water content causes the fibres to swell. Fig. 4 shows that changes in water content are parallel

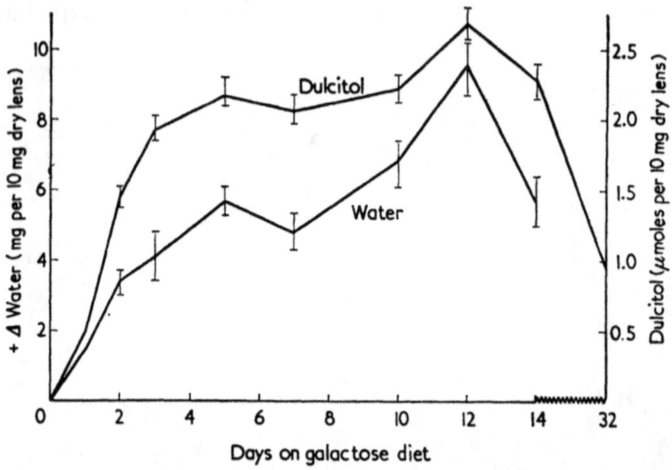

Fig. 4. The accumulation of dulcitol and water in galactose cataract. (From Kinoshita, Merola and Dikmak, 1962)

to those in dulcitol content. A similar mechanism must be postulated for xylose cataract since the gross and microscopic appearance of the cataract, as well as other biochemical changes, are similar. Lens swelling is less marked (van Heyningen, 1963; Lerman *et al.*, 1965), but Kinoshita (unpublished) has found a 15% increase in the lens water of the rat fed 50% xylose for 10 days.

Patterson and Bunting (1966) found that it was necessary to feed a diet containing 50% xylose for the same degree of lens swelling as that obtained by feeding 35% galactose. On these two diets, the average level of lens polyols on days 7 to 14 was similar (xylitol 60 mM, and galactitol 70 mM).

Metabolism of Xylose by the Lens

In spite of the high level of xylitol (up to 40 μmoles/g) that is found in the lens of the xylose-fed rat, no xylitol was found in other organs (blood, kidney, lung, liver, heart, muscle, retina, testis, seminal vesicles and spleen) of the xylose-fed rat (van Heyningen, 1959a). Hayman, Lou, Merola and Kinoshita (1966) have recently examined the aldose-reducing activity of the lens and other tissues of the rabbit. Table I shows their findings (data from their Tables I and II). They had earlier shown that the lens contains only one aldose-reducing enzyme (Hayman and Kinoshita, 1965), aldose reductase (E.C. 1.1.1.21). From a study of the relative rates of reduction (or oxidation in the case of gulonate) of a number of substrates, they conclude that this is also the main enzyme with aldose-reducing activity in the adrenal and skeletal muscle of the rabbit.

Table I. *The distribution of enzymatic activity in the rabbit*

Organ	Rate of reduction of DL-glyceraldehyde (μmoles/min per g tissue)	Relative rate of reaction of other substrates			
		D-glucuronate	D-glucuronolactone	D-xylose	L-gulonate
Lens	0.53	87	102	55	34
Adrenal	1.20	133	91	58	18
Muscle	0.069	220	88	56	22
Kidney	0.24	1315	197	25	283
Liver	0.36	993	131	7	141
Heart	0.012	810	115	45	240
Spinal cord	0.038	618	140	18	145
Brain	0.031	908	180	29	207

The relative rates are expressed as percentages of those obtained with DL-glyceraldehyde as a substrate.

Hayman *et al.* (1966).

The activity in the other organs examined is largely due to the activity of $NADP^+$ L-hexonate dehydrogenases (E.C. 1.1.1.19 and 20) (Mano *et al.*, 1961; York *et al.*, 1961). Thus all these tissues are able to reduce xylose, though the rate was somewhat more rapid in the lens, adrenal and skeletal muscle. The fact that xylitol could not be detected in organs other than the lens suggests either that xylitol may leak out from these tissues more readily than it does from the encapsulated lens, or that it is much more rapidly removed in these tissues by oxidation to xylulose followed by phosphorylation, or both. The presence of L-iditol dehydrogenase has, however, been shown in the lens (van Heyningen, 1959a), and the formation of both labelled xylulose and lactic acid has been shown to occur *in vitro* in a rat lens incubated in a medium containing [U-^{14}C]xylose (Fig. 5). In this radioautograph 0.5% of the counts are in xylulose, 4.5% in lactic acid and 2.5% in the spot labelled "mixture including phosphates", 53% in xylitol and 39% in xylose. (A sample of authentic xylulose was kindly given by Dr. O. Touster.)

Aldose reductase is the only enzyme in *rat* lens which uses xylose as substrate. In this xylose differs from galactose (which can also be

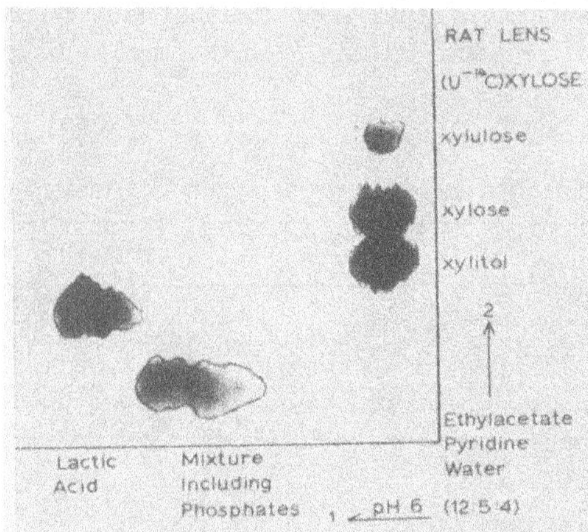

Fig. 5. Radioautogram of the labelled products of [U-^{14}C]xylose metabolism by the lens of the rat. Rat lenses were incubated in a medium (5 ml) containing [U-^{14}C]xylose (5 mg) and unlabelled glucose (9 mg). An amount equivalent to 25 mg of lens was subjected to 2-way electrophoresis and chromatography, followed by radioautography (van Heyningen, 1965)

phosphorylated by a kinase, though the activity of this enzyme is weak in the lens; Hayman *et al.*, 1966), and from glucose which is metabolised by three enzymes; aldose reductase, hexokinase and a glucose dehydrogenase. Pottinger (1967) has examined these three enzymes in extracts of *cow* lens and shown that the K_m (glucose) is 28 mM, 0.12 mM and 690 mM respectively. The relative importance of the metabolic route glucose→sorbitol→fructose in the normal lens is not known. Under abnormal conditions (high blood glucose in diabetes, or high blood levels of xylose or galactose in xylose- or galactose-feeding) there is accumulation of polyols in the lens. The K_m of purified aldose reductase for glucose (70 mM) is more than ten times that for xylose (5 mM) (Hayman and Kinoshita, 1965) and the V_{max} for the two substrates is the same; therefore when levels of the order of 5 mM both glucose and xylose occur in the blood (and aqueous humour) the formation of xylitol in preference to sorbitol is to be expected. (The K_m for NADPH with the two substrates is however unknown.)

Impermanence of Xylose Cataract

An intriguing problem, specific to xylose cataract, is the accumulation of sorbitol, which occurs together with that of xylitol, albeit at a much lower concentration. Another problem, also specific to xylose cataract, is its impermanence. Fig. 6 shows both these phenomena.

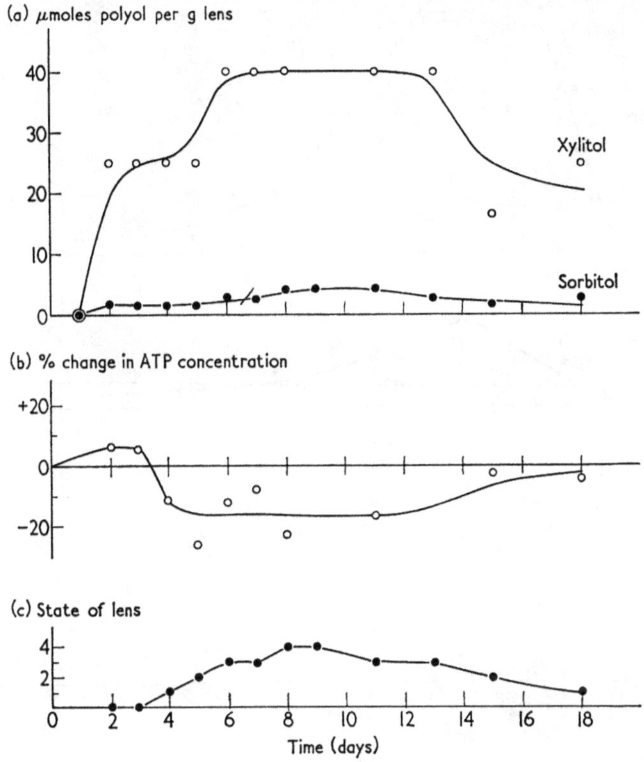

Fig. 6a—c. Daily changes in lens of xylose-fed rat. a Xylitol and sorbitol concentrations, b % change in ATP concentration

$$\left[\left(100 \times \frac{\mu\text{moles per g lens, xylose-fed rat}}{\mu\text{moles per g lens, glucose-fed rat}}\right) - 100\right]$$

and c state of lens (*1* trace of opacity, *2* about $^1/_3$ opaque, *3* about $^3/_3$ opaque, *4* almost completely opaque) (van Heyningen, 1963)

As the lens becomes more opaque the xylitol reaches a level of 40 μmoles/g lens (8th—11th day) and sorbitol reaches a level of 4 μmoles/g lens. After the 11th day of feeding the levels of xylitol and sorbitol both fall. The ratio of lens xylitol/lens sorbitol fell continuously after the first two days (Fig. 7). This figure includes data from another series of experiments to be described below; it shows that sorbitol in the lens of the xylose-fed rat reaches, on the 10th day, a concentration five times that in the normal lens (Kuck, 1963) and thereafter it falls.

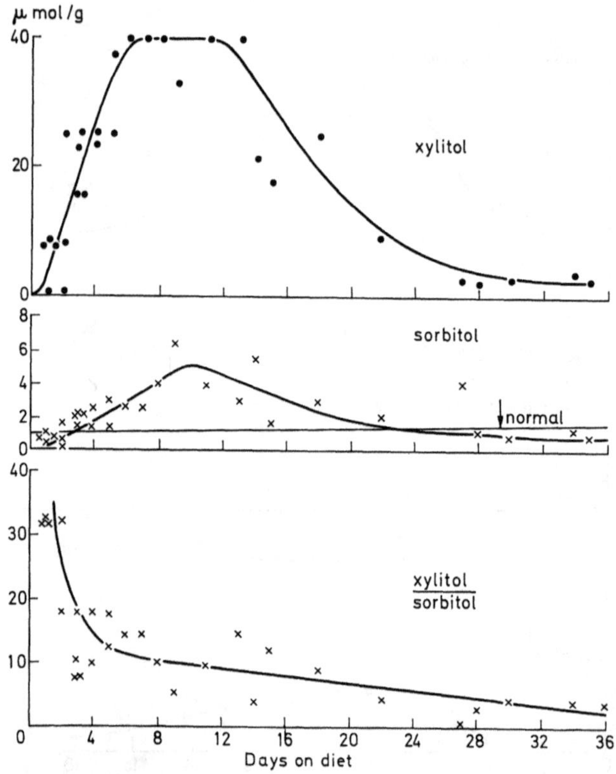

Fig. 7. Xylitol, sorbitol and $\frac{\text{xylitol}}{\text{sorbitol}}$ in the lens of the xylose-fed rat. Weanling rats were fed continuously on a diet containing 35% xylose and killed at various times from 3—35 days after the start of the diet. Normal sorbitol levels from Kuck (1963)

This impermanence is peculiar to xylose cataract and quite different from the sequence of events in diabetic or galactose cataracts, which proceed to completion and final disintegration of the lens if the level of the blood glucose or galactose remains elevated. Mature xylose cataracts are, however, not unknown; they have been produced in Sprague-Dawley rats fed a diet of 50% xylose (Patterson and Bunting, 1966).

Table II. *Blood sugars and lens polyols in the xylose-fed rat*

	Xylose (μmol/ml)	Glucose (μmol/ml)	Xylose / Glucose	Xylitol (μmol/g)	Sorbitol (μmol/g)	Xylitol / Sorbitol
2	2.5 ± 1.2 (1.5 to 4.2)	5.5 ± 0.8 (4.6 to 6.6)	0.45 ± 0.15 (0.33 to 0.70)	8.15 ± 8.3 (0 to 25)	0.2 ± 0.5 (0 to 1.4)	
	3.9 ± 1.5 (1.9 to 7.3)	5.4 ± 1.0 (3.9 to 7.2)	0.74 ± 0.28 (0.31 to 1.40)	25.5 ± 8.2 (16 to 40)	2.0 ± 0.6 (1.4 to 3.0)	13.0 ± 3.3 (8 to 18)
3	—	—	—	38.6 ± 3.1 (33 to 40)	4.0 ± 1.1 (2.8 to 6.5)	10.7 ± 4.2 5—14
35	2.0 ± 0.1 (1.9 to 2.2)	6.0 ± 4.9 (3.6 to 8.2)	0.36 ± 0.14 (0.24 to 0.53)	2.9 (2 to 4)	1.7 (0.8 to 4.0)	3.4 (1 to 4)
al	—	6.7 ± 3.3	—	—	1.0 to 1.3	—

[a] Kuck (1966, 1963).

Results are given as mean value of 5 to 10 determinations ± standard error (Range).

The impermanence of xylose cataract does not seem to be due to a decrease in aldose reductase activity as the lens ages, as Kinoshita *et al.* (1963) could find no difference between the lens of a 100 g rat and a 300 g rat.

A possible explanation is that the changes in the lens reflect changes in the levels of xylose and glucose in the blood of the rat after continued feeding on the diet. This possibility was tested (van Heyningen, unpublished) and a summary of the results is shown in Table II; the difference between days 3 to 6 and days 27 to 35 for blood xylose concentrations (3.8 and 2.0 μmoles/ml) and for the ratio blood xylose/blood glucose (0.74 and 0.36) are both significant at the 1% level. The Table II also shows the rise and fall in the level of lens polyols and the precipitous fall in the ratio xylitol/sorbitol.

The question is, can the decrease in the level of blood xylose, and the even slightly greater decrease in the ratio blood xylose/blood glucose, be the cause of the drastic changes in the levels of these polyols and hence in the state of clarity of the lens? A suggested sequence of events is shown in Fig. 8.

Aldose reductase has an affinity for xylose more than 10 times that for glucose; when both glucose and xylose are at a concentration of 10 mM the rate of reduction of xylose is six times that of glucose (Hayman and Kinoshita, 1965). These findings refer to the purified enzyme, with adequate amounts of NADPH, but they lead one to expect that xylitol would be formed more rapidly than sorbitol during days 3 to 6

Days on diet	Blood			Lens	
	Xylose (mM)	Glucose (mM)	Xylose / Glucose	Aldose reductase K_m xylose 5 mM K_m glucose 70 mM	L-iditol dehydrogenase K_m xylitol 1.7 mM K_m sorbitol 5.2 mM
3— 6	3.9	5.4	0.74	*Xylitol* formed more rapidly than sorbitol	High concentrations of *xylitol* suppress its own oxidation
	↓		↓	*Sorbitol* formation suppressed by presence of xylose	*Sorbitol* oxidation suppressed by presence of xylitol
27—35	2.0	6.0	0.36	*Xylitol* formed less rapidly	Lower concentration of *xylitol* is less of an inhibition to its own oxidation
				Sorbitol formation less suppressed	Lower concentration of xylitol allows greater rate of oxidation of *sorbitol*

Fig. 8. Suggested sequence of events leading to the development of and recovery from xylose cataract. Weanling rats were fed continuously on a diet containing 35% xylose and killed at various times from 3—35 days after the start of the diet

when the concentration of xylose in the blood is not much less than that of glucose. The rate of formation of sorbitol from glucose in the presence of xylose would probably be less than that formed in its absence.

The level of polyols in the lens depends not only on their rate of formation by aldose reductase but also on the rate of removal by L-iditol dehydrogenase. This enzyme in the lens (Kinoshita, 1963) is like that of liver (Smith, 1962), in that its affinity for xylitol is greater than that for sorbitol at pH 7.4. The high ratio xylitol/sorbitol in the lens will therefore reduce the availability of sorbitol to the enzyme and suppress its oxidation; this could account for the accumulation of this substance above its normal value even though its rate of formation may be depressed. When xylitol is present at a high concentration ($>$ 10 mM) its oxidation by L-iditol dehydrogenase is itself suppressed (Smith, 1962); this would tend to prevent its oxidation and to keep up its concentration in the lens.

After the diet has been fed for some time the blood xylose and ratio xylose/glucose in the blood falls; the rate of formation of xylitol by aldose reductase would therefore be gradually decreased; at a lower concentration it would be less of an inhibitor to its own removal by L-iditol dehydrogenase and would be more rapidly oxidised. Also as its concentration fell, nearer to that of sorbitol, the sorbitol formation would be less suppressed but it also would be more readily oxidised by L-iditol dehydrogenase and its level would gradually return to within normal limits.

The above "explanation" greatly oversimplifies the situation. In particular it does not take into account the concentration of coenzymes (reduced and oxidised) within the lens, or the fact that the K_m's of the sugars and polyols depend upon these concentrations. Nevertheless the finding (Fig. 9) that xylose/glucose in blood bears a straight-line relationship with xylitol/sorbitol in lens indicates that the explanation for lens changes may be found in the changes in blood xylose and glucose.

BÄSSLER and co-workers (see elsewhere in this volume) have shown that adaptation by rats to the feeding of *xylitol* involves an increased activity of L-iditol (or polyol) dehydrogenase in the liver. Adaptation to *xylose*, manifested by a lowering of the level of xylose in the blood, could similarly involve an increased activity of enzymes which metabolise xylose, for example, the aldose-reducing enzymes found in various organs (Table I). It is possible that this adaptation takes place more rapidly in the older rat, and this, together with the fact that in general the younger the animal the more susceptible it is to experimental cataract, could explain why xylose cataract develops only in the weanling rat.

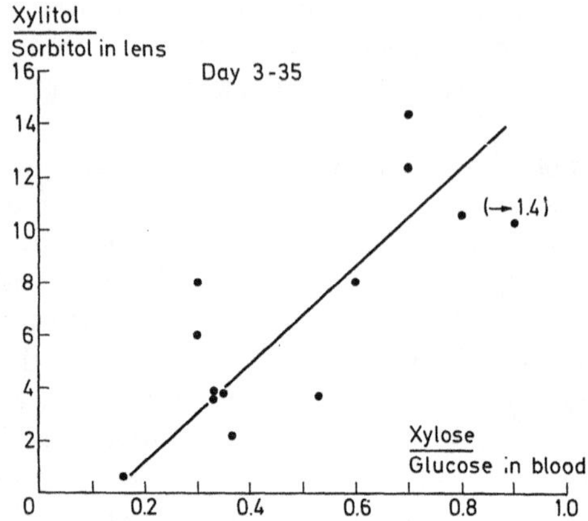

Fig. 9. Relationship between $\frac{\text{xylitol}}{\text{sorbitol}}$ in the lens and $\frac{\text{xylose}}{\text{glucose}}$ in the blood of the xylose-fed rat. Weanling rats were fed continuously on a diet containing 35% xylose and killed at various times from 3—35 days after the start of the diet

Oxidation of Xylose by the Lens

The labelled compounds found in a rat lens incubated in [U-^{14}C]xylose are, besides xylose itself, xylitol and small amounts of xylulose, lactic acid and phosphate esters (Fig. 5). The calf lens, on the other hand, incubated in labelled xylose, forms xylonic acid and xylitol in equivalent amounts (Table III). The glucose dehydrogenase that brings about the

Table III. *Radioactivity in the labelled products of [U-^{14}C]xylose metabolism by the lens of the rat and the calf*

	Rat (% of total)	Calf (% of total)
Xylitol	53.5	26.8
Xylose	39	32.3
Xylulose	0.5	—
Lactic acid	4.5	0.2
Xylonic acid	—	38.2
Mixture including phosphates	2.5	—
Unknown	—	2.5

Rat and calf lenses were incubated in a medium (5 ml) containing [U-^{14}C]xylose (5 mg) and unlabelled glucose (9 mg). An amount equivalent to 25 mg of lens was subjected to 2-way electrophoresis and chromatography, followed by radioautography (van Heyningen, 1965). Radioactivity was counted on both sides of spots on paper, the background subtracted and the values averaged.

oxidation of xylose (van Heyningen, 1958, 1964) has been partly characterised by Pottinger (1967). She could find virtually no trace of the enzyme in the *rat* lens, whereas in the *bovine* lens it was twice as active (measured under optimal conditions) as the aldose reductase. The difference between species, in the metabolism of xylose by the lens, appears therefore to be due to the difference in content of these two enzymes, aldose reductase and glucose dehydrogenase.

Acknowledgement. I thank the Royal National Institute for the Blind for a grant for technical assistance.

References

Darby, W. J., and P. L. Day: J. biol. Chem. **133**, 503 (1940).

Hayman, S., and J. H. Kinoshita: J. biol. Chem. **240**, 877 (1965).

— M. F. Lou, L. O. Merola, and J. H. Kinoshita: Biochim. biophys. Acta (Amst.) **128**, 474 (1966).

Heyningen, R. van: Biochem. J. **69**, 481 (1958).

— Biochem. J. **73**, 197 (1959a).

— Nature (Lond.) **184**, 194 (1959b).

— Chap. The lens, in: The eye, vol. I, ed. H. Davson, p. 256. London: Academic Press 1962.

— Exp. Eye Res. **2**, 362 (1963).

— Biochem. J. **92**, 14C (1964).

— Biochem. J. **96**, 419 (1965).

Kinoshita, J. H.: Invest. Ophthal. **4**, 786 (1965).

— S. Futterman, K. Satoh, and L. O. Merola: Biochim. biophys. Acta (Amst.) **74**, 340 (1963).

— L. O. Merola, and E. Dikmak: Exp. Eye Res. **1**, 405 (1962).

Kuck, J. F. R.: Invest. Ophthal. **2**, 607 (1963).

— Invest. Ophthal. **5**, 65 (1966).

Lerman, S., M. Burton, S. Zigman, and M. Pitel: Acta ophthal. (Kbh.) **43**, 764 (1965).

—, and F. W. Heggeness: Biochem. J. **79**, 224 (1961).

Mano, Y., K. Suzuki, K. Yamada, and W. Shimazono: J. Biochem. (Tokyo) **49**, 618 (1961).

Mitchell, H. S., and W. M. Dodge: J. Nutr. **9**, 37 (1935).

Patterson, J. W.: Arch. Biochem. **58**, 23 (1955).

—, and K. W. Bunting: Docum. ophthal. (Den Haag) **20**, 64 (1966).

Pottinger, P. K.: Biochem. J. **104**, 663 (1967).

Smith, M. G.: Biochem. J. **83**, 135 (1962).

York, J. L., A. R. Grollman, and C. Bublitz: Biochim. biophys. Acta (Amst.) **47**, 298 (1961).

Discussion

Dr. TOUSTER: Since adaptive enzyme changes are being found in liver, isn't it possible that enzyme levels are changing in the lens during xylose feeding which might explain your results?

Dr. VAN HEYNINGEN: It is possible, although no one has yet found adaptive enzyme changes of this type in the lens.

Dr. TOUSTER: Has the xylitol inhibition of the lens L-iditol dehydrogenase actually been demonstrated?

Dr. VAN HEYNINGEN: No.

Dr. KUZUYA: What do you think is the causative relationship of polyol accumulation with the occurrence of cataract? Is the accumulation of water the primary mechanism or does the accumulation of polyol damage the lens in a more specific manner?

Dr. VAN HEYNINGEN: The accumulation of water takes place rapidly, and the water clefts which are formed constitute, or contribute to, the opacity. Metabolic changes, such as the abnormally high rate of oxidation of NADPH also occur and presumably account for the cataractogenic effect of xylose feeding.

Dr. BÄSSLER: Is it possible to produce cataract by the administration of polyols?

Dr. VAN HEYNINGEN: Not as far as I know.

Dr. BÄSSLER: The oxidation of xylitol in the lens is slow at high xylose/glucose ratio in the blood as you have shown. You attribute this to the fact that xylitol in high concentration inhibits its own oxidation; I would suggest an additional mechanism. We found that xylose is a competitive inhibitor for transketolase. By this means high xylose concentration could inhibit xylitol oxidation too.

Dr. VAN HEYNINGEN: This is interesting and I agree that it could be an additional mechanism.

Dr. OPITZ: Did you ever investigate the influence of dimethyl sulfoxide (DMSO) on these polyol-cataracts?

Dr. VAN HEYNINGEN: No.

Dr. POGELL: Do you mean by xylitol suppression of its own oxidation that you are getting substrate inhibition of the oxidase?

Dr. VAN HEYNINGEN: Yes; I am assuming that the enzyme in lens behaves the same way as that in liver, isolated by Smith.

Dr. LANG: In our feeding experiments with xylitol we studied carefully if lens opacities would occur in our rats. We never found any cataractogenic action of xylitol and the same with all feeding experiments conducted in other laboratories. There must be a difference in lens metabolism between endogenously formed xylitol and exogenous xylitol. The differences may be due to permeability or differences in metabolism. There is some literature concerning rat strains not susceptible to xylose cataracts. Is the metabolism of xylose in the lens of such nonsusceptible strains modified? Have you any experience in this field?

Dr. VAN HEYNINGEN: Our rats are Wistar albino strain; I have no experience of other strains. To produce cataract, it is necessary for the xylitol to be formed within the lens from xylose occurring in the blood and aqueous humour.

Pentoses in Mucopolysaccharides and Glycoproteins

LENNART RODÉN

Departments of Pediatrics and Biochemistry,
and the La Rabida-University of Chicago Institute,
University of Chicago, Chicago, Illinois, U.S.A.

Only one of the four aldopentoses, ribose, has previously been firmly established as a component of macromolecules of animal origin. From time to time, however, the finding of other pentoses, particularly xylose, in mammalian tissues has been reported. For a number of reasons little attention has generally been paid to such findings: the quantities observed in any one tissue or purified product have been small, and in no case was the work carried beyond the stage of simple qualitative analysis. It has also often been assumed that xylose may arise by decarboxylation of glucuronic acid — a component of the acid mucopolysaccharides — and such a trivial explanation for the presence of pentose in polysaccharide hydrolysates has not been conducive to a deeper penetration of the subject. Yet another explanation of the finding of xylose in animal tissues, such as intestinal mucus and mucosa from the pig, has been brought forward, i.e., the possibility of contamination of the tissue with plant material which often contains xylose [1].

In recent years it has become clear, however, that D-xylose does indeed occur as a structural component of animal tissues and, in particular, of the acid mucopolysaccharides. Xylose also appears to be a component of several animal glycoproteins, although information concerning structural details is still lacking. Furthermore, L-arabinose has been found in hyaluronic acid preparations from brain [2] and it may be assumed that the two pentoses play an important role in the structure or metabolism of these macromolecules.

At the present time the role of xylose as a component of animal macromolecules has been clearly defined only with regard to the acid mucopolysaccharides in which the pentose serves as the point of linkage between carbohydrate and protein. This paper will therefore deal mainly with this group of substances, and in order to view the role of xylose from a somewhat broader perspective, the main structural features of the mucopolysaccharides will be briefly considered. Details of the structure and metabolism of these compounds have been covered in several recent reviews [3—13].

The acid mucopolysaccharides of connective tissues occur in their native state as covalently bound complexes with protein. Whereas little is known about the structures of the protein moieties, the polysaccharide components have been studied for many years, and with the exception of certain details their structures have been largely elucidated. Compared to many glycoproteins the acid mucopolysaccharides exhibit an almost monotonous regularity in structure. Each polysaccharide chain contains a large number of identical, repeating disaccharide units, composed of one uronic acid and one hexosamine residue (Table I). The uronic acid may be either D-glucuronic acid or L-iduronic acid. The former is the exclusive uronic acid component of hyaluronic acid and chondroitin 4- and 6-sulfate. L-Iduronic acid predominates in dermatan sulfate (chondroitin sulfate B), but small amounts of D-glucuronic acid are also present in this polysaccharide [14, 15]. Heparin and heparitin sulfate (heparan sulfate) contain mainly D-glucuronic acid but also some L-iduronic acid. The hexosamine moiety of the chondroitin sulfates and dermatan sulfate is galactosamine, whereas hyaluronic acid, heparin and heparitin sulfate contain glucosamine. One of the substances classified as an acid mucopolysaccharide, i.e. keratosulfate, does not conform to the pattern outlined above for the composition of the repeating disaccharide units, inasmuch as it does not have a uronic acid component. Instead, the disaccharide unit contains galactose, with glucosamine as the hexosamine moiety. In addition keratosulfate contains small and varying amounts of sialic acid and fucose, monosaccharides which are generally regarded as typical constituents of glycoproteins. Keratosulfate may therefore in a sense be considered as a compound

Table I. *Monosaccharide composition of acid mucopolysaccharides*

Mucopolysaccharide	Monosaccharide components of repeating disaccharide unit	
	hexosamine	uronic acid
Chondroitin 4- and 6-sulfate	D-Galactosamine	D-Glucuronic acid
Dermatan sulfate (chondroitin sulfate B)	D-Galactosamine	L-Iduronic acid D-Glucuronic acid
Heparin	D-Glucosamine	D-Glucuronic acid L-Iduronic acid
Heparitin sulfate	D-Glucosamine	D-Glucuronic acid L-Iduronic acid
Hyaluronic acid	D-Glucosamine	D-Glucuronic acid
Keratosulfate	D-Glucosamine	(D-Galactose)

on the borderline between glycoproteins and the typical, uronic acid-containing mucopolysaccharides. It may be noted that sulfated glycoproteins other than keratosulfate have recently been isolated from epithelial mucins [16, 17] and that the distinction between glycoproteins and" acid mucopolysaccharides" is therefore becoming somewhat diffuse.

The structure of the chondroitin 4-sulfate-protein complex of cartilage has been studied more extensively than that of any other mucopolysaccharide-protein complex. On the basis of physico-chemical studies it has been suggested that the complex consists of a protein core, to which many individual polysaccharide chains are attached [18, 19]. Evidence concerning the nature of the carbohydrate-protein linkage in this complex was first obtained by Muir [20]. She determined the amino acid composition of the complex and the chondroitin sulfate prepared from the complex by digestion with papain. After the proteolytic treatment, half of the original amount of serine was still bound to the polysaccharide, whereas the quantities of the other remaining amino acids were much less. Serine was the only amino acid present in a quantity corresponding to one mole per mole of polysaccharide, a finding which strongly indicated serine as the point of linkage between the protein and carbohydrate moieties.

More conclusive and detailed knowledge of this problem was obtained by degrading the complex from bovine nasal septum enzymatically to small glycopeptides, amenable to structural studies [21]. Degradation of the complex with testicular hyaluronidase yielded a product with a molecular weight of approximately 250,000. This material consists of a protein core to which many small oligosaccharide chains are attached as would be expected on the basis of the structure proposed by Mathews and Lozaityte [18]. Further degradation of the hyaluronidase-treated complex with papain, leucine aminopeptidase and carboxypeptidase yielded a glycopeptide fraction which contained serine as its only amino acid. The linkage to serine suggested by Muir has thus been confirmed. Additional evidence for the involvement of serine and, in particular, of its hydroxyl group has been obtained by the elegant alkali degradation studies of Anderson, Hoffman and Meyer [22]. These authors demonstrated that alkali treatment of peptide-containing chondroitin sulfate preparations resulted in destruction of serine with concomitant production of alanine under reducing conditions, indicating that the hydroxyl group of serine is involved in a linkage which may be cleaved by alkali in a β-elimination reaction.

At the time our studies on the carbohydrate-protein linkage of the chondroitin 4-sulfate-protein complex were initiated, it was naturally believed that the polysaccharide chains were linked to protein via either glucuronic acid or galactosamine, which were the only known monosaccharide constituents of chondroitin sulfate. Analysis of the carbohydrate composition of the purified glycopeptide fraction demonstrated the expected presence of glucuronic acid and galactosamine, and the molar ratio, 2:1, indicated that the chondroitin sulfate chains proper terminated with a glucuronic acid residue. However, in addition to the two characteristic components of chondroitin sulfate, the glycopeptide fraction also contained galactose and xylose, in a molar ratio of 2:1. The presence of these atypical monosaccharides suggested that they were involved in the linkage of the chondroitin sulfate chains to protein, and this assumption was substantiated by further studies of the detailed structure of the glycopeptide. Following partial acid hydrolysis a number of "linkage region" fragments were isolated [23—25], the structures of which are shown in Table II.

Table II. *Fragments isolated from chondroitin 4-sulfate glycopeptides by partial acid hydrolysis*

1. O-β-D-xylosyl-L-serine
2. 4-O-β-D-galactosyl-D-xylose
3. 4-O-β-D-galactosyl-O-β-D-xylosyl-L-serine
4. 3-O-β-D-galactosyl-D-galactose
5. O-β-D-galactosyl-(1 → 3)-O-β-D-galactosyl-(1 → 4)-D-xylose
6. 3-O-β-D-glucuronosyl-D-galactose

The characterization of Fragment 1 which was composed of xylose and serine, demonstrated that the carbohydrate-protein linkage is a β-glycosidic linkage between D-xylose (pyranose form) and the hydroxyl group of L-serine. The structure of xylosylserine has been confirmed by comparison with authentic material obtained by chemical synthesis [26]. It is of interest to note that xylosylserine has been found in normal urine [27] and it seems likely that the urinary compound represents a degradation product arising from the catabolism of chondroitin 4-sulfate and other acid mucopolysaccharides.

The characterization of the galactose-containing fragments shown in Table II now permits a detailed formulation of the structure of the entire carbohydrate sequence in the carbohydrate-protein linkage region (Fig. 1).

The presence of a single xylose residue in a key position in the chondroitin 4-sulfate-protein complex naturally suggests some specific structural or metabolical function which is of importance for the complex as a whole. It may be hypothesized that the relatively small xylose residue constitutes a more suitable connecting link than the bulkier uronic acid or N-acetylgalactosamine sulfate residues. However, it seems

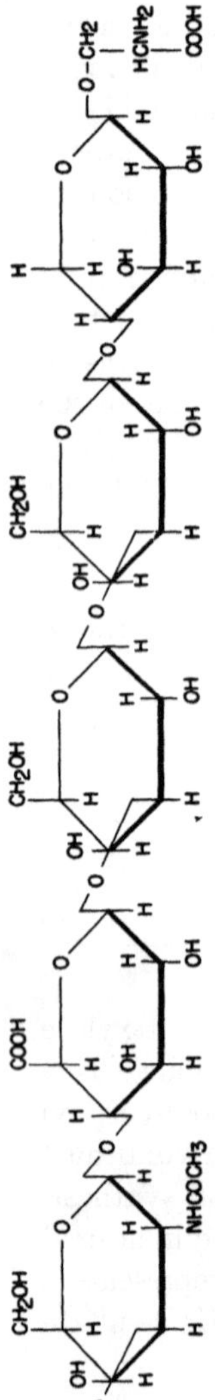

Fig. 1. Structure of the carbohydrate sequence in the polysaccharide-protein linkage region of the chondroitin 4-sulfate-protein complex

more likely that the solution to the problem resides in the biosynthesis of the complex. In order to clarify this point some pertinent aspects of this process will be briefly considered. Previous studies in several laboratories have indicated that the biosynthesis of glycoproteins in animal tissues in general proceeds by sequential addition of the sugar residues to a preformed protein acceptor, possibly the entire protein moiety of the glycoprotein. If this were also the mechanism of formation of the chondroitin sulfate-protein complex, the first reaction in the synthesis of the carbohydrate chain would be the addition of xylose to serine residues of the protein acceptor. The incorporation of xylose from UDP-D-xylose into material precipitable with trichloroacetic acid has indeed been demonstrated in several cell-free tissue systems: hen's oviduct [28], mast cell tumor [29], embryonic chick cartilage [30] and embryonic chick brain [31]. Xylosylserine was isolated from the reaction products after proteolytic treatment [28, 30] and, although the nature of the protein acceptors has not yet been investigated in detail, it seems reasonable to assume that they represent the protein moieties of the polysaccharide-protein complexes.

The specificity of UDP-xylose as glycosyl donor was investigated by Neufeld [32] and it was found that three other xylose-containing nucleotides yielded only minimal xylose incorporation. UDP-Xylose is formed in animal tissues [33, 34] as well as in plants [35] and yeast [36, 37] by decarboxylation of UDP-glucuronic acid which, in turn, originates from UDP-glucose:

$$\text{UDP-glucose} \xrightarrow{\text{DPN}} \text{UDP-glucuronic acid} \quad (1)$$

$$\text{UDP-glucuronic acid} \longrightarrow \text{UDP-xylose} + CO_2 \quad (2)$$

It has been observed by Neufeld and Hall [38] that UDP-xylose is a potent and specific inhibitor of UDP-glucose dehydrogenase, the enzyme catalyzing Reaction (1). Consequently, UDP-xylose may function as a feed-back regulator, or synchronizer, of the synthesis of the chondroitin 4-sulfate-protein complex. Thus, in case the rate of synthesis of the protein moiety of the polysaccharide-protein complex would decrease, the resultant accumulation of UDP-xylose would inhibit Reaction (1), and the rate of formation of UDP-xylose and its precursor, UDP-glucuronic aicd, would decrease. The reduction of the UDP-glucuronic acid pool would then result in a diminished synthesis of the chondroitin sulfate chains proper to balance the reduced protein syntheiss. In this

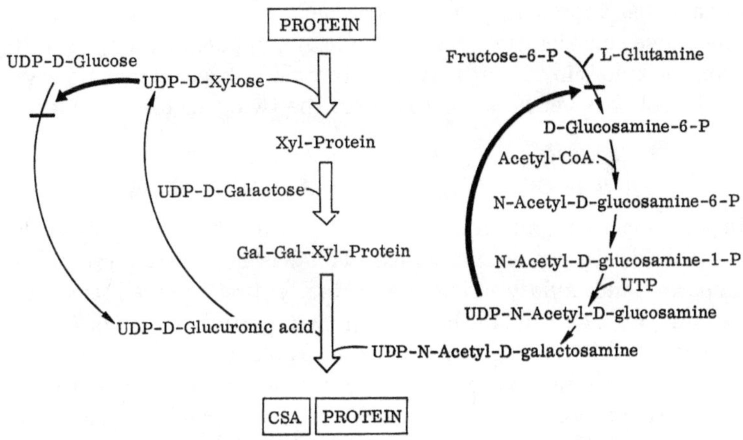

Fig. 2. Pathways of biosynthesis of the chondroitin 4-sulfate-protein complex. Reactions subjected to feed-back regulation are indicated by heavy arrows

context it is of interest to note that the synthesis of the hexosamine moiety of chondroitin sulfate is also under feed-back regulation via an inhibitory effect of UDP-N-acetylglucosamine on the first enzyme of the hexosamine pathway, fructose 6-phosphate: L-glutamine transamidase [39]. A summary of some of the reactions involved in the synthesis of the chondroitin sulfate protein-complex is shown in Fig. 2.

In addition to what has been discussed above, a few comments with regard to the biosynthesis of the galactose moieties are in order. It has been shown that galactose is incorporated from UDP-galactose into the trichloroacetic acid-insoluble fraction of the chick cartilage system [30] and, following treatment of this material with alkali in the presence of borohydride, galactosylxylitol was obtained as a reaction product. In preliminary experiments evidence has also been obtained for the formation of the second galactose residue of the carbohydrate-protein linkage region [40]. It is not clear, however, whether the system contains one or two separate galactosyltransferases, and the answer to this problem will have to await further progress in the attempts to solubilize and fractionate the enzymes involved.

Since the first glucuronic acid residue of the chondroitin sulfate chains proper is linked to galactose rather than to galactosamine the question also arises whether the addition of this glucuronic acid moiety — which has been demonstrated in preliminary experiments in our laboratory — requires a separate enzyme or if the specificity of the "polymerizing" glucuronosyltransferase is broad enough to permit transfer to galactose as well as to galactosamine.

The above considerations serve to indicate, in part at least, a specific function for xylose in the carbohydrate-protein linkage region. However, no special role has yet been suggested for the two galactose residues.

Xylose in Other Polysaccharide-Protein Complexes

In addition to chondroitin 4-sulfate, several other mucopolysaccharides are linked to protein via xylose. This has been firmly established for heparin, and xylosylserine was actually first isolated from this compound [41, 42]. Lindahl has shown that partial acid hydrolysis of heparin yields all the fragments listed in Table II and concludes that there is complete identity between the carbohydrate sequences in the polysaccharide-protein linkage regions of heparin and chondroitin 4-sulfate [43—45]. Similarly, chondroitin 6-sulfate has an identical carbohydrate-protein linkage region, since (a) alkali treatment cleaves the linkage and destroys serine [22], (b) glycopeptides containing galactose and xylose in a molar ratio of 2:1 may be obtained by degradation with hyaluronidase and proteolytic enzymes [46] and (c) the typical linkage region oligosaccharides, 3-O-β-D-glucuronosyl-D-galactose, 4-O-β-D-galactosyl-D-xylose, 3-O-β-D-galactosyl-D-galactose and O-β-D-galactosyl-$(1 \rightarrow 3)$-O-β-D-galactosyl-$(1 \rightarrow 4)$-D-xylose have been isolated from chondroitin 6-sulfate [47].

Evidence of a similar nature has been obtained for dermatan sulfate (chondroitin sulfate B), including the isolation of the three typical galactose-containing oligosaccharides [48, 49].

Some evidence has been presented to indicate that the carbohydrate-protein linkage in heparitin sulfate is the same as in heparin and the chondroitin sulfates. Thus, Jacobs and Muir found that serine was the major amino acid in heparifin sulfate isolated from aorta by proteolytic digestion [50]. The presence of xylose and galactose in heparitin sulfate from the same source as well as from tissues of patients with Hurler's syndrome was demonstrated by Knecht, Cifonelli and Dorfman [51, 52].

Arabinose in Hyaluronic Acid

Whereas the existence of covalently bound complexes between protein and the sulfated mucopolysaccharides has been well known for a long time, the situation with regard to hyaluronic acid has been less clear. Only in the last few years has it become reasonably certain that hyaluronic acid does indeed occur in the native state as a covalently linked protein complex. However, the mode of linkage between protein and carbohydrate is still unknown. Some interesting observations with possible bearing on this problem have made been by Stary and collaborators. In the analysis of hyaluronic acid from brain tissue, it was discovered that such preparations contained L-arabinose, galactose and glucose [2]. The two latter sugars have also been observed by Hamerman, Rojkind and Sandson [53] in hyaluronic acid-protein samples from synovial fluid. Arabinose was not found in these preparations, nor could it be detected except possibly in trace amounts in hyaluronic acid from umbilical cord [54]. It is tempting to speculate that L-arabinose may represent the point of linkage between carbohydrate and protein in much the same way as xylose joins the two moieties in the chondroitin sulfate-protein complex. The present experimental data are still incomplete, however, and do not permit any definite conclusions regarding this problem.

Xylose in Glycoproteins

Information concerning the occurrence of xylose in carbohydrate-protein compounds other than the acid mucopolysaccharides is at present rather vague. It is quite clear that xylose is a constituent of several glycoproteins of plant as well as of animal origin. These include bile and urinary glycoproteins [55], a glycoprotein from placenta [56], orosomucoid and ribonuclease [57], the plant enzymes Taka-amylase [58] and pineapple-stem bromelain [59] as well as others. Structural studies are in progress in several laboratories and it is anticipated that a more complete picture of at least some of these compounds will emerge in the near future.

Concluding Remarks

Throughout the history of xylose the appearance of this monosaccharide in substances of animal origin has been regarded with disbelief and often been interpreted as an artifact. It is of interest to note that — to the best of the author's knowledge — the presence of xylose in animal tissues was first reported in 1910 by the Japanese biochemist Kura Kondo from Tokyo [60]. Kondo found that a compound with the same melting point (143°) as the osazone of xylose was formed when he treated chondroitin sulfate (!) with phenylhydrazine and hydrochloric acid. His discovery was soon forgotten, but on the occasion of this symposium on pentoses it seems appropriate to draw attention to this remarkable achievement and conclude by paraphrasing an old saying: Nothing is new under the sun of Nippon!

References

1. Werner, I.: Acta Soc. Med. upsalien. **58**, 1 (1953).
2. Wardi, A. H., W. S. Allen, D. L. Turner, and Z. Stary: Arch. Biochem. **117**, 44 (1966).
3. Brimacombe, J. S., and J. M. Webber: Mucopolysaccharides. Amsterdam: B. B. A. Library, Elsevier 1964.
4. Muir, H.: Int. Rev. Connective Tissue Res. **2**, 101 (1964).
5. Ashwell, G.: Ann. Rev. Biochem. **33**, 101 (1964).
6. Fitton Jackson, S., R. D. Harkness, S. M. Partridge, and G. R. Tristram (eds.): Structure and function of connective and skeletal tissues. London: Butterworths 1965.
7. Gottschalk, A.: Glycoproteins. Amsterdam: B. B. A. Library, Elsevier 1965.
8. Neufeld, E. F., and V. Ginsburg: Ann. Rev. Biochem. **34**, 297 (1965).
9. Quintarelli, G.: The chemical physiology of mucopolysaccharides. Boston: Little, Brown & Co. 1968.
10. Sharon, N.: Ann. Rev. Biochem. **35**, 485 (1966).
11. Schiller, S.: Ann. Rev. Physiol. **28**, 137 (1966).
12. Roseman, S. (ed.): 4th Intern. Conference on Cystic Fibrosis of the Pancreas (Mucoviscidosis), Part II. Basel and New York: Karger 1968.
13. Rodén, L., in: Third Nuffield Conference on Rheumatism, Ditchley Park, 1967.
14. Fransson, L.-Å., and L. Rodén: J. biol. Chem. **242**, 4161 (1967).
15. — — J. biol. Chem. **242**, 4170 (1967).
16. Kent, P. W., and J. C. Marsden: Biochem. J. **87**, 38P (1963).
17. Inoue, S., and Z. Yosizawa: Arch. Biochem. **117**, 257 (1966).
18. Mathews, M. B., and I. Lozaityte: Arch. Biochem. **74**, 158 (1958).
19. Partridge, S. M., H. F. Davis, and G. S. Adair: Biochem. J. **79**, 15 (1961).
20. Muir, H.: Biochem. J. **69**, 195 (1958).
21. Gregory, J. D., T. C. Laurent, and L. Rodén: J. biol. Chem. **239**, 3312 (1964).
22. Anderson, B., P. Hoffman, and K. Meyer: J. biol. Chem. **240**, 156 (1965).
23. Rodén, L., and G. Armand: J. biol. Chem. **241**, 65 (1966).
24. Lindahl, U., and L. Rodén: J. biol. Chem. **241**, 2113 (1966).
25. Rodén, L., and R. Smith: J. biol. Chem. **241**, 5949 (1966).
26. Lindberg, B., and B.-G. Silvander: Acta chem. scand. **19**, 530 (1965).

27. Tominaga, F., K. Oka, and H. Yoshida: J. Biochem. **57**, 717 (1965).
28. Grebner, E. E., C. W. Hall, and E. F. Neufeld: Biochem. biophys. Res. Commun. **22**, 672 (1966).
29. — — — Arch. Biochem. **116**, 391 (1966).
30. Robinson, H. C., A. Telser, and A. Dorfman: Proc. nat. Acad. Sci. (Wash.) **56**, 1859 (1966).
31. Rodén, L., and B. Togias: Unpublished results.
32. Neufeld, E. F.: International Symposium on glycosaminoglycans, glycoproteins and glycolipids. Korin Kaku, Tokyo, Japan, 1967.
33. Bdolah, A., and D. S. Feingold: Biochem. biophys. Res. Commun. **21**, 543 (1965).
34. Castellani, A. A., in: G. Quintarelli (ed.), The chemical physiology of mucopolysaccharides. Boston: Little, Brown & Co. 1968.
35. Feingold, D. S., E. F. Neufeld, and W. Z. Hassid: J. biol. Chem. **235**, 910 (1960).
36. Ankel, H., and D. S. Feingold: Biochemistry **4**, 2468 (1965).
37. — — Biochemistry **5**, 182 (1966).
38. Neufeld, E. F., and C. W. Hall: Biochem. biophys. Res. Commun. **19**, 456 (1965).
39. Kornfeld, S., R. Kornfeld, E. F. Neufeld, and P. J. O'Brien: Proc. nat. Acad. Sci. (Wash.) **52**, 371 (1964).
40. Helting, T., and L. Rodén: J. biol. Chem. **244**, 2790 (1969).
41. Lindahl, U., J. A. Cifonelli, B. Lindahl, and L. Rodén: J. biol. Chem. **240**, 2817 (1965).
42. —, and L. Rodén: J. biol. Chem. **240**, 2821 (1965).
43. — Biochim. biophys. Acta (Amst.) **130**, 361 (1966).
44. — Biochim. biophys. Acta (Amst.) **130**, 368 (1966).
45. — Arkiv Kemi **26**, 101 (1966).
46. Schmidt, M., A. Dmochowski, and B. Wierzbowska: Biochim. biophys. Acta (Amst.) **117**, 258 (1966).
47. Helting, T., and L. Rodén: Biochim. biophys. Acta (Amst.) **170**, 301 (1968).
48. Stern, E. L., J. A. Cifonelli, L.-Å. Fransson, B. Lindahl, L. Rodén, S. Schiller, and M. L. Spach: Arkiv Kemi **30**, 583 (1969).
49. Stern, E.: Fed. Proc. **27**, 596 (1968).
50. Jacobs, S., and H. Muir: Biochem. J. **87**, 38P (1963).
51. Knecht, J., and A. Dorfman: Biochem. biophys. Res. Commun. **21**, 509 (1965).
52. — J. A. Cifonelli, and A. Dorfman: J. biol. Chem. **242**, 4652 (1967).
53. Hamerman, D., M. Rojkind, and J. Sandson: Fed. Proc. **25**, 1040 (1966).
54. Allen, W. S., A. H. Wardi, and Z. Stary: Fed. Proc. **26**, 282 (1967).
55. Weicker, H., and D. Grässlin: Nature (Lond.) **212**, 715 (1966).
56. Tomoda, M., and K. Murayama: Jap. J. exp. Med. **35**, 159 (1965).
57. Weicker, H., and R. W. Jeanloz: Fed. Proc. **26**, 607 (1967).
58. Tsugita, A., and S. Akabori: J. Biochem. **46**, 695 (1959).
59. Murachi, T., A. Suzuki, and N. Takahashi: Biochemistry **6**, 3730 (1967).
60. Kondo, K.: Biochem. Z. **26**, 116 (1910).

Discussion

Dr. Touster: Couldn't UDP-xylose arise from UDP-L-iduronic acid? Perhaps this is the reason that the production of UDP-xylose from UDPGA has not been found in mammals as yet.

Dr. Rodén: This is certainly possible. However, formation of UDP-xylose via UDP-L-iduronic acid would seem to represent an unnecessary detour in the metabolism. Furthermore, L-iduronic acid is not found in cartilage, whereas UDP-xylose is utilized in this tissue for the formation of the chondroitin 4-sulfate-protein complex. Also, since UDP-xylose is formed from UDPGA in bacteria, plants, and avian tissues it would seem most reasonable to assume that the same pathway occurs in mammals, even if it has not yet been conclusively demonstrated.

Dr. Horecker: D-xylose and L-arabinose have been isolated from brain. Is anything known with respect to the nature of the mucopolysaccharide from which they are derived?

Dr. Rodén: In the fractionation of the mucopolysaccharide mixture from brain tissue, Stary and his collaborators found that xylose followed the chondroitin sulfate fraction, whereas the arabinose was considered to be a component of the hyaluronic acid. However, the exact location of arabinose in the hyaluronic acid molecule has not yet been determined, and in view of the difficulties in purifying the acid mucopolysaccharides there is even a possibility that arabinose may be part of a macromolecule other than hyaluronic acid.

Dr. Ninomiya: You did these experiments with cartilage and skin. Is the xylose content high in these tissues?

Dr. Rodén: We have not made any direct determinations of the xylose content of these tissues. However, if xylose is present exclusively in the protein complexes of chondroitin 4-sulfate and dermatan sulfate, the xylose content could be expected to be in the order of 0.1% (of dry weight) for cartilage and much less in skin, since the latter tissue has a much lower content of polysaccharide.

Dr. Yoshikawa: Is there any specific enzyme to split the bond between pentose and serine?

Dr. Rodén: A xylosidase has been isolated from liver which cleaves xylosylserine. However, no enzyme has been found which can effect the cleavage of the xylose-serine bond in the polysaccharide-protein complex.

Formation of a New Family of Aminopentose Phosphates by Extracts of Streptomyces Alboniger *

BURTON M. POGELL[1], PETER F. REBELLO, and PATIT P. MUKHERJEE

Department of Biochemistry, Albany Medical College,
Albany, New York, U.S.A.

Introduction. The only known naturally-occurring aminopentose is 3-amino-3-deoxy-D-ribose, which has been found in three different microorganisms as part of adenine nucleosides. Fig. 1 shows the structure of two of these derivatives. 3′-Amino-3′-deoxyadenosine (Fig. 1a) was

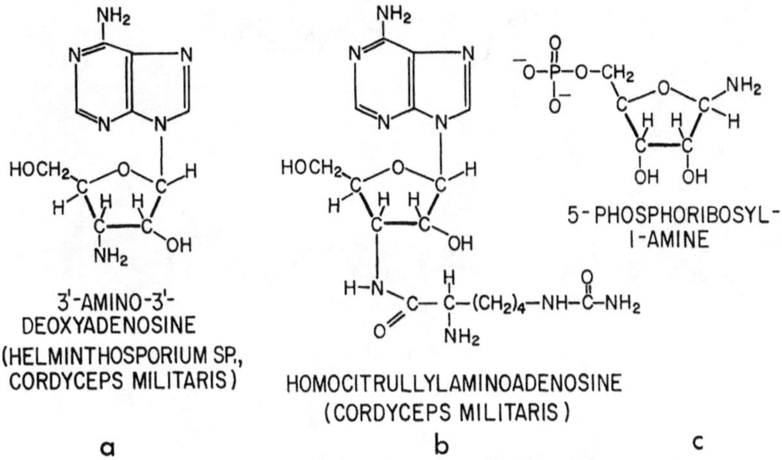

Fig. 1. Natural compounds which contain aminopentoses

isolated from *Helminthosporium sp.* [2]. It is identical with adenosine except that the 3′-hydroxyl group has been replaced by an amino group. This compound was also found in *Cordyceps militaris*, along with its derivative, homocitrullylaminoadenosine (Fig. 1b), which is 3′-amino-

---

* This work was supported by grants from the National Institutes of Health, U.S.A. (GM-12888) and the National Science Foundation, U.S.A. (GB-2441). A preliminary report of these studies has been presented [1].

1. Research Career Development Awardee of the United States Public Health Service, No. GM-K3-3033. Present address: Department of Microbiology, St. Louis University School of Medicine, St. Louis, Missouri, U.S.A.

3′-deoxyadenosine with the amino acid, homocitrulline, linked in amide linkage to the 3′-amino group [3, 4]. The first compound containing 3-aminoribose isolated from microorganisms was the antibiotic and antitumor agent, puromycin, whose structure is shown in Fig. 2 [5]. It is composed of a dimethyladenine moiety, a 3-aminoribose group, and an O-methyl-L-tyrosine, which, here again, is attached to the amino group of ribosamine in amide linkage. No normal metabolic function in microorganisms has been discovered for any of these compounds. Puromycin,

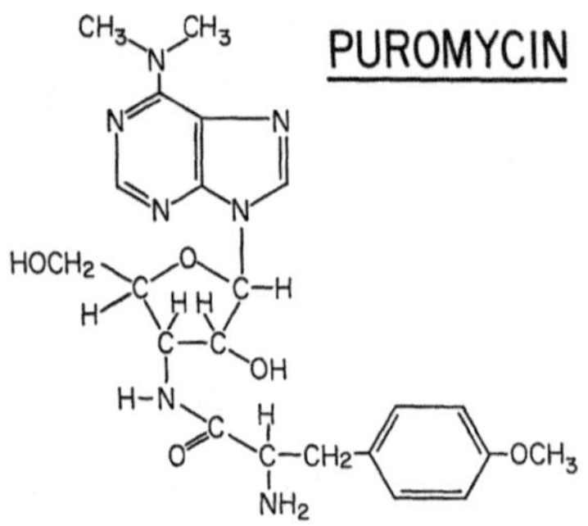

Fig. 2. Structure of puromycin

of course, has proved to be very useful to biochemists, primarily as an inhibitor of protein synthesis.

Another aminopentose, 5-phosphoribosyl-1-amine (Fig. 1c) is known for its role as an intermediate in the biosynthesis of purines. This compound is very labile and has not been isolated from natural sources.

Initially, in beginning our work on puromycin biosynthesis, we chose to study the enzymatic mechanism of 3-aminoribose formation. The cultures used for our current studies are shown in Fig. 3. The upper portion shows the first isolated culture which produced puromycin, namely *Streptomyces alboniger*. The name becomes quite obvious when you see the picture. This actinomycete produces beautiful white spores and a pigment is excreted into the medium by the cells, which at first is green and then after further growth turns black. The lower portion of Fig. 3 shows the mutant isolated by Lederle Laboratories which produced higher amounts of puromycin. This strain has the unique feature of not forming any black pigment. The chemical nature of this pigment, whether it is of a melanin-type formed by tyrosine oxidation or of some other structure, is still unknown.

Preliminary Studies on Aminopentose Formation. The enzymatic formation of the 3-aminoribose moiety of puromycin was followed using the Exley-Elson-Morgan colorimetric test for aminosugars [6]. 3-Aminoribose gives a purple color with the Elson-Morgan reaction for hexosamines, with about 10% of the color yield of glucosamine and a slightly different absorption maximum, at 550 mμ instead of 530—540 mμ.

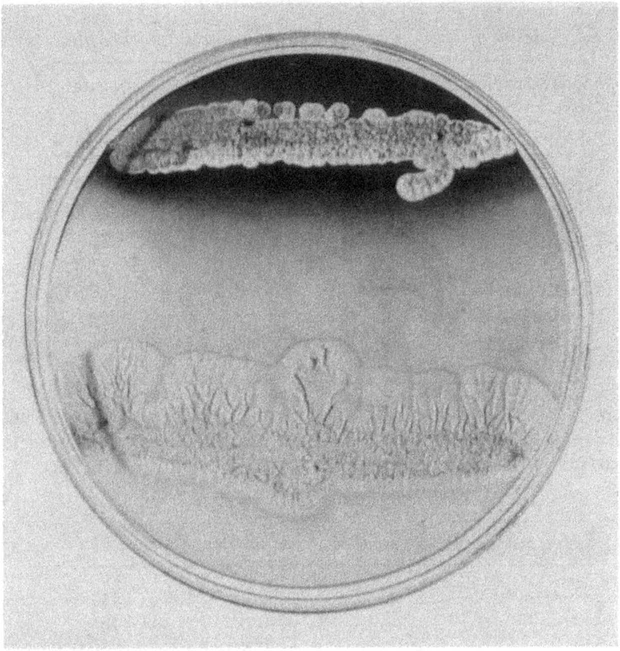

Fig. 3. Cultures of *Streptomyces alboniger* on glucose-yeast extract-malt extract agar. Upper: ATCC No. 12461; lower: ATCC No. 12462

Mutant cell cultures were grown to the point where maximum antibiotic production occurred and then harvested, washed, and a cell-free extract made by sonication. In our preliminary experiments, we found that the soluble fraction from these extracts would produce Exley-positive color when incubated with D-ribose 5-phosphate, ammonium chloride, and phosphate buffer at pH 7.5 and 38°. Full enzyme activity was retained after dialysis, no other cofactors were required, and the enzyme activity was destroyed upon boiling.

We then examined the substrate requirements of this reaction. The carbon donor specificity studies are summarized in Table I. It may be seen that ribose 5-phosphate was a better carbon source for aminosugar formation than either glucose 6-phosphate or fructose 6-phosphate. Ribose alone gave no synthesis; with added ATP, a small increase in color was observed, presumably from phosphorylated ribose. No color formation was found in the absence of an added carbon source.

Specificity studies with regard to nitrogen source are summarized in Table II. In the first experiment with undialyzed crude supernatant, NH_4^+ appeared to be a better source of the nitrogen group than either L-glutamine or L-asparagine. In the second experiment, we used partially purified enzyme, a thoroughly dialyzed 45 to 65% saturation ammonium sulfate precipitate. A much larger increase in differential color yield was found. NH_4^+ was now clearly the best nitrogen donor, and yields from either glutamine or asparagine were not much above that observed

Table I. *Specificity of carbon donor for aminopentose phosphate synthesis*

Carbon Source	Aminosugar formed mμmoles per ml Exp. 1	Exp. 2
Ribose-5-P	100	190
Glucose-6-P (5 mM)	—	0
Glucose-6-P (10 mM)	36	18
Glucose-6-P (20 mM)	—	57
Fructose-6-P (10 mM)	—	0
Ribose (10 mM)	0	—
Ribose + ATP (10 mM)	17	—
None	—	0

In Exp. 1, 10 mM ribose-5-P and 15 mM NH_4Cl were used; in Exp. 2, 15 mM ribose-5-P and 20 mM NH_4Cl. Undialyzed crude supernatant of *S. alboniger* was present in Exp. 1, dialyzed supernatant in Exp. 2.

Table II. *Specificity of nitrogen donor for aminopentose phosphate synthesis*

Nitrogen Source	Aminosugar formed mμmoles per ml Exp. 1	Exp. 2
NH_4Cl	86	224
L-Glutamine	59	78
L-Asparagine	53	73
None	33	41

In Exp. 1, 10 mM ribose-5-P, 15 mM nitrogen source, and undialyzed crude extract were present; in Exp. 2, 20 mM ribose-5-P, 30 mM nitrogen source, and partially-purified enzyme.

without addition of any nitrogen source. One problem encountered here is the difficulty in removal of all residual ammonium ions bound to enzyme protein. Linearity of product formation with time was observed with the partially purified enzyme (Fig. 4). On the basis of these experiments, we suspected we were observing aminopentose formation as follows:

$$\text{D-ribose-5-P} \rightleftharpoons \text{D-ribulose-5-P} + NH_4^+ \rightarrow \text{aminopentose-5-P} + H_2O.$$

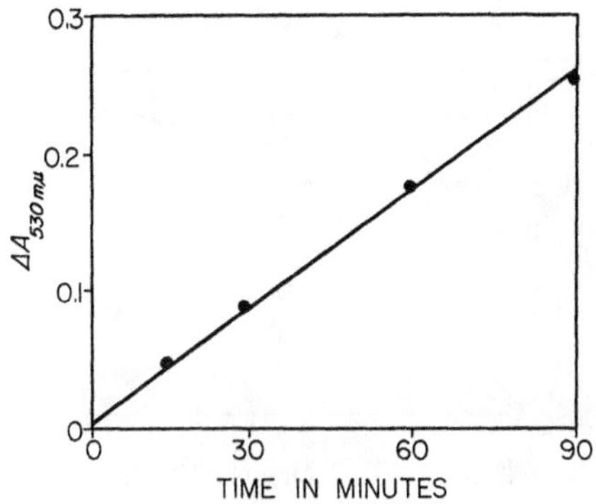

Fig. 4. Effect of time on aminosugar formation with partially-purified enzyme. 20 mM ribose-5-P, 30 mM NH_4Cl

A very active phosphoriboisomerase activity is present in these extracts, about 100-fold greater than the rate of aminosugar formation, and we wanted to make as certain as possible that we were indeed getting enzymatic formation of aminosugar. The following experiments were done to establish this point. There was no question that enzyme was required for aminosugar formation from ribose 5-phosphate and NH_4^+, since no product was formed in the absence of added extract. However, we were concerned about nonenzymatic product formation from ribulose 5-phosphate and possibly from other pentose phosphates. When ribose 5-phosphate was first incubated with enzyme in the absence of ammonium ions, this mixture then boiled and NH_4Cl added, only a very small increase in Exley color was observed upon further incubation. We also incubated ribose 5-phosphate with enzyme in the absence of ammonium ions, removed protein with perchloric acid, isolated the mixture of unknown pentose phosphates, and then incubated this mixture with NH_4Cl with and without the further addition of fresh extract. These results are shown in Fig. 5. A rather large color differential was observed in the absence and presence of enzyme. In this experiment, ribose 5-phosphate was preincubated for 5 mins with enzyme before isolation of the pentose phosphate mixture. Similar results were observed with a 30 min preincubation period. In a third experiment, equilibrium mixtures of ribose 5-phosphate and ribulose 5-phosphate were prepared by incubating ribose 5-phosphate with a partially purified spinach phosphoriboisomerase, and here again 70—80% of the aminosugar color formed from this mixture required the presence of enzyme. It thus appeared established that a *Streptomyces* enzyme was required for aminosugar formation from either ribose 5-phosphate, ribulose 5-phosphate,

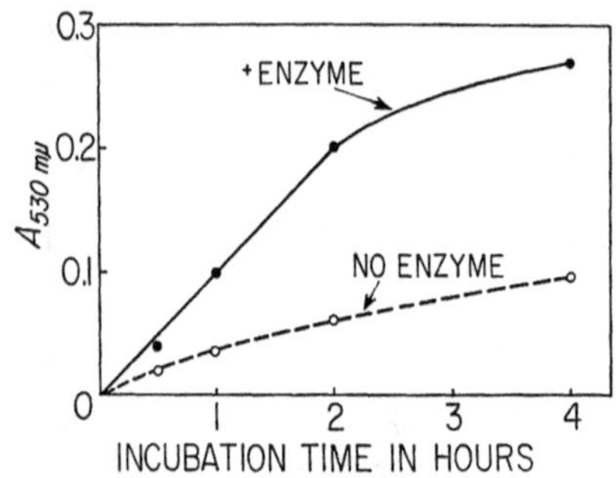

Fig. 5. Demonstration of enzyme requirement for aminosugar formation from enzymatically-formed pentose phosphate mixture

or any other pentose phosphates formed by the extract. However, a small amount of aminosugar product was formed nonenzymatically from ribulose 5-phosphate and other pentose phosphates and NH_4Cl.

Nature of Aminosugar Product. A large scale preparation of aminosugar phosphates was isolated after incubating ribose 5-phosphate, NH_4Cl, and dialyzed *S. alboniger* extract for 2 hrs at pH 7.5 and 38°. We were not able to increase the yield of product from ribose 5-phosphate to much more than 2%, so that presumably we are studying an aminopentose phosphate deaminase and the reaction is being measured in the direction of aminosugar synthesis. Purification of the product was attained by adsorption and elution from a Dowex 50 (H^+) column (Fig. 6a). Aminosugar phosphates form "zwitterions" and therefore stick very loosely to Dowex 50. By elution with slightly acidic water (pH 4.5), the aminosugar phosphates, but not free aminosugars, were removed. Glucosamine 6-phosphate previously had been shown to behave in this manner on cation exchange resins [7]. The unknown mixture was then treated with potato acid phosphatase. Most of the organic phosphate was released with little loss of aminosugar color.

The dephosphorylated material upon rechromatography behaved as shown in Fig. 6b. Aminosugars were eluted only after treatment with 0.2N HCl.

Paper chromatography of the purified aminosugar phosphates in isopropanol: HCl clearly established that our product was not glucosamine

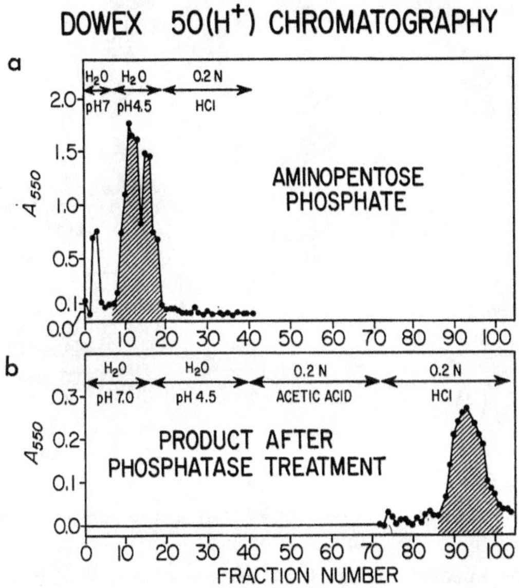

Fig. 6. Chromatography of unknown aminosugar products on Dowex 50 (H^+) resin before and after dephosphorylation

6-phosphate (Fig. 7). Evidence for the existence of at least two new compounds was found, both easily separable from glucosamine 6-phosphate. Similar results were obtained by paper chromatography in phenol. We also checked the behavior of the free aminosugars obtained after dephosphorylation in three different chromatographic solvents and found mobilities clearly different from either D-glucosamine or D-galactosamine and very similar to 3-aminoribose. However, the color spectrum of the aminosugar phosphate in the Exley reaction, shown in Fig. 8, was very similar to glucosamine 6-phosphate, and quite distinct from 3-aminoribose. The color obtained with the free aminosugars was identical to that found with the aminosugar phosphate. So, on the basis of these experiments, we assumed that our product was a mixture of 2-amino-2-deoxy-pentose phosphates.

Upon more refined Dowex 50 chromatography by shifting the pH of elution to 5.5, we were able to resolve the aminophosphate mixture

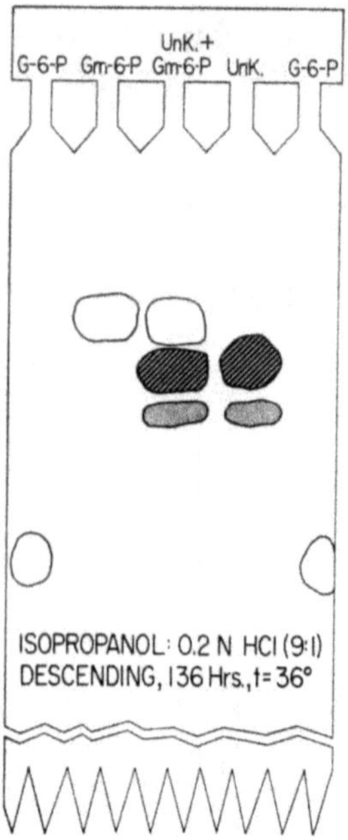

Fig. 7

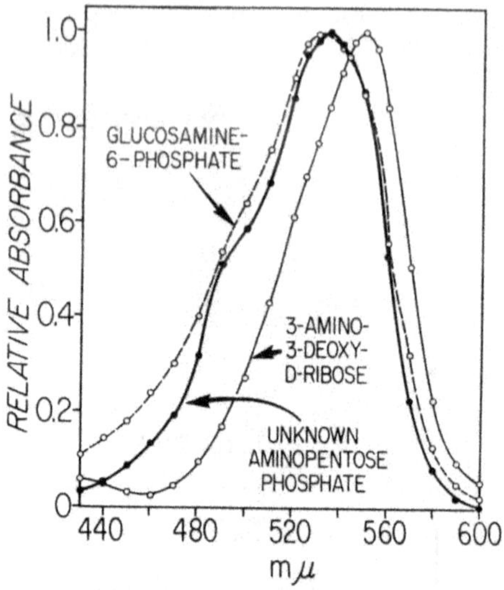

Fig. 8

Fig. 7. Paper chromatography of unknown aminosugar phosphates in isopropanol: HCl

Fig. 8. Absorption spectra in Exley procedure of aminopentose phosphate product, glucosamine-6-P, and 3-aminoribose

into three different fractions (Fig. 9). Each fraction was concentrated and rechromatographed on the same size column. From fraction 1, we obtained two different peaks, which are referred to as 1 a and 1 b. From fractions 2 and 3, we obtained two rather homogeneous large peaks. The small minor peaks were discarded. These results indicated that our mixture contained at least three different aminopentose phosphates. The concentrated fraction 3 crystallized as fine needles upon freezing and thawing.

Each of these rechromatographed products was then separately dephosphorylated with potato acid phosphatase and chromatographed on Dowex 50 (H^+) with 0.2N HCl as eluant (Fig. 10). From fraction 1 a, we obtained two peaks. The first compound corresponded to glucosamine. The second peak appeared to contain two components. Fractions 1 b, 2, and 3 gave largely homogeneous peaks.

The purity of the aminopentoses from these latter fractions was further confirmed by both paper chromatography and by quantitative

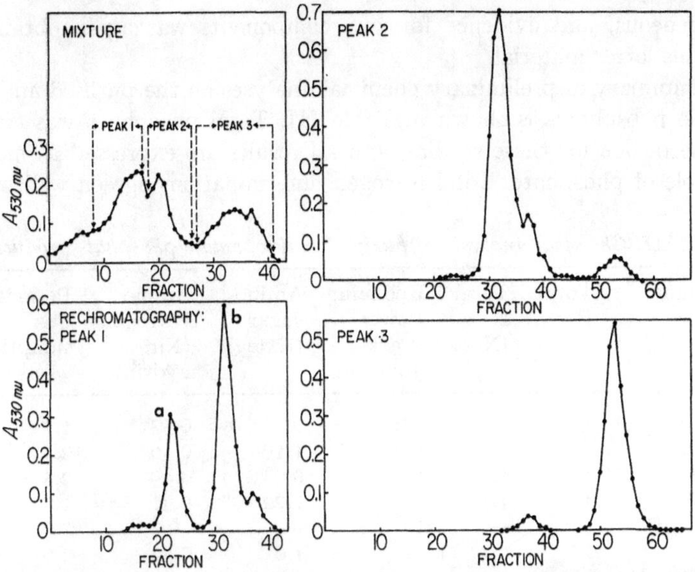

Fig. 9. Dowex 50 (H^+) chromatography of aminopentose phosphates. Elution with water adjusted to pH 5.5. Products isolated after 5 hr incubation

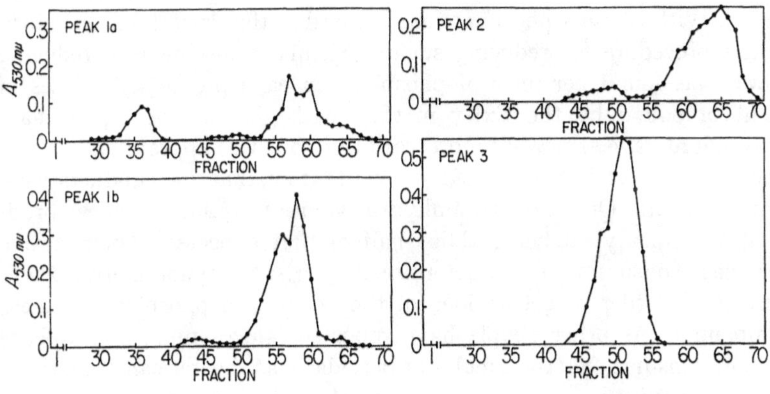

Fig. 10. Dowex 50 (H^+) chromatography of free aminopentoses after dephosphorylation of purified aminopentose phosphate products. Elution with 0.2 N HCl

measurement in the amino acid analyzer. The aminopentoses appeared after glucosamine and galactosamine on an Aminex-H column with 0.35 M Na-citrate, pH 5.28, as eluant and were determined by reaction with ninhydrin[2]. All fractions except the second peak from 1a appeared

2. T. H. Plummer Jr., New York State Health Research Laboratories, kindly performed these determinations.

homogeneous, and evidence for two components was again obtained with this latter material.

A summary of preliminary chemical analyses on the purified aminopentose phosphates is shown in Table III. Total phosphate was given a value of one for these studies, and all results are expressed as moles per mole of phosphate. Total nitrogen determinations agreed well with

Table III. *Chemical analysis of purified aminopentose phosphate products*

Compound	Total P	Total N (Nessler)	Reducing Power (Park-Johnson)	Amino-sugar (Exley)	Amino Nitrogen (Ninhydrin)	Periodate Consumption
Peak 1a	1	1.2	1.2	0.74	0.53	1.7
Peak 1b	1	1.3	1.0	1.10	0.46	2.4
Peak 2	1	1.1	1.0	0.94	0.49	2.5
Peak 3	1	1.1	0.8	1.22	0.31	1.9
Ribose-5-P	1	—	1	—	—	3.0
Glucos-amine-6-P	1	1.1	—	1.0	0.72	3.6

the presence of one amino group per more of phosphate in each compound. With ribose 5-phosphate as standard in the Park-Johnson colorimetric procedure for reducing sugars [8], about one mole of reducing power was found per mole of phosphate in each product. Analyses of each compound by the Exley method with glucosamine 6-phosphate as standard gave somewhat more variation, but was approximately 1:1. These results further supported the contention that the products were 2-aminosugars. Quantitative ninhydrin values were low for these products, the color yield being about half of that expected. Quantitative periodate consumption was determined by a spectrophotometric procedure [9] in order to get an idea of the position of phosphate in these compounds. As other people have reported, glucosamine 6-phosphate did not consume exactly 4 moles of periodate, as theoretically expected. If this behavior is assumed to be general for the unknown aminosugar phosphates, then the results obtained agree with phosphate being on carbon 5 in compounds 1b and 2, and also probably in 3.

Ninhydrin degradation was utilized for further characterization of the free aminosugars. This procedure splits off carbon-1 and the 2-amino group, giving the corresponding nitrogen-free tetrose from the aminopentose. Each free aminosugar was first heated with excess ninhydrin in pyridine, the mixture then passed through a mixed-bed ion exchange column, and the effluent concentrated and compared with authentic threose and erythrose, the two possible tetroses that are formed by

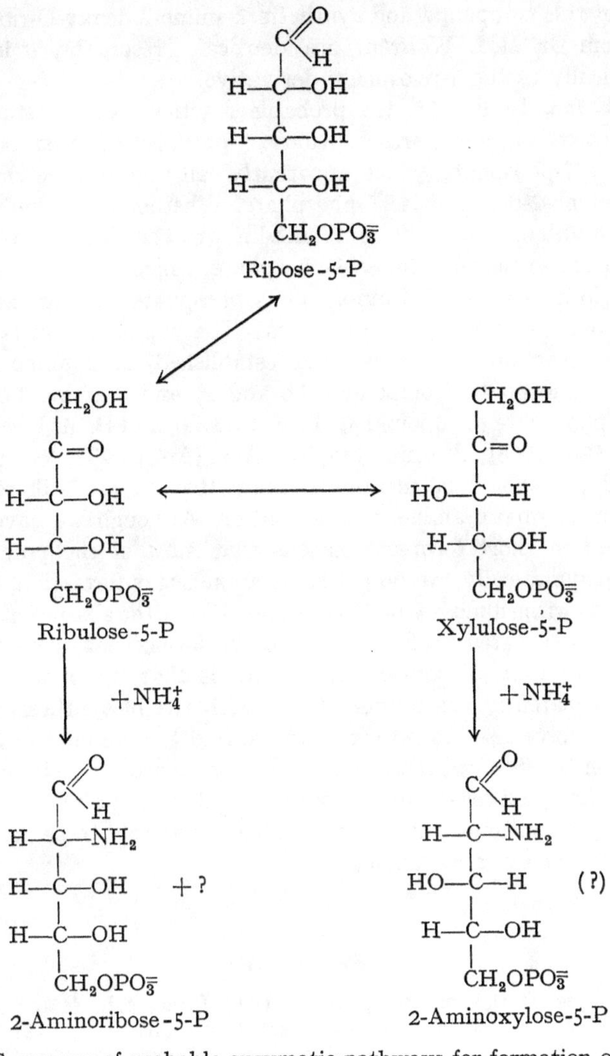

Fig. 11. Summary of probable enzymatic pathways for formation of aminopentose phosphates

removal of carbon-1. The two tetroses are clearly separated by electrophoresis in molybdate [10] and also by paper chromatography in phenol. Compounds 1a, 1b, and 2 gave only erythrose and compound 3, only threose. On the basis of these results, compounds 1a, 1b, and 2 have to be identical in configuration with either ribose or arabinose and compound 3, with either xylose or lyxose. The one compound for which we have definitely established identity is aminosugar 2. Microinfrared spectra of this compound and synthetic 2-amino-2-deoxy-D-ribose-HCl, a gift from Dr. M. L. Wolfrom, are identical. Presumably it is formed enzymatically as the 5-phosphate derivative.

Conclusion. In Fig. 11, the probable reactions we are studying in these extracts are summarized. Ribose 5-phosphate is first isomerized to ribulose 5-phosphate, which apparently can then react with NH_4^+ to form 2-amino-2-deoxyribose 5-phosphate. Whether compound 1b is a ribose or arabinose derivative is still not known. The ribulose 5-phosphate is further converted to xylulose 5-phosphate, which can then react with NH_4^+ to form 2-amino-2-deoxyxylose 5-phosphate (and/or 2-amino-2-deoxylyxose 5-phosphate). Addendum: The major products of the enzymatic reaction have now been established as 2-amino-2-deoxy-D-ribose 5-phosphate (compounds 1b and 2) and 2-amino-2-deoxy-D-lyxose 5-phosphate (compound 3) [P. F. Rebello, and B. M. Pogell, Bact. Proc., p. 126 (1968); Biochim. biophys. Acta (Amst.) **177**, 468 (1969)].

Finally, it should be mentioned again that we are still concerned about nonenzymatic aminosugar formation. Although we have clearly established in short term incubations that most of the products are formed enzymatically, we do get some aminosugar formation in longer incubations with ribulose 5-phosphate and NH_4^+ in the absence of enzyme. So all of our isolated products may not be formed enzymatically. The other matter that, of course, concerns us is that this new enzymatic reaction apparently has nothing to do with the biosynthesis of puromycin. We have also looked very hard to find the presence of 2-aminopentoses in *S. alboniger*, but so far have no evidence for their occurrence in acid-hydrolyzed cells; only glucosamine and muramic acid were found. We are now continuing our studies to see what role these compounds play in *Streptomyces* metabolism.

References

1. Mukherjee, P. P., and B. M. Pogell: Bact. Proc., p. 83 (1965).
2. Gerber, N. N., and H. A. Lechavalier: J. org. Chem. **27**, 1731 (1962).
3. Kredich, N. M., and A. J. Guarino: J. biol. Chem. **236**, 3300 (1961).
4. Guarino, A. J., and N. M. Kredich: Biochim. biophys. Acta (Amst.) **68**, 317 (1963).
5. Fryth, P. W., C. W. Waller, B. L. Hutchings, and J. H. Williams: J. Amer. chem. Soc. **80**, 2736 (1958).
6. Exley, D.: Biochem. J. **67**, 52 (1957).
7. Pogell, B. M., and R. M. Gryder: J. biol. Chem. **228**, 701 (1957).
8. Park, J. T., and M. J. Johnson: J. biol. Chem. **181**, 149 (1949).
9. Horecker, B. L., J. Hurwitz, and A. Weissbach: J. biol. Chem. **218**, 785 (1956).
10. Bourne, E. J., D. H. Hutson, and H. Weigel: J. chem. Soc. **163**, 4252 (1960).

Discussion

Dr. TAKAGI: Have you found some relationship between aminopentose phosphate formation and the growth of the cells?

Dr. POGELL: No, we have not done such studies.

Dr. TOUSTER: Couldn't your enzyme have a hexose derivative as its normal substrate? Competition studies between glucose 6-phosphate and ribose and ribulose 5-phosphates would be required.

Dr. POGELL: This is a possibility, although from our carbon source specificity studies, I would predict that this enzyme is not identical with glucosamine-6-phosphate deaminase.

Dr. TOUSTER: Is there a color to your enzyme that might suggest the presence of riboflavin?

Dr. POGELL: Our partially-purified ammonium sulfate precipitate is still yellow.

Dr. HORECKER: The fact that the addition of NH_3 is not stereospecific suggests that it may not be an enzyme-catalyzed reaction. Is it possible that the enzyme preparation is contributing a metal, or other non-protein, catalyst?

Dr. POGELL: Our enzyme preparations were always dialyzed against EDTA. No aminosugar was formed from ribose-5-phosphate and NH_4Cl with boiled enzyme. Also, if ribose-5-phosphate was first incubated with enzyme, the mixture boiled and then NH_4Cl added, only a small increase in aminosugar content was observed upon further incubation.

Dr. HORECKER: Have you tried ribulose diphosphate?

Dr. POGELL: No.

Session II

Physiology of Pentoses and Pentitols

Utilization of Xylitol in Animals and Man

KONRAD LANG

Physiologisch-Chemisches Institut der Johannes-Gutenberg-Universität
Mainz, West Germany

The discovery of the pentose phosphate cycle and the glucuronic acid xylulose cycle as important metabolic pathways raised the question of the nutritive and dietetic significance of the metabolites involved in the mentioned cycles or related substances and also the problem of their clinical use in disorders of the carbohydrate metabolism. The most interesting substance in this respect seems to be xylitol for many reasons:

1. Preliminary experiments in my laboratory and some data in the literature showed that the tested pentoses can not be used efficiently as a source of calories. 1957 and 1959 Wyngaarden, Segal and Foley [1, 2] reported experiments on the metabolism of tracer doses of infused ^{14}C labelled pentoses in man. With the exception of D-ribose, the pentoses given were not oxidized to $^{14}CO_2$ in a larger extent. Most of the activity was detected in the urine.

Table I. *Disposition of ^{14}C after infusion of tracer doses of labelled pentoses*

Pentose	% of the doses administered after 6 hours	
	expired as $^{14}CO_2$	excreted in the urine
D-xylose	16	35
D-lyxose	14	72
D-arabinose	19	57
L-arabinose	0.8	85
D-ribose	48	10

Experiments in which larger amounts of unlabelled pentoses ranging from 5 to 20 g were infused, the urinary excretion of the pentoses amounted to 27—60% of the dose given and averaged 44%.

In similar experiments with rats and guinea pigs, where ^{14}C labelled D-xylose was used, McCormick and Touster [3] found the same results:

only 11—15% of the activity given were oxidized to $^{14}CO_2$ and most of the activity excreted in the urine. The same could be shown in experiments using the corresponding pentitols. D- and L-arabitol were not oxidized to $^{14}CO_2$ in a larger extent, same as D-ribitol.

Table II. *Infusion of 10—20 g pentoses in normal subjects*

Pentose	*n*	% of the administered dose found in the urine
D-xylose	13	27—55
D-lyxose	2	57—86
D-arabinose	3	36—60
L-arabinose	4	38—60

Table III. *Utilization of pentitols by the rat*

Pentitol	% of the administered ^{14}C activity		
	expired as $^{14}CO_2$	in the urine	as liver glycogen
D-arabitol	14	36	0.2— 0.3
L-arabitol	13	38	0.9— 3.2
D-ribitol	31	27	17 —22

In contrast, tracer doses of xylitol were oxidized to a larger extent and more rapidly to $^{14}CO_2$ and this substance was efficiently utilized as a precursor of glycogen. Besides xylitol, only D-ribose was also well utilized in the metabolism by rats and guinea pigs.

2. On larger scale, xylitol can be prepared in a relatively easy and cheap way.

3. Xylitol is a very stable substance and gives no Maillard reaction when heating together with amino acids and related substances. This is a very important point regarding the practical use of this substance in the infusion therapy. And last not least, xylitol cristallizes very well, so that it is easy to obtain pure preparations needed for the parenteral use.

4. The metabolism of xylitol is independent of insulin. We found in eviscerated rats the xylitol space to be in the average 40% of the body weight [4]. The same space was found by us also in man and also in alloxandiabetic and eviscerated rats, and the space was not changed at all by administration of insulin.

Regarding the mentioned facts, we were encouraged to perform experiments on a larger scale for elucidating the biological properties of xylitol.

The first problem to be cleared was the acute and chronic toxicity and the compatibility of the substance. Xylitol was proved in our experiments as very non toxic. The LD_{50} in mice was estimated 25.7 g/kg body weight when given per os. In long-term feeding experiments on groups of each 30—60 rats, we found that 10% and 30% of xylitol in the diet were well tolerated and produced no adverse effects regarding growth, protein efficiency, reproduction and histopathology of the main organs. Lens opacities did not occur. In the first days of the feeding experiments, a part of the animals had diarrhea which ceased in the second week and was no more observed during the rest of the feeding period [5]. Besides, the feeding experiments gave clear evidence that the amounts of xylitol fed were utilized efficiently in the metabolism and could cover the caloric requirements of the animals.

In other experiments, the compatibility of parenteral administration of xylitol was tested in rats. Repeated intraperitoneal or i.v. infusions of large amounts of xylitol did not cause any disturbances or histopathological changes in liver, kidneys and heart [6].

Next we studied the problem of the capacity of man and animals to metabolize xylitol. We found the capacity considerable. Experiments in rats with xylitol-U-^{14}C using the continuous i.v. infusion technique showed that about 60% of the dose were oxidized to CO_2 and 30—35% were retained in the body. Liver glycogen accounted for 4—6% [7]. The rest was excreted in the urine. Feeding experiments on a larger scale on rats gave the same results. The total xylitol turnover of man measured by the rate of disappearance from the blood after i.v. infusion was estimated in my laboratory by Bässler [8]. He found that the xylitol turnover in man is in the same range as that of fructose. Giving 0.4 g xylitol/kg body weight, the total turnover averaged in adults 12 mg/kg/min, 11.3 mg being metabolized and 0.7 mg (corresponding to 6.1% of the dose) being excreted in the urine. The data of the elimination of xylitol from the blood as well as those of urinary excretion were the same in adults, children and even in premature infants, indicating that the premature infant possesses the full capacity of metabolizing xylitol. Figures in the same order were obtained also by Bässler in similar experiments on rats [9].

Kinetic studies on the decrease of the xylitol level in the blood after i.v. infusion of xylitol showed, that the process can be described as a reaction of first order. The decrease of the blood level was synchronous with the decrease of the xylitol level in the tissues. Because there was no significant excretion of xylitol in the urine (in these short-time experi-

Table IV. *Xylitol turnover.* (Data from Bässler)

	Adults	Children	Premature children	Rat
Xylitol administered g/kg	0.4	0.2	0.2	0.5
Total turnover mg/min/kg	12.0	7.2	7.2	10—20
Xylitol metabolized mg/min/kg	11.3	6.5	6.5	—
Xylitol excreted in the urine % of the dose administered	6.1	9.2	9.2	—

Table V. *Decrease of the xylitol level in the blood and tissues (*Bässler *et al., 1966) [9]. Rats were given 0.5 g/kg body weight of xylitol i.v.*

Time (min)	Level mg-%			
	Blood	Liver	Muscle	Adipose tissue
5	768	381	204	72
15	392	272	125	108
30	301	157	107	77
45	95	13	33	17
60	67	9	18	10
180	0.6	1.6	11	1.5

ments), the decrease of the blood level must be directly related to the turnover rate of the substance in the animal.

Using the continuous infusion technique Strack *et al.* found also the same figures for the xylitol turnover in rabbits, xylitol in a dose of 0.5 g/kg body weight/hour was metabolized by normal and alloxan-diabetic rabbits as well as fructose [10].

Table VI. *Xylitol clearance in the rabbit.* (Data from Strack *et al.*)

	Clearance ml/min/kg body weight	
	Normal animal	Diabetic animal
xylitol 0.50 g/kg	46.3	43.6
sorbitol 0.50 g/kg	32.3	20.2
D-fructose 0.50 g/kg	32.0	29.2

In the experiments of Strack *et al.* the turnover of xylitol in the rabbit averaged 8.33 mg/kg body weight, 13.2% being excreted in the urine.

Experiments on young and old normal persons and on patients with diabetes mellitus or liver diseases performed by Mehnert *et al.* showed that i.v. given xylitol could be utilized at the same extent by all persons tested. The urinary loss after a dose of 0.5 g of xylitol/kg body weight amounted in these experiments to 8—12% of the dose [11]. In my laboratory we also found no difference in the xylitol utilization in normal and diabetic subjects [12, 13].

Studies on the tissue specifity for xylitol utilization showed that the liver is the main organ in xylitol metabolism. Besides the liver only the kidney can metabolize xylitol to a significant extent. Red blood cells produce very efficiently lactic acid from xylitol [14]. The relation of the capacity of the liver to the capacity of the remainder of the organism was estimated by Bässler to about 10:1 for the rat, measuring the activity of xylitol dehydrogenation by the different organs in vitro [9].

Table VII. *Dehydrogenation of xylitol in vitro* (Bässler, 1966)

Rat organs, homogenates	µmol. dehydrogenated	
	mg Protein	whole organ
Liver	24.7	27,000
Kidney	6.5	2,150
Testes	1.5	240
Small intestine	0.26	75
Heart	0.40	25
Lung	0.13	20
Adipose tissue	0.56	15
Brain	0.10	10

In liver perfusion studies and perfusion of the whole animal Strack *et al.* found that in the rabbit about 15% of the xylitol were utilized by the extrahepatic tissues and 85% by the liver [10].

It seems to be remarkable that in man, rat and rabbit all parameters measured of xylitol metabolism and utilization are in such a close relation.

The turnover of xylitol is in a large scale influenced by the diet. Prof. Bässler's paper will discuss in detail the adaptation mechanism involved in the enzyme induction of xylitol metabolizing enzymes by prefeeding xylitol. Prefeeding the animals with xylitol will significantly increase absorption, turnover and oxidation of the substance. This point is very important e.g. for the utilization of xylitol when the substance is given per os.

The absorption of xylitol is very slow. In rats using xylitol-U-^{14}C, we found $t/2$ of the absorption to be 7 ± 1 hour [7]. This corresponds to an absorption rate of about 20% of that of glucose. Mehnert *et al.* found in man the same absorption rate [11]. The absorption of xylitol is a passive process. The slow rate of absorption causes diarrhea (for osmotic reasons) after feeding higher doses of xylitol in man and all animals tested. As mentioned above, an adaptive increase of absorption occurs to that extent that in our animal feeding experiments the diarrhea disappears in a few days (even feeding 30% of xylitol in the diet). However the slow absorption rate may limit the tolerable amount of xylitol when given per os to a certain extent.

After oral or parenteral administration of xylitol we never found increased excretion of pentoses in the urine. This is in contrast to D-xylose, which by blocking the pentose phosphate shunt at the transketolase level raises the excretion of ribulose and xylulose about 100 fold above the normal level [15, 16].

The ability of man to metabolize xylitol to a large extent and the compatibility of the substance, together with its useful dietetic properties like independence of its insulin metabolism, its strong antiketogenic effect and its depressing action on the blood level of non-esterified fatty-acids suggest that xylitol can be useful in treatment of metabolic disorders e.g. in diabetes mellitus. In this respect it should be mentioned that the sweet taste of xylitol is practically the same as that of sucrose.

References

1. Wyngaarden, J. B., S. Segal, and J. B. Folley: J. clin. Invest. **36**, 1395 (1957).
2. Segal, S., and J. B. Folley: J. clin. Invest. **38**, 407 (1959).
3. McCormick, D. B., and O. Touster: J. biol. Chem. **229**, 451 (1957).
4. Bässler, K. H., u. W. Prellwitz: Klin. Wschr. **42**, 94 (1964).
5. Kieckebusch, W., W. Griem u. K. Lang: Klin. Wschr. **39**, 447 (1961).
6. Griem, W., u. K. Lang: Klin. Wschr. **40**, 801 (1962).
7. Schmidt, B., M. Fingerhut u. K. Lang: Klin. Wschr. **42**, 1073 (1964).
8. Bässler, K. H., W. Toussaint u. G. Stein: Klin. Wschr. **44**, 212 (1966).
9. — G. Stein u. W. Belzer: Biochem. Z. **346**, 171 (1966).
10. Müller, F., E. Strack, E. Kuhfahl u. D. Dettmer: Z. ges. exp. Med. **142**, 338 (1967).
11. Mehnert, H., J. D. Summa u. H. Förster: Klin. Wschr. **42**, 382 (1964).
12. Bässler, K. H., W. Prellwitz, V. Unbehaun u. K. Lang: Klin. Wschr. **40**, 791 (1962).
13. Prellwitz, W., u. K. H. Bässler: Klin. Wschr. **41**, 196 (1963).
14. Bässler, K. H., u. W. V. Reimold: Klin. Wschr. **43**, 169 (1965).
15. — H. Holzmann, W. Prellwitz u. J. Stechert: Klin. Wschr. **43**, 27 (1965).
16. —, u. J. Stechert: Klin. Wschr. **43**, 31 (1965).

Discussion

Dr. Horecker: Did you analyze for changes in blood levels of glucose and pyruvate during glucose infusion?

Dr. Lang: We found no changes in the glucose level. The pyruvate level was not checked by us, but I think in other papers of this symposium figures will be given.

Dr. Kumagai: Do you think younger children excrete more xylitol in urine than do adults?

Dr. Lang: It is not the case. Dr. Erdmann will discuss this problem in detail.

Dr. MacDonald: Did you find any difference in the lipid metabolism between xylitol and glucose?

Dr. Lang: We did not make experiments in this field.

Dr. Touster: Dr. Lang, I believe you stated that the absorption of xylitol is a passive process, yet adaptation occurs which increases the rate of absorption. Will you amplify this point please?

Dr. Lang: The absorption is a passive process. Dr. Bässler will show in his paper that by adaptation phenomena the metabolism of xylitol will be enhanced on the cellular level, so that the blood level decreases. Therefore by smaller concentration gradient the absorption is favored.

Utilization of Xylitol in Human Erythrocytes

TOSHIO ASAKURA, HARUTADA NINOMIYA, SHIGEKI MINAKAMI, and HARUHISA YOSHIKAWA

Department of Physiological Chemistry and Nutrition, Faculty of Medicine, University of Tokyo, Tokyo, Japan

Mature human erythrocytes synthesize no DNA, RNA, heme or protein. They possess no citric acid cycle or electron transfer system, and obtain the energy mainly from carbohydrate metabolism through the Embden-Meyerhof and Warburg-Dickens pathways.

Of several sugar alcohols metabolized by erythrocytes, the unusual behavior of xylitol prompted the present investigation of its effects upon the reduction of methemoglobin and oxidized glutathione, and the protection of the cells from hemolysis. The relationship of these effects to the metabolism of xylitol was also studied.

Reduction of Methemoglobin

The reduction of methemoglobin to hemoglobin is an extremely important process because the former does not combine with oxygen. It is established that hemoglobin is oxidized spontaneously to methemoglobin *in vivo*, and that the methemoglobin is reduced in erythrocytes. Since xylitol was found to be an efficient reductant of methemoglobin in these cells [1], the mechanism of this reduction as well as the metabolism of this sugar alcohol have been investigated.

In Table I is shown the effect of various sugar alcohols and glucose on the rate of methemoglobin reduction. Each substance was added at a final concentration of 100 mM. In this experiment, erythrocytes were treated with sodium nitrite to oxidize hemoglobin to methemoglobin, and the cells were incubated with or without substrate after the excess nitrite was washed out. Methemoglobin content was measured by the method of Evelyn and Malloy [2]. As can be seen in the table, xylitol, sorbitol and glucose are effective reducing agents, whereas the other sugar alcohols are less effective. As shown in Table II, 10 mM xylitol is still active, whereas sorbitol had low activity at this concentration; xylitol is the most effective polyol for the reduction of methemoglobin.

Table I. *Effect of various sugars on methemoglobin reduction*

Substrate	met-Hb reduction (%/4 hrs)
None	−10.4
Glucose	26.8
Glycerol	4.9
Erythritol	2.9
Dulcitol	4.1
iso-Dulcitol	7.6
Xylitol	48.0
Inositol	4.7
Mannitol	12.4
Sorbitol	37.8

Table II. *Effect of sugar concentration on the rate of methemoglobin reduction*

Substrate	Concentration	met-Hb reduction (%/4 hrs)
—	—	−10.4
Xylitol	10	34.5
	100	48.0
Sorbitol	10	5.4
	100	37.8
Glucose	10	26.1
	100	26.8

Although the reduction rate is dependent upon polyol concentration, the rate of methemoglobin reduction remains constant even when glucose concentration is increased from 10 to 500 mM. The relationship between the rate of methemoglobin reduction and xylitol concentration is shown in Fig. 1. At a level of 150 mM, xylitol is approximately twice as effective as glucose.

It is known that in blood stored for more than 4 weeks glucose is not utilized probably because of the low ATP level in the cells. We found that xylitol was effective for the methemoglobin reduction even in the long stored blood (Fig. 2). The lower curve represents the rate of methemoglobin reduction by xylitol and the upper curve that by glucose. These results indicate that xylitol is actively metabolized in mature erythrocytes, although probably by a mechanism different from that of glucose (see below).

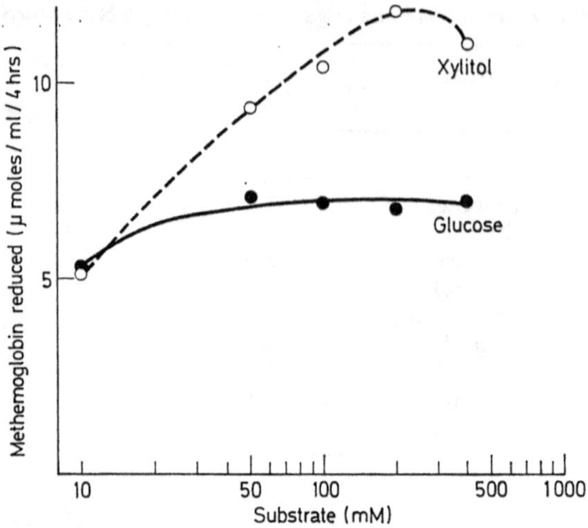

Fig. 1. Effect of substrate concentration on the rate of methemoglobin reduction [1]

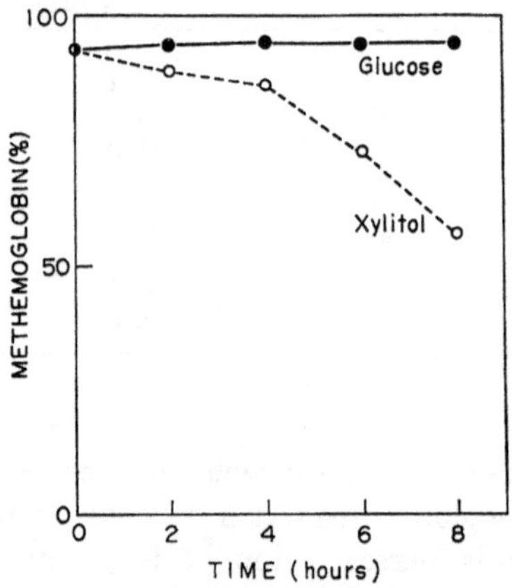

Fig. 2. Comparison of the rate of methemoglobin reduction by glucose and xylitol in stored blood (ACD medium, 4 W). Nitrite treated red cells were suspended in a solution containing 100 mM NaCl, 10 mM KCl, 30 mM Triethanolamine, and 100 mM substrate, at pH 7.4, and incubated at 37° C

Reduction of Oxidized Glutathione

Another important function of xylitol in erythrocytes is the reduction of oxidized glutathione, a substance which is thought to exert a protective effect on the integrity of hemoglobin and certain enzymes in the cells. As shown in Fig. 3, either xylitol or glucose can reduce oxidized glutathione in erythrocytes. The upper curve represents reduction by xylitol and the middle one that by glucose. The reduction of oxidized glutathione by xylitol is also observed in stored blood, whereas glucose is ineffective (Fig. 4). Reduced glutathione was measured by the method of Beutler [3].

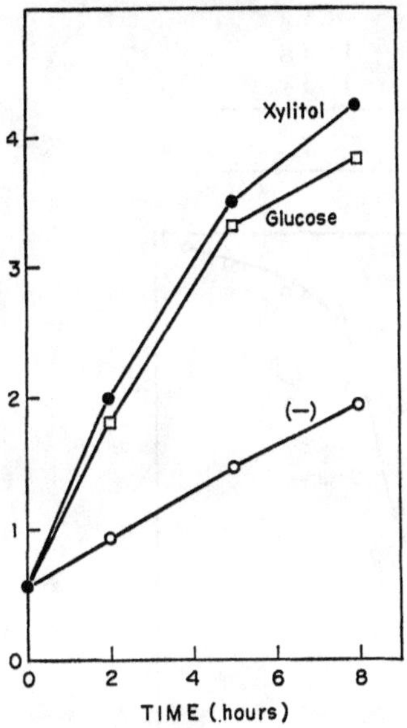

Fig. 3

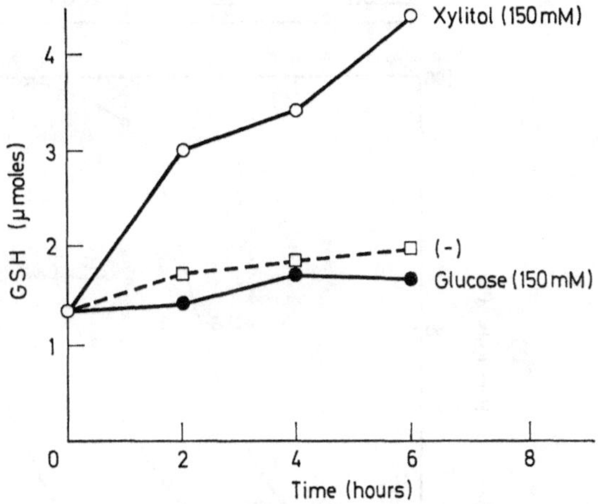

Fig. 4

Fig. 3. Reduction of GSSG by glucose or xylitol in fresh blood. Erythrocytes were treated with sodium nitrite in the absence of glucose and incubated with glucose or xylitol

Fig. 4. Reduction of GSSG by glucose or xylitol in stored blood

Protection of Cells from Hemolysis

It is known that a variety of hemolytic process is associated with diminished sulfhydryl activity in the red cell. Thus as a result of our experiments on the reduction of oxidized glutathione, we also have investigated the effect of xylitol on the osmotic fragility of stored blood.

Components of the blood storage system are shown in Table III. In system A, glucose was added and in B xylitol was added to the usual ACD mixture. The former conveniently will be referred to as ACDG and the latter as ACDX. In both systems the osmolarity are identical.

Table III. *Composition of blood storage system*

Additions	Usual	Experimental system	
		A (ACDG)	B (ACDX)
Na-Citrate	13.2 g	13.2 g	13.2 g
Citrate	4.8 g	4.8 g	4.8 g
Glucose	13.3 g	26.6 g	13.3 g
Xylitol	—	—	11.3 g
Blood	200 g	200 g	200 g

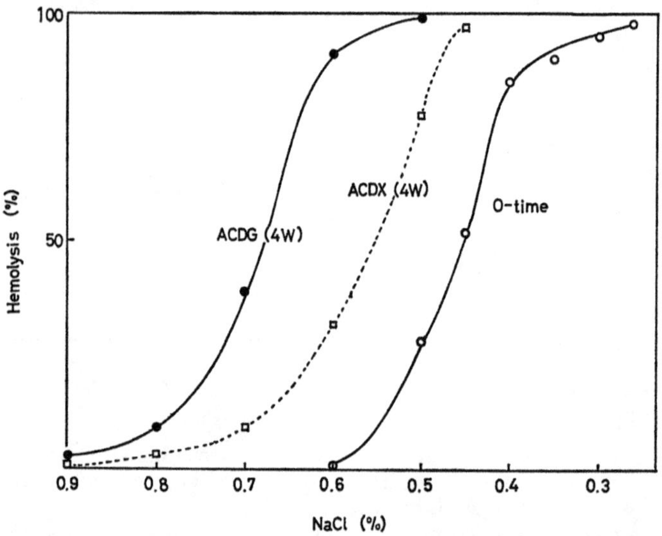

Fig. 5. Effect of xylitol on the osmotic fragility during storage

Typical results of an osmotic fragility test are shown in Fig. 5. Addition of xylitol clearly protects against hemolysis during storage. In Fig. 6, sodium chloride concentration at 50% hemolysis is plotted against refrigerated storage time. Although hemolysis increases both in the ACDG and ACDX systems during storage, the rate is decreased in the presence of xylitol. In order to determine the mechanism of protective action of xylitol against hemolysis, ATP and GSH levels of the stored blood were measured in the presence or absence of xylitol. Table IV shows the ATP level of stored blood. System A indicates ACDG and B ACDX. The medium containing xylitol exhibits a slightly higher ATP level. Although the level of reduced glutathione was also measured in

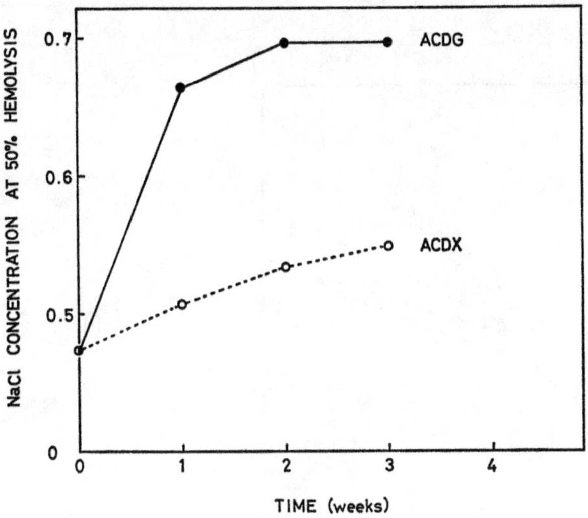

Fig. 6. Effect of xylitol on osmotic fragility during storage

Table IV. *ATP content in the stored blood (mμmoles/mμmoles heme)*

		0	1	2	3	4
		weeks				
I	A	51.5	30.8	24.5	18.9	9.5
	B	49.2	30.8	30.0	21.2	20.5
II	A	49.1	27.8	22.2	18.5	16.7
	B	50.0	—	23.6	23.6	19.2

A = ACDG; B = ACDX.

stored blood, there was no significant difference in the presence or absence of xylitol within 4 weeks.

Lactate formation was also observed during storage. The results are shown in Fig. 7. The dotted line represents lactate formation in the presence of xylitol and the solid lines that with glucose. It is apparent from this figure that the addition of xylitol promotes lactate formation during storage. This probably is due to the higher ATP lebel and also to the increase of NADH/NAD ratio in the presence of xylitol. Further investigation is necessary in order to find the optimal conditions for the use of xylitol in the blood storage system, since hemolysis occurs in the presence of sugars at higher concentrations.

As presented above, xylitol reduces methomoglobin and oxidized glutathione in red cells, and also protects against hemolysis. In order

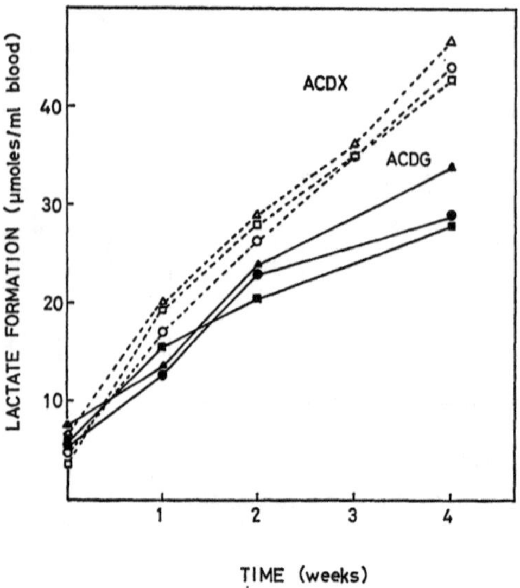

Fig. 7

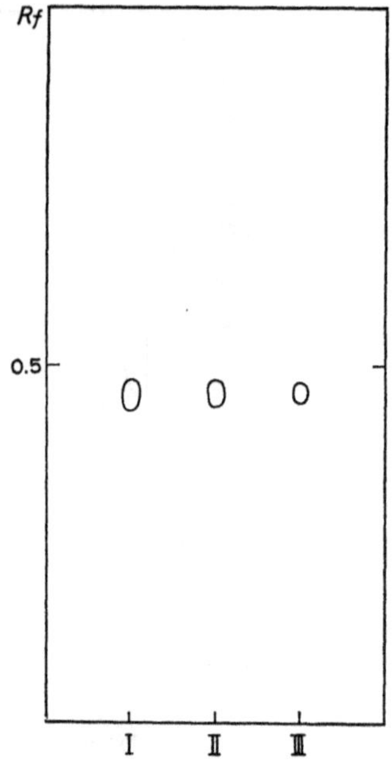

Fig. 8

Fig. 7. Rate of lactate formation during blood storage in the presence or absence of xylitol in ACD medium

Fig. 8. Paper chromatogram of the isolated ketopentose. Butanol-acetone-water (50:50:20). *I* and *III* authentic xylulose; *II* isolated ketopentose

to learn the mechanism of these effects of xylitol on erythrocytes, the metabolism of this sugar alcohol in mature human erythrocytes has been investigated also.

The reaction product of xylitol in erythrocytes was at first identified as xylulose. This was done by paper chromatography, employing either color reactions with cysteine carbazole and orcinol or gas chromatography. Fig. 8 shows the paper chromatogram of the isolated ketopentose, which was separated from the incubation mixture of red cells with xylitol. Only one spot can be detected as shown in this figure, and it has a same R_f value as the authentic xylulose has. In Fig. 9 is shown the spectra of the cysteine carbazole reaction products with the isolated ketosugar and authentic xylulose. The spectrum of the isolated ketopentose is quite similar to that of synthetic xylulose. Similar results were obtained with spectra of the orcinol reaction products (Fig. 10). Gas chromatography of the isolated ketopentose and authentic xylulose was carried out after trimethysilylation of the sugars. The retention time of both sugars were also in good agreement [1].

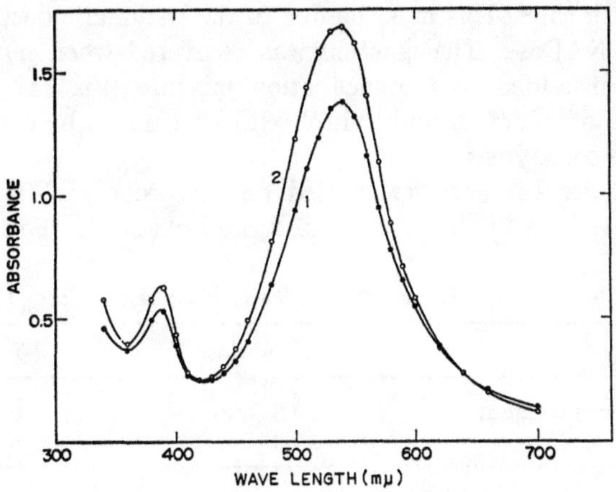

Fig. 9. Cysteine carbazole reaction products of the isolated ketopentose. *1* authentic D-xylulose; *2* isolated ketopentose

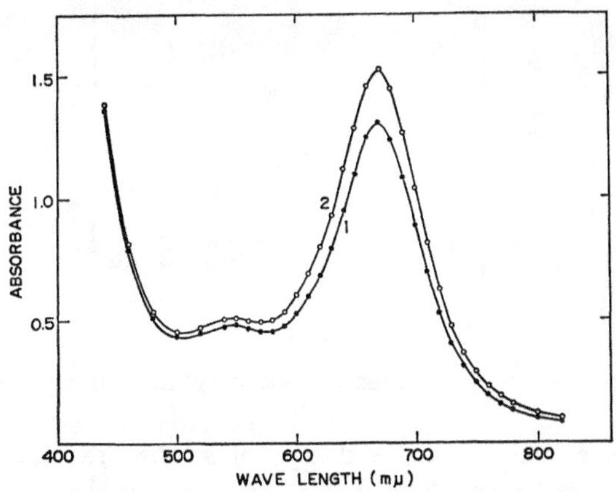

Fig. 10. Orcinol reaction product of the isolated ketopentose. *1* authentic D-xylulose; *2* isolated ketopentose

The following studies were carried out to determine whether the ketopentose was in the D- or L-form. This was done either by the use of an enzyme specific for D-xylulose or by the measurement of optical rotation. The results are shown in Table V, which indicate that 77% of the xylulose was in the D-form. In this experiment, L-xylulose content was calculated by subtraction of D-xylulose from total xylulose which was measured by color reaction of cysteine carbazole.

Further studies were made using hemolysate. In the hemolysate experiment, the rate of methemoglobin reduction by xylitol markedly decreased with time. This must be due to the enzymatic decomposition of NAD by NADase. The reaction was recovered when nicotinamide coenzyme was added to the incubation mixture (Fig. 11). Thus the methemoglobin reduction and xylitol oxidation might be coupled with nicotinamide coenzymes.

The next experiment was carried out in order to learn whether NADP can be used in the reaction of xylitol oxidation. This was done

Table V. D-*xylulose content in the isolated ketopentose* [1]

	D-xylulose	L-xylulose
Enzymatic measurement	19.8 (mg)	6.8 (mg)
Optical rotation	20.5	6.1

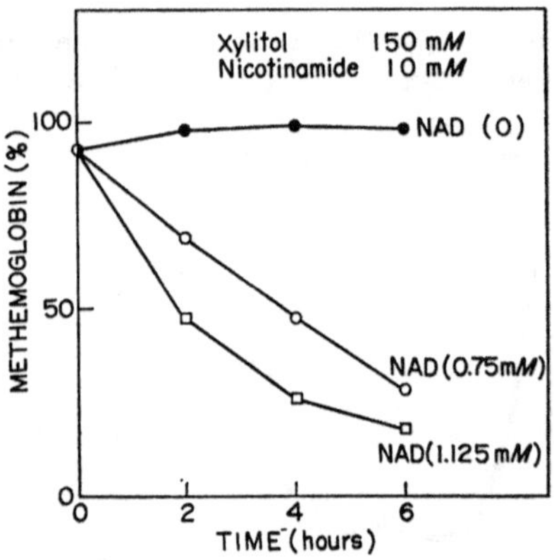

Fig. 11. Effect of NAD on the methemoglobin reduction rate in hemolysate

by incubation of hemolysate with xylitol in the presence of NAD or NADP. After incubation, ketopentose was separated and measured as the D- or L-form of xylulose. As shown in Table VI, in the presence of NAD the D-form was predominant, but in the case of NADP about 50% was the L-form.

What is the Further Metabolism of Xylulose? Bässler and Reimold observed lactate formation when xylitol was incubated with human erythrocytes [4]. This, however, does not necessarily indicate direct metabolic change of xylitol to lactate, since the conversion of xylitol to xylulose is associated with the increase of NADH/NAD ratio in the cells. As a result, the lactate/pyruvate equilibrium may be shifted toward lactate. This possibility can not be neglected. In Table VII, levels of glycolytic intermediates are shown after incubation of erythrocytes with xylitol, glucose or both substrates. As can be seen in the table, lactate and dihydroxyacetone phosphate accumulate in the presence of xylitol. This increase of lactate and triosephosphate can be explained by the change of NADH/NAD ratio in the cells when xylitol is metabolized to xylulose with formation of NADH.

The ATP level in erythrocytes decreases when cells are incubated with xylitol. This may also be due to inhibition of triosephosphate dehydrogenase by the increased NADH/NAD ratio. As a result, ATP

Table VI. *Effect of nicotinamide coenzyme on the formation of* D-*xylulose in hemolysate system*

Pyridine nucleotide	D-xylulose	L-xylulose
NAD	4.5 (mg)	0.2 (mg)
NADP	2.0	1.9

Table VII. *Effect of xylitol metabolism on glycolytic intermediates*

Substrate		Lactate	Pyruvate	DHAP	ATP	2,3-DPG
Glucose mM	Xylitol mM	μmoles/ml	mμmoles ml/cells	mμmoles ml/cells	mμmoles ml/cells	mμmoles ml/cells
0	0	2.25	2,320	10.1	240	2,930
10	0	5.46	856	15.2	750	3,540
0	10	4.54	180	30.4	462	3,580
10	10	7.06	45	709.0	725	3,940

production is inhibited. On the other hand, when glucose is added to the xylitol system, ATP is maintained at the initial level (Fig. 12). The time courses of the ATP level and lactate formation are shown in Fig. 12. The ATP level decreases when the cells are incubated only with xylitol as mentioned above. But in the presence of glucose, the ATP level is maintained at the initial level. Under this condition, lactate formation can be observed to continue constantly in spite of the accumulation of triosephosphates caused by the insufficient supply of NAD. The results indicate that the NADH/NAD ratio in the cells may not play an important role in the regulation of glycolysis. For these reasons, the measurement of lactate is not sufficient to evaluate lactate formation from xylitol; it is necessary to use radioactive xylitol in order to determine the true feature of xylitol metabolism in the cells.

Fig. 13 shows the chromatogram with Dowex-1-formate column after incubation of erythrocytes with ^{14}C-xylitol. Peak I was identified as

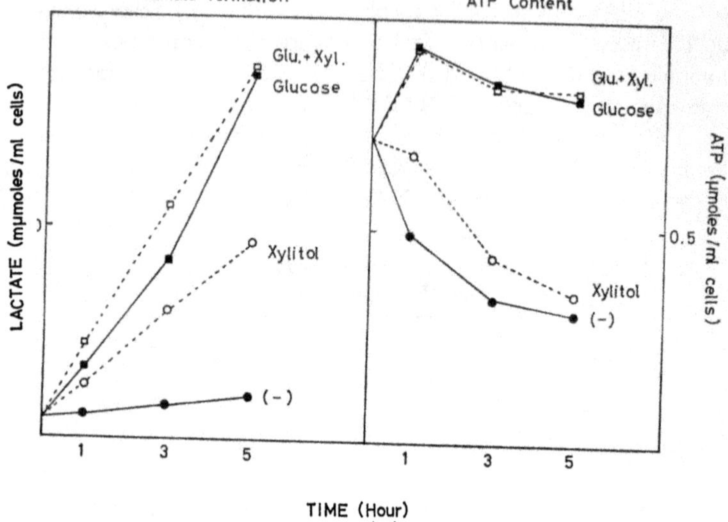

Fig. 12. Time courses of ATP level and lactate formation after incubation of red cells with xylitol, glucose or both substrate

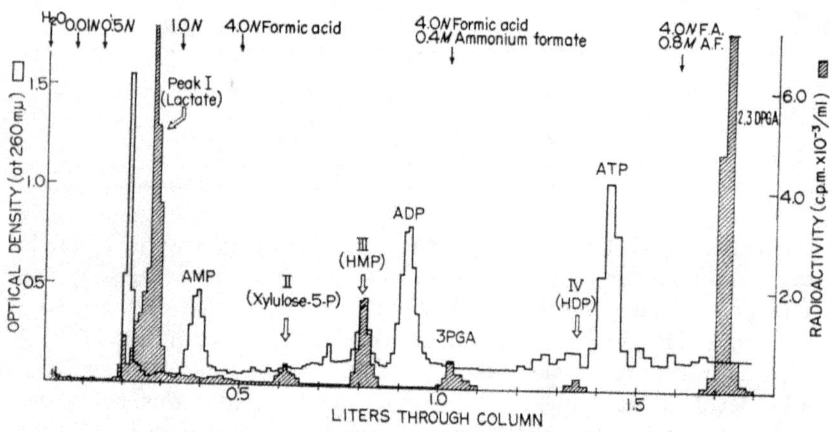

Fig. 13. Column chromatography of acid soluble compounds after incubation of red cells with radioactive xylitol

lactate, Peak II and III were sugar monophosphates and Peak IV was sugar diphosphate. Thus, radioactivity was shown to be incorporated into some glycolytic intermediates, proving that xylitol is metabolized to lactate in erythrocytes.

The suggested route of xylitol metabolism to lactate is shown in Fig. 14. Xylitol is first oxidized to D-xylulose, which probably is phos-

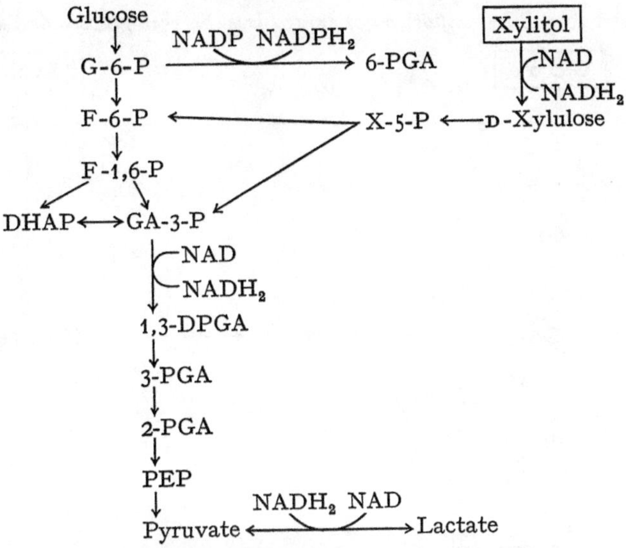

Fig. 14. Scheme of xylitol metabolism in red cells

Table VIII. *Lactate formation from radioactive xylitol in red cells*

State of Hb	Xylitol	Lactate formation (μmole/ml cell/3 hr)
HbO_2	10 mM	0.42
met-Hb	10 mM	1.09

phorylated to D-xylulose 5-phosphate in the presence of ATP. The enzyme, xylulo-kinase, has been found in liver by Hickman and Ashwell. D-xylulose 5-phosphate, a well known intermediate of the pentose phosphate pathway, is metabolized to glyceraldehyde 3-phosphate and to fructose 6-phosphate. The incorporation of radioactivity of C^{14}-xylitol into these sugar-phosphates supports the above-mentioned metabolic route from xylitol to lactate.

The rate of lactate formation from xylitol can be measured accordingly, only when radioactive xylitol is used as the substrate. The results obtaimed by this method are shown in Table VIII. The rate of lactate formation from xylitol depends on whether or not the intracellular hemoglobin is in the reduced state. When the hemoglobin first is oxidized to methemoglobin by treatment with sodium nitrite, lactate is formed at a rate of 1.07 μmoles/ml cells/3 hours at 37° C. This may be due to

Table IX. *Comparison of energy level between glucose and xylitol*

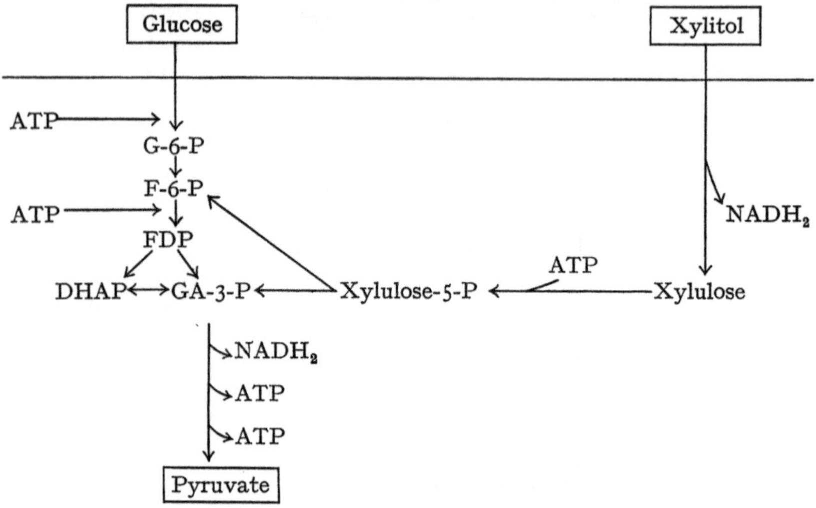

	Glucose	Xylitol
ATP	0.33	0.33
$NADH_2$	0.33 (1.0 ATP)	0.53 (1.6 ATP)

(mole/C)

the high supply of NAD resulting from the reduction of methemoglobin by NADH. According to the results, xylitol oxidation to xylulose increases, and further metablism to lactate ensues.

We have demonstrated that various effects of xylitol on erythrocytes are related to the metabolism of xylitol in red cells. It appears that there are two reasons for these remarkable effects.

One is that xylitol is metabolized to D-xylulose with formation of NADH before it is phosphorylated to D-xylulose 5-phosphate. But in the case when glucose is utilized, it is necessary to be phosphorylated prior to further utilization in the cells. As shown in Table IX, when glucose is metabolized to pyruvate, 0.33 mole ATP and 0.33 mole NADH are produced per mole of glucose. On the other hand, 0.33 mole ATP and 0.53 mole NADH are formed per mole of xylitol.

Another reason is the high permeability of xylitol with respect to the membrane. This permeability of xylitol was investigated using radioactive xylitol. This experiment was carefully performed. As presented in Fig. 15, xylitol passes the membrane easily and equilibrium was attained less than one minute. Since the rate of incorporation is not effected even when the experiments are carried out in an ice bath, xylitol may pass through the membrane by a simple diffusion mechanism.

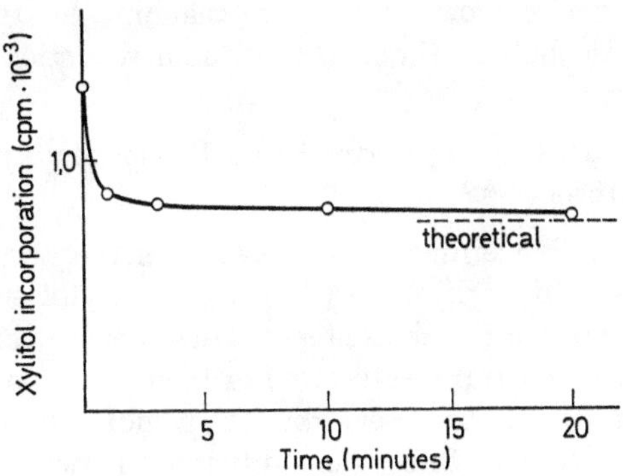

Fig. 15. Permeability of xylitol through red cell membrane

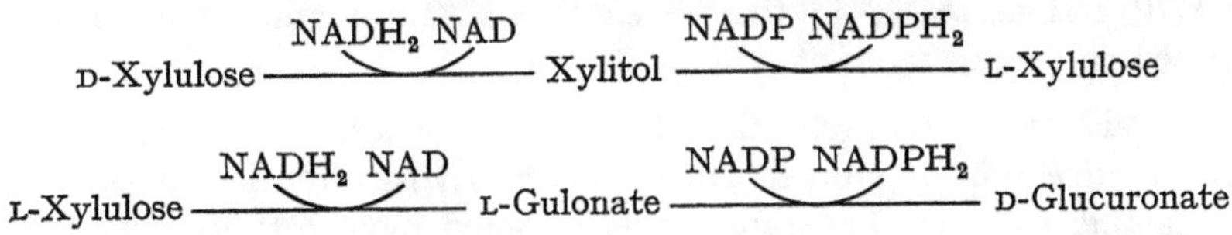

Fig. 16. Conversion of D- and L-xylulose

This seems to be important for the utilization of xylitol. For example, sodium gulonate, which is located in a similar metabolic position (Fig. 16), is oxidized to L-xylulose in the presence of NAD and to glucuronate in the presence of NADP in liver. When sodium gulonate was incubated with erythrocytes containing methemoglobin, the reduction of methemoglobin could not be detected, but in the experiments with hemolysate the sugar can be utilized for the reduction. This may be due to the low permeability of the sugar through the red cell membrane.

References

1. Asakura, T., K. Adachi, S. Minakami, and H. Yoshikawa: J. Biochem. **62**, 184 (1967).
2. Evelyn, K. A., and H. T. Malloy: J. biol. Chem. **126**, 655 (1938).
3. Reutler, E.: J. Lab. clin. Med. **49**, 84 (1957).
4. Bässler, K. H., u. W. V. Reymold: Klin. Wschr. **43**, 169 (1965).

Discussion

Dr. Horecker: Have you had an opportunity to test xylitol in congenital idiopathic methemoglobinemia?

Dr. Asakura: No, I have not. I am planning to study the effect of several sugar alcohols on the methemoglobin reduction in congenital methemoglobinemia.

Dr. Bässler: Have you prepared a NADP-specific xylitol dehydrogenase from erythrocytes?

Dr. Asakura: Concerning the problem whether or not red cells contain NADP-specific xylitol dehydrogenase (L-xylulose forming), we have no direct evidence for its existence. However, we can mention the following two experimental results. One of them is the fact that xylitol oxidation to xylulose can be observed in the hemolysate by the addition of NADP. Another is that there was a difference between total xylulose measured by the cysteine carbazole reaction and that measured enzymatically by NAD enzyme. The difference increased in the presence of NADP and decreased in the presence of NAD. Further experiments will be performed to clarify this problem.

Dr. Erdman: As a clinician, I'd like to ask a question concerning the practice of blood storage. Have you already used routinously xylitol containing solutions for stabilization of sored blood? If so, what are the results you have got ten?

Dr. Asakura: No, we have not used xylitol for the practice of blood storage yet. Further studies are necessary, for example, concerning the amount of xylitol to be added to ACD medium.

Dr. Touster: Do you know whether the normal blood levels of xylitol are such as to make your observations physiologically significant under normal conditions? What is the xylitol concentration of blood?

Dr. Asakura: We have not measured xylitol concentration in normal blood, but it is supposed to be low. Under physiological conditions this sugar alcohol may not have significant meaning for the reduction of methemoglobin because of its low content in blood.

Dr. Horecker: You observed the reduction of GSSG by added xylitol. Do you attribute this to the presence of an NAD-glutathione reductase, or was it due to the reduction of NADP by the second xylitol dehydrogenase?

Dr. Asakura: I think erythrocytes contain NAD-glutathione reductase on the basis of the following experimental results. Hemolysate was incubated at 37° C for 5 minutes to decompose NAD and NADP enzymatically, and was passed through charcoal as suggested by Dr. S. Rapoport. The solution was incubated with xylitol and NAD. GSSG was reduced in this system.

Dr. Winegrad: Dr. Frank Oski in Dr. Irwin Rose's laboratory was prompted by your observation to determine whether xylitol added *in vitro* would protect red cells from patients with congenital glucose 6-phosphate dehydrogenase deficiency from the hemolytic effect of acetyl phenylhydrazine. He was unable to observe a protective effect. This may be related to Dr. Horecker's comment concerning the unlikelihood that glutathione reduction is accomplished by a DPNH linked enzyme.

Experimental Findings on the Utilization of Xylitol in the Heart Muscle of the Guinea Pig

H. Bickel and W. Fekl

From the Scientific Department of J. Pfrimmer & Co. Erlangen, West Germany

The extrahepatic utilization of xylitol is measured as the difference between the administration on the one hand and the turnover in the liver + excretion on the other hand. It was indicated by Mueller and co-workers with 15.9% of the quantity infused. As can be concluded from the enzyme activities of the dehydration of xylitol in the individual tissues, apart from the kidneys, testes, lungs and adipose tissues, the heart muscle can also take part in this turnover.

In order to solve this problem, the heart-lung preparation of the guinea pig was used as a model for this experiment.

After urethane narcosis, animals of a body weight of about 400 gms are tracheotomized and thoracotomized and appropriate cannulas are introduced into the aorta, the venae cavae superior and inferior. Then, an extra-corporal circulation is established which has a constant temperature of 39° C by means of a double-wall system. Thus, the total amount of blood discharged by the left ventricle (less the coronary amount) is pushed through the aorta-cannula, passes a peripheral resistance device and is recorded by a bubble-flow meter. The blood is collected in a container from which it flows back through the vena cava inferior into the right atrium. By this preparation, cardiac output, pressure in the right atrium and arterial pressure could be recorded.

If the heart preparations while perfused with the blood of guinea pigs showed sufficient reactions when loaded with volume, we changed over to a suspension of blood cells having a defined substrate content. For this artificial perfusate frequently washed bovine erythrocytes were suspended in Ringer's solution at a hematocrit of 30. For maintaining the colloid-osmotic pressure, this suspension contained 3.5% of gelatine. The perfusate was gradually added from a temper container within 15 minutes.

When a perfusion without any substrates took place, the hearts of guinea pigs showed already after a short period the typical symptoms of energy deficit, i.e. a decrease of the original tonus and of the contractility occurs. About 30 minutes after the substrate-free perfusion, the intracellular substrate reserves are also exhausted and a cardiac arrest results.

When, however, the heart muscle is perfused with a perfusate containing xylitol (xylitol content: 90 mg-%), the cardiac activity is completely maintained for $1^1/_2$ to 2 hours, i.e. hearts react upon a volume load with a physiological increase of the stroke volume and of the blood pressure, whereas the increase of the pressure in the right atrium is only unimportant.

During this period the cardiac output amounts to 3 to 4 liters. The same situation results when the hearts are perfused with a perfusate containing glucose. In this case, the survival time of the heart is also scarcely 2 hours with the same output capacity.

These findings show that the heart muscle is really able to meet its energy requirements for a certain period by exclusively giving xylitol. Since according to Gercken and Huerter markedly insufficient phenomena result when myocardial ATP-content is less than 70% of the initial value, it can be supposed that the ATP-content can be maintained for a certain period above this threshold by xylitol as the sole energy source and that the restitution of ATP takes place at sufficient speed.

In comparison to the perfusion containing glucose, it can furthermore be concluded from the hemodynamic situation that the substrate xylitol is — under these special experimental conditions — as well utilizable as glucose.

Experiments on the heart-lung preparation of the rabbit which were carried out in a similar way by Gercken and Huerter showed in contrast to this that fructose and mannose cannot maintain the functioning of heart muscle. When these two substances were given, the hearts reacted as in the case of a perfusion free of any substrates.

We can confirm these findings for fructose. Our preparations, too, did not present any considerably longer periods of activity when perfused with a suspension containing fructose than with a substrate-free perfusion.

In these model experiments, it cannot be excluded, however, that an indirect energy supply of the heart muscle by metabolites of xylitol (e.g. lactic acid) formed in the lungs takes place. This problem is under investigation.

References

Gercken, G., u. P. Hürter: Stationäre Metabolitkonzentrationen im insuffizienten Säugetierherzen nach Monojodacetat- und Natriumfluoridvergiftung. Pflügers Arch. ges. Physiol. **292**, 100—117 (1966).

— — Substratversorgung und Stoffwechselhemmung in situ perfundierter Kaninchenherzen. Angew. Chem. **77**, 1081 (1965).

Müller, F., E. Strack, E. Kuhfahl u. D. Dettmer: Der Stoffwechsel von Xylit bei normalen und alloxandiabetischen Kaninchen. Z. ges. exp. Med. **142**, 338—350 (1967).

Discussion

Dr. Horecker: The effects with xylitol in contrast to fructose are very interesting. I wonder if the ability of xylitol to pass the blood-brain barrier has been studied, and what effects of xylitol on brain might be expected?

Dr. Fekl: There are some differences between fructose and sorbitol on the one side and xylitol on the other side, and I suppose that xylitol offers some advantages compared to sorbitol and fructose. We are continuing our work in this field. Of course, brain metabolism of xylitol would be a very interesting problem, but as far as I know, this question has not yet been checked.

Dr. Lang: I could imagine that in the case of your heart-lung-preparations lactic acid will not be utilized efficiently by the heart. In contrast, I suppose that in such a preparation the arteriovenous difference of lactic acid is rather inverse. So I believe that you need not worry about metabolites originating from the erythrocytes or the lungs. What is your opinion?

Dr. Fekl: That is our opinion, too, but of course we have to check this. It will be necessary to perfuse separate heart preparations, as e.g. the Langendorff heart, as a model, and to measure arteriovenous differences of metabolites and to carry out such perfusions with other substrates, too. The perfusion of isolated lung preparations may also be of great interest.

Dr. Bässler: Couldn't you perfuse your preparations with lactic acid to find out whether it is used under your conditions?

Dr. Fekl: Such perfusions are under trial together with measurement of arteriovenous difference.

Absorption of Xylitol

K. H. Dehmel, H. Förster, and H. Mehnert

IIIrd Medical Department of the Städtische Krankenhaus
München-Schwabing, West Germany
Medizinische Poliklinik, University of München,
West Germany

I. Introduction

The diet of diabetic patients should not contain sugars with insulin-dependent utilization and rapid absorption. For this reason, the diabetic patient is, in general, not allowed to take glucose, sucrose or maltose. Up to now, we used fructose or the polyol sorbitol as sugar substitutes. Both substances have an almost identical metabolic pathway, since sorbitol after dehydrogenation by sorbitol-dehydrogenase is converted to fructose and is now subject to the same metabolism as fructose. It is well known that this catabolic pathway is insulin-independent. Several groups could demonstrate that the breakdown of xylitol is also independent of insulin, for this metabolite appears in the glucuronic acid-xylose-cycle and enters the Horecker-shunt at the D-xylose 5-phosphate step. It was now of interest to find out whether this property which recommends xylitol as a sugar substitute for diabetic subjects adds to the advantage of a relatively slow absorption of the polyol. We decided, therefore, to examine the absorption of xylitol in man and rat and compare the results with those obtained with other kinds of sugars and polyols.

II. Methods

In order to examine the absorption from the intestine we used the so-called Collidon-clearance of Mehnert and Förster by which method the absorption of certain substances in man and animal can be investigated more accurately. In this test, collidon (poly-vinyl-pyrrolidon, molecular weight approx. 30,000), a substance which is non-absorbable, water-soluble, of high molecular weight and easily to be assessed is administered together with the substances to be tested, i.e. sugars. Thus the extent of the absorption at any site of the gastrointestinal tract is measured by evaluating the change in the ratio between the absorbable test substance and the non-absorbable collidon.

By this method we examined the absorption of the monosaccharides glucose, galactose and fructose, the disaccharides maltose, sucrose and lactose as well as the polyols sorbitol and xylitol in man and rat.

The investigation in man was restricted to the examination of duodenal absorption by means of a double balloon sound (Bartelheimer) the balloons of which blocked the duodenum distally of the pylorus and behind the flexura duodenojejunalis. The solution containing the test substance and the control substance was injected into the duodenum

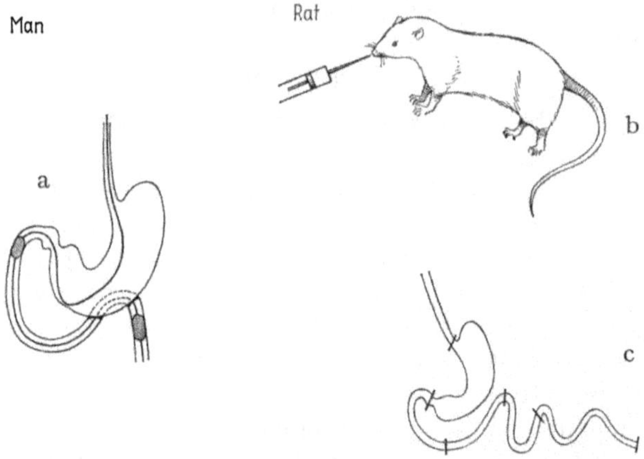

Fig. 1 a—c. Examination of absorption in man and rat. a, Examination of duodenal absorption by the double balloon sound; b, Application of substances into the stomach by tube; c, Ligation of gastrointestinal tract in various sections; and determination of their contents

through this tube. In 5, 15 and 30 minute samples the extent of the absorption was examined. Small losses of the test substance caused by an incomplete occlusion of the examined intestinal section could be corrected since the water-soluble control substance would always pass through the balloon as well and thus the calculation of the absorption was practically not impaired (Fig. 1).

Experiments in rats were carried out by injecting the test substance into the stomach by a stomach tube. After 15, 30 or 60 minutes the animals were sacrificed and the whole intestine was examined for the presence of test substance and collidon in different sections. From the presence or absence of collidon in the different intestinal sections it could be seen whether and to which extent the test substance — determined at the same time — was absorbed or whether the test substance had reached the examined intestinal section at all. The collidon was assessed according to Schubert, xylitol according to Rapoport and West.

III. Results

Fig. 2 shows the results obtained in man. It can be seen that the absorption of galactose and glucose is the fastest and that of sorbitol and xylitol is the slowest. Fructose and the disaccharides showed intermediate rates of absorption. We want to mention particularly that the validity of the results with this technique is limited as far as the disaccharides are concerned because the enzyme lactase is localized in the mucous membrane of the duodenum, the enzymes for the hydrolysis

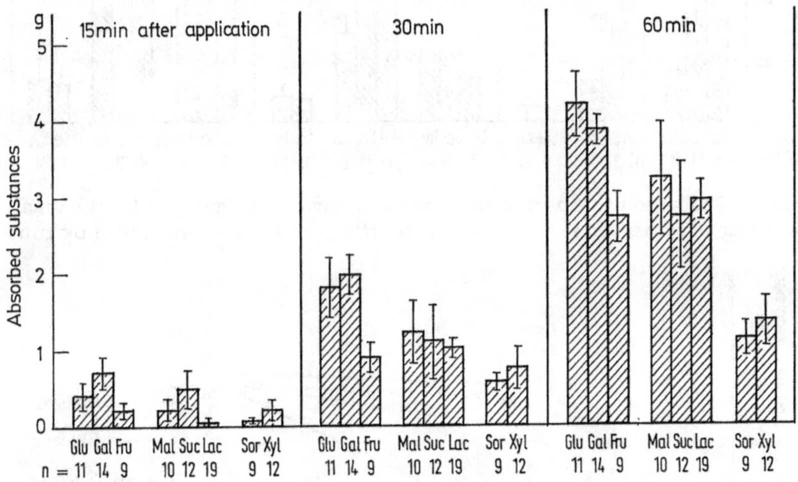

Fig. 2. The absorption of various sugars and polyols from the human duodenum after obstruction to both ends by double balloon sound. 5 g of the substances were given (concentrations started with 5%)

of maltose and sucrose, however, are to be found particularly in the lower parts of the small intestine. To say something about the absorption of the total intestine, experimental investigations in animals were necessary, the results of which are shown in Figs. 3 and 4. In comparison with the results obtained by investigations of the human duodenum, it could be shown that the disaccharides maltose and sucrose are absorbed significantly faster than lactose and — during short test periods — also faster than fructose. As in man, the absorption of galactose and glucose in the rat is also the fastest, that of sorbitol and xylitol is the slowest. Furthermore, our investigations show that the amount of the substances released by the stomach increases with the duration of the test, with the absorption rate of the test substance and with the reduction in concentration of the solutions used. That means that the mechanism of stomach emptying is of great influence on the absorption.

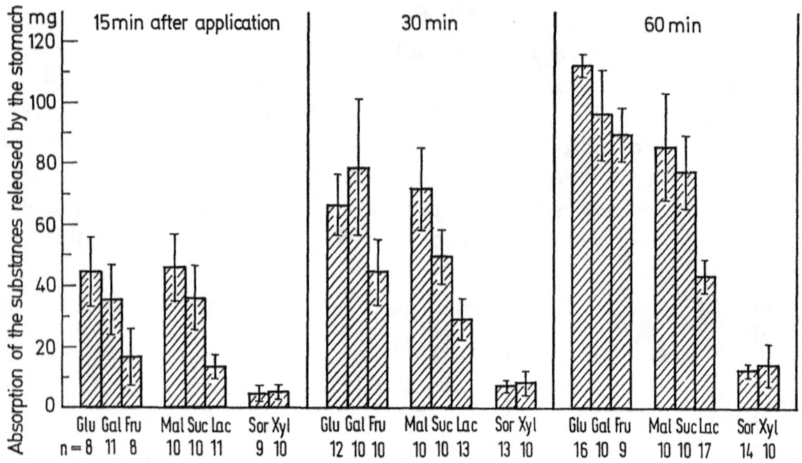

Fig. 3. The absolute absorption of various sugars and polyols by the small intestine of rats after 1.3 ml of a 10% solution (= 130 mg substance) by tube

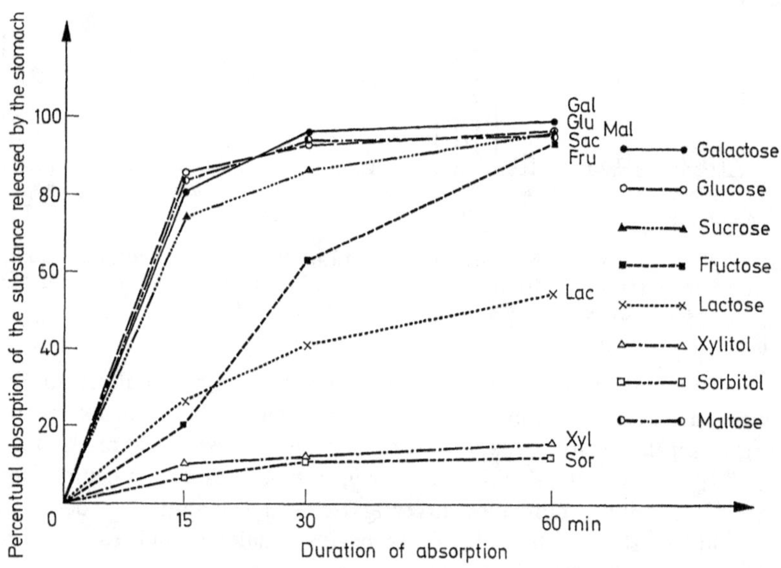

Fig. 4. Percentual absorption of 10% solutions (1.3 ml) of galactose, glucose, fructose, maltose, sucrose, lactose, xylitol and sorbitol in rats

IV. Conclusions

From the comparative investigations of the absorption of sugars and polyols we conclude that xylitol is absorbed very slowly. This property of xylitol has many advantages and renders it particularly useful for the diet of diabetic persons, because the diabetic organism does not get a too immediate load. Provided that the conversion of xylitol to blood glucose is of any significant extent this slow absorption would certainly restrict the process. A certain disadvantage of the slow absorption is the fact that xylitol just as sorbitol may cause osmotic diarrhea. According to our own experiences, 40 gms. of xylitol per day should therefore not be exceeded in diabetic patients. In this connection the results of Bässler *et al.* who observed an adaptation of the xylitol absorption are very interesting: diarrhea was reduced because xylitol absorption was gradually increased.

It should be mentioned, however, that in diabetic patients, particularly in obese persons, xylitol is of course a source of calories in contrast to saccharine. An uncontrolled intake of xylitol in diabetics, especially of old age, does practically not produce a hyperglycemia but may cause an increase in body weight because of its caloric value.

References

Bässler, K. H., W. Prellwitz, V. Unbehaun u. K. Lang: Xylitstoffwechsel beim Menschen. Zur Frage der Eignung von Xylit als Zucker-Ersatz beim Diabetiker. Klin. Wschr. **40**, 791 (1962).

— G. Stein u. W. Belzer: Xylitstoffwechsel und Xylitresorption. Stoffwechseladaptation als Ursache für Resorptionsbeschleunigung. Biochem. Z. **346**, 171 (1966).

Dehmel, K. H.: Untersuchungen zur Resorption des Zuckeralkohols Xylit bei der Ratte. Inaug.-Diss. München 1965.

Griem, W., u. K. Lang: Pathologisch-histologische Untersuchungen an Rattenorganen nach parenteraler Verabreichung von Xylit. Klin. Wschr. **40**, 801 (1962).

Hollmann, S.: Nicht-glykolytische Stoffwechselwege der Glucose. Stuttgart: Georg Thieme 1961.

Lang, K.: Xylit als Nahrungskohlenhydrat. Med. u. Ernähr. **4**, 45 (1963).

Mehnert, H., u. H. Förster: Die Prüfung der Resorption von Glucose, Fruktose und Sorbit mit einer neuen Methode. Klin. Wschr. **39**, 596 (1961).

— K. Stuhlfauth, B. Mehnert, R. Lausch u. W. Seitz: Vergleichende Untersuchungen zur Resorption von Glukose, Fruktose und Sorbit beim Menschen. Klin. Wschr. **37**, 1138 (1959).

Mellinghoff, C. H.: Über die Verwertbarkeit des Xylit als Ersatz-Zucker bei Diabetikern. Klin. Wschr. **39**, 447 (1961).

West, C. D., and S. Rapoport: Modification of calorimetric method of determination of mannitol and sorbitol in plasma and urine. Proc. Soc. exp. Biol. (N.Y.) **70**, 141 (1949).

Effect of Xylitol on the Metabolic Rate of Glucose in Alloxan Diabetic Rat Liver Slices

NORIMASA HOSOYA, TAKASHI SAKURAI, HIDEO TAKAGI, and TADAHIKO MACHIYA

Nutritional Laboratory, School of Health Sciences, Faculty of Medicine, University of Tokyo, Tokyo, Japan

Glycogenic and glycolytic enzymes function systematically in the cell to establish the regulated flow of each of the metabolic pathways. Chaikoff *et. al.* [1] reported that enzyme activities of glycogenesis and glycolysis are markedly disturbed in the liver of alloxan diabetic rats. In order to elucidate the effect of xylitol on glucose metabolism [2], we have measured intermediates of glycogenesis and glycolysis in diabetic liver, in the presence or absence of added xylitol.

Xylitol Effect on Uptake and Output of Glucose. Glucose levels in alloxan diabetic liver are 30% higher than in normal liver (Table I).

Table I. *Changes of glucose content in normal or diabetic rat liver slices (μmoles/g wet weight; 10 determinations)*

		Endogenous original content	After 1.5 hr incubation		
			without	with insulin	with xylitol
Normal	Tissue	3.370 ±0.302	−1.470 ±0.124	−1.790 ±0.157	+0.440 ±0.061
	Medium		+3.466 ±0.305	+3.240 ±0.312	+1.592 ±0.151
Diabetic	Tissue	4.417 ±0.415	+2.183 ±0.206	−0.547 ±0.063	+1.328 ±0.146
	Medium		+5.394 ±0.514	+2.622 ±0.252	+1.811 ±0.175

Female rats (80—100 g) were injected i.p. with alloxan (150 mg/kg), and after two weeks the rats having over 300 mg/ml blood sugar level were used as the diabetic group. The rats were fasted for one day before sacrifice.

Rat liver slices weighing 500 mg were suspended in a Krebs-Ringer phosphate buffer (final volume 3.0 ml, pH 7.2), containing 20 μmoles of glucose, and 20 μmoles of xylitol or 2 m units of insulin. After 1.5 hours of incubation the reaction vessel was immersed into crushed ice, 0.3 ml of ice cold 0.3 N perchloric acid was added, and the acid-insoluble fraction was removed by centrifugation. Glucose was measured colorimetrically with glucose oxidase [7].

During incubation of diabetic liver for 1.5 hours, glucose content as well as output of glucose from the tissue increases. Addition of insulin or xylitol to diabetic liver slices causes a decrease of glucose output. Glucose content of the diabetic tissue is decreased by the addition of insulin, whereas an increase of glucose is not so large in the presence of xylitol.

Table II. *Glucose-U-C^{14} Incorporation into glucose and glycogen of rat liver slices (mμmoles/g wet weight)*

Rat	Conversion of glucose-U-C^{14} to	Incorporation of glucose-U-C^{14}		
		without	with insulin	with xylitol
Normal	glucose	162.4 ±12.2	155.3 ±13.5	134.2 ±14.8
	glycogen	267.0 ±22.6	283.0 ±25.4	293.0 ±26.3
Diabetic (blood sugar 300 mg/dl)	glucose	142.0 ±11.7	136.3 ±12.2	132.3 ±13.6
	glycogen	260.0 ±25.5	328.0 ±33.3	371.7 ±42.1

The reaction mixture contained rat liver slices, 500 mg; Krebs-Ringer phosphate buffer (pH 7.2), 2.5 ml; glucose (0.25 μc/μmoles), 20 μmoles; xylitol, 20 μmoles; and insulin, 2.000 microunits. Incubation was performed for 1.5 hr, at 37° C. The isolation of glucose or glycogen from the tissue slice for C^{14} analysis was carried out by the method of Villee *et al.* [6].

On the other hand, incorporation of glucose-U-C^{14} into tissue glucose is identical in normal and diabetic liver (Table II). The results shown in Tables I and II suggest that insulin stimulates the utilization of glucose, and that xylitol withholds glucose in the tissue.

Xylitol Effect on Glucogenesis. Incorporation of glucose-U-C^{14} into glycogen is identical in normal and diabetic rat liver slices, as indicated in Table II. Lang *et al.* [3] have observed that xylitol administration to diabetic rats causes a stimulation of liver glycogen synthesis. Our experiment also demonstrated that glycogen synthesis of the diabetic tissue is stimulated by the addition of either insulin or xylitol (Table II). Accordingly to elucidate the xylitol effect on glycogen synthesis the intermediates of glycogenesis were assayed by the enzymatic micromethods of Bergmeyer [4]. Pyridine nucleotide oxidation or reduction was assessed from the change in fluorescence using a 365 mμ excitation light and a secondary emission filter at 460 mμ (Hitachi Fluorophotometer was modified and attached to a recorder).

In diabetic slices glucose 6-phosphate content increases two-fold, and uridine-diphosphate glucose decreases to about a third by the addition of xylitol. Since glucose 6-phosphate is an essential cofactor of glycogen synthetase, it seems that stimulation of glycogen synthesis by the addition of xylitol may be due to the increase of glucose 6-phosphate (Table III).

Table III. *Intermediates of glycogenesis in normal or diabetic rat liver slices (mμmoles/g wet weight)*

Intermediates	Normal rat		Diabetic rat	
	without xylitol	with xylitol	without xylitol	with xylitol
Glycogen (μmoles/g)	212.0 ±16.3	223.0 ±19.6	196.0 ±20.2	234.0 ±24.4
UDP-Glucose (mμmoles/g)	10.7 ±1.1	5.3 ±0.7	32.0 ±4.0	10.7 ±9.8
Glucose 1-P (mμmoles/g)	50.4 ±5.2	50.2 ±5.0	87.6 ±7.3	63.0 ±7.1
Glucose 6-P (mμmoles/g)	46.5 ±4.4	42.9 ±3.0	20.8 ±3.5	41.6 ±3.2
Glucose (mμmoles/g)	1,900.0 ±178.0	3,810.0 ±363.0	6,600.0 ±621.0	5,745.0 ±561.0

After the incubation as described in Table II the tissue slice was homogenized in 3.0 ml of ice cold 0.3 N perchloric acid. The supernatant obtained from the homogenate by centrifugation was adjusted to pH 5—6 with 5 N potassium hydroxide. The precipitated potassium perchlorate was removed, and the clear supernatant was used for fluorometric analysis.

Methods of enzymatic assay were slightly modified from those of Bergmeyer [4]. The standard assay mixture contained 90 μmoles of triethanolamine buffer, 100 μl of perchlorate extract, 0.5—2.0 μmoles of pyridine nucleotide and 5—30 μl of enzyme solution, in a total volume of 2.0 ml.

Glycogen was determined according to the method of Walaas and Walaas [8].

Xylitol Effect on Glycolysis. The amounts of the intermediates of glycolysis in a liver slices were also measured by means of the fluorophotometer. In the diabetic tissue the concentrations of glycolytic intermediates and ATP, ADP, and NAD are lower than those in the normal (Tables IV and V). Thus in the diabetics the flux of glycolysis probably decreases compared to the normal.

Addition of xylitol or insulin to the diabetic tissue causes an increase of the intermediates of glycolysis, and particularly, those after the step of triose-phosphate. Addition of xylitol to the diabetic tissue causes

Table IV. *Intermediates of glycolysis in normal rat liver slices (mμmoles/g wet weight)*

Inter-mediates	Before incubation	After incubation without insulin	After incubation with insulin	After incubation with xylitol
G-6-P	136.8 ± 5.7	46.5 ± 4.4	60.2 ± 5.3	42.9 ± 3.0
F-6-P	18.6 ± 2.4	7.5 ± 1.8	12.8 ± 3.6	11.7 ± 1.3
FDP	27.2 ± 3.0	17.3 ± 2.8	20.4 ± 3.1	33.1 ± 3.9
DHAP	186.1 ± 9.1	128.7 ± 8.4	147.3 ± 9.7	38.1 ± 6.0
GAP	39.4 ± 4.9	44.7 ± 4.9	49.6 ± 6.5	32.7 ± 2.5
3-PGA	46.6 ± 6.1	60.3 ± 6.5	68.5 ± 6.7	32.4 ± 3.3
2-PGA	60.5 ± 3.8	52.5 ± 5.1	43.6 ± 5.6	45.4 ± 3.6
PEP	56.9 ± 7.3	53.7 ± 5.9	54.9 ± 6.3	45.0 ± 3.3
Pyruvate	79.4 ± 5.3	70.5 ± 4.3	67.9 ± 5.2	83.2 ± 5.0
ATP	354.5 ± 11.1	130.6 ± 9.6	134.8 ± 8.3	101.8 ± 9.9
ADP	284.6 ± 10.7	138.0 ± 7.8	140.5 ± 5.9	146.5 ± 8.8
DPN	189.9 ± 7. 8	67.6 ± 5.7	53.5 ± 6.3	48.7 ± 6.1

Table V. *Intermediates of glycolysis in diabetic rat liver slices (mμmoles/g wet weight)*

Inter-mediates	Before incubation	After incubation without insulin	After incubation with insulin	After incubation with xylitol
G-6-P	155.8 ± 6.5	20.8 ± 3.5	40.8 ± 5.4	41.6 ± 3.2
F-6-P	33.6 ± 4.9	6.8 ± 1.2	8.7 ± 2.4	12.9 ± 1.8
FDP	34.0 ± 3.6	15.8 ± 2.1	18.0 ± 3.3	16.5 ± 1.3
DHAP	121.3 ± 9.3	43.7 ± 5.9	90.6 ± 8.7	24.8 ± 3.2
GAP	49.0 ± 6.1	19.8 ± 3.5	30.2 ± 5.3	31.6 ± 4.0
3-PGA	49.5 ± 3.6	20.8 ± 2.6	67.2 ± 7.4	65.0 ± 7.7
2-PGA	51.9 ± 7.8	16.3 ± 2.0	44.7 ± 4.8	46.2 ± 4.4
PEP	54.7 ± 2.9	22.1 ± 3.6	70.5 ± 5.9	69.3 ± 4.9
Pyruvate	62.6 ± 5.5	23.6 ± 2.8	75.3 ± 4.8	97.3 ± 9.5
ATP	271.0 ± 11.3	87.1 ± 7.1	100.0 ± 5.2	132.4 ± 9.1
ADP	328.1 ± 9.71	30.7 ± 8.7	103.4 ± 6.3	107.3 ± 8.2
DPN	174.8 ± 9.3	50.5 ± 6.4	42.5 ± 5.8	38.5 ± 4.7

the amount of dihydroxyacetone-phosphate to decrease one half. Xylitol probably is metabolized via the Pentose Phosphate Cycle and produces reduced NADP which in turn results in stimulation of triglyceride synthesis from dihydroxyacetone-phosphate. ATP and ADP in the diabetic tissue are also increased by the addition of insulin or xylitol.

Table VI. *Free energy changes in the system of Glucose → Pyruvate in liver slices*

	Without	With insulin	With xylitol
Normal	−15.7	−15.5	−15.6
Diabetes	−17.8	−16.2	−16.6

ΔF (Kcal).
Glucose + 2NAD + 2ADP + 2Pi → 2 Pyruvate + 2NADH + 2ATP

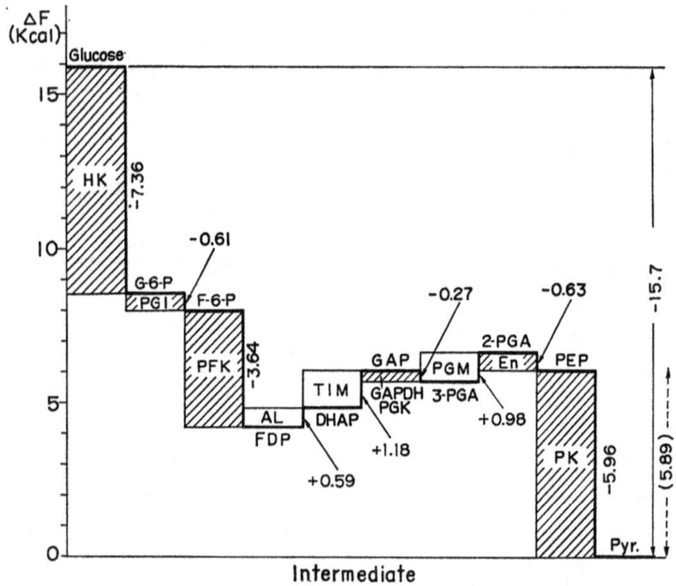

Fig. 1. Free energy change in glycolysis in normal rat liver slice with glucose

From the observed concentrations of intermediates the free energy changes in glycolysis were calculated according to the method of Burton and Krebs [5]. The calculations are based on standard free energy data at pH 7.0 at 25°, and determined by means of the equation $\Delta G = \Delta G_0 + 1{,}42 \log K$. The change of free energy due to the temperature difference (25° and 37°) is neglected, because it is not so large. It is assumed for convenience in the calculation that the concentration of inorganic phosphate is 10^{-3} M and that the ratio of NAD to $NADH_2$ is 10.

The free energy change in glycolysis in normal rat liver slices incubated with glucose is −15.7 Kcal/mole (Fig. 1). When glycolysis is divided into its nine steps, the results shown in the figure indicate that glucokinase, phosphofructokinase and pyruvate kinase are rate-limiting in liver glycolysis. The free energy change in glycolysis in normal rat liver slices is —15.5 and —15.6 Kcal/mole in the presence of insulin or xylitol, respectively. The pattern of the free energy change in normal liver shows almost no difference in the presence of insulin or xylitol.

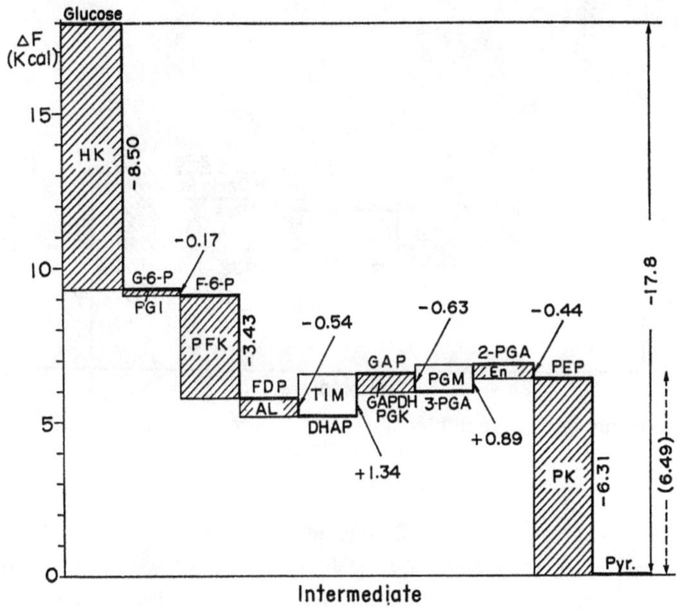

Fig. 2. Free energy change in glycolysis in diabetic rat liver slice with glucose

The free energy change in glycolysis in diabetic rat liver slices incubated with glucose is —17.8 Kcal/mole (Fig. 2). The flow of glycolysis was impaired in the diabetics. Upon addition of insulin to the diabetic tissue the pattern of glycolysis resembles the pattern of the normal rat (—16.2 Kcal/mole). Insulin probably stimulates the glycolytic enzyme activities. The free energy change in glycolysis in diabetic rat liver slices in the presence of xylitol is —16.6 Kcal/mole (Fig. 3). By addition of xylitol to diabetic tissue, the free energy change from glyceraldehyde-phosphate to pyruvate decreases (—6.49 → —4.92 Kcal/mole). These results suggest that xylitol is metabolized via the Pentose Phosphate Cycle and that the triose-phosphate is metabolized via the Embden-Meyerhof pathway. Xylitol induces the enzyme activities from glyceraldehyde-phosphate to pyruvate and this part of the glycolytic pathway is stimulated.

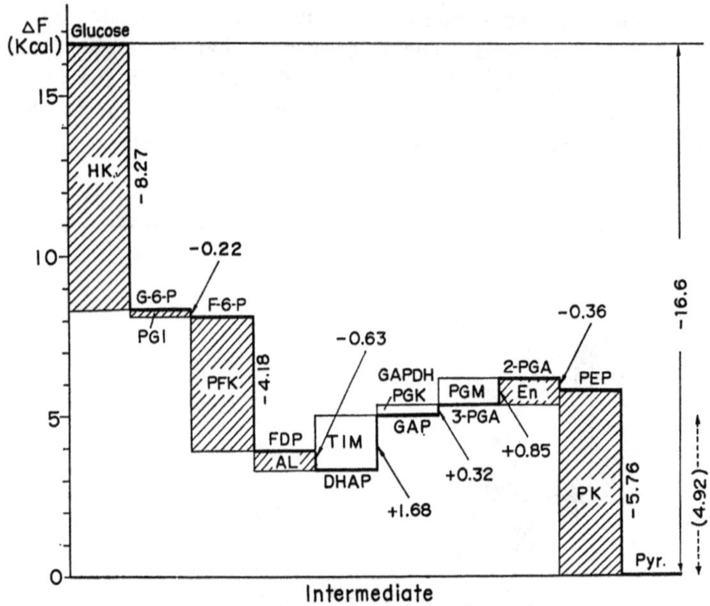

Fig. 3. Free energy change in glycolysis in diabetic rat liver slice with glucose and xylitol

Conclusion

The impaired glycolytic activities in diabetic rat liver slices are recovered upon addition of insulin or xylitol. Xylitol penetrates the tissue without any relation to insulin, and exerts a positive feedback on glycolysis via the Pentose Phosphate Cycle.

References

1. Fitch, W. M., and I. L. Chaikoff: Biochim. biophys. Acta (Amst.) **57**, 588 (1962).
2. Hosoya, N., T. Sakurai, and N. Shimazono: Proceedings of VII. Intern. Congr. of Nutrition.
3. Schmidt, B., M. Fingerhut u. K. Lang: Klin. Wschr. **42**, 1073 (1964).
4. Bergmeyer, H. U.: Methoden der Enzymatischen Analyse. Weinheim: Verlag Chemie 1962.
5. Burton, K., and H. Krebs: Biochem. J. **54**, 94 (1953). — Burton, K.: Ergebn. Physiol. **49**, 275 (1957).
6. Villee, C. A., and A. B. Hastings: J. biol. Chem. **179**, 675 (1949).
7. Huggett, A. St. G., and D. A. Nixon: Biochem. J. **66**, 129 (1957); — Lancet **1957 II**, 368.
8. Walaas, O., and E. Walaas: J. biol. Chem. **187**, 769 (1950).

Discussion

Dr. Touster: What is the cause of the UDPG decrease?

Dr. Hosoya: It might be considered that the reason is stimulation of glycogen synthetase. Addition of xylitol to the diabetic liver tissue causes an increase of glucose-6-phosphate, and this glucose-6-phosphate would increase glycogen synthetase activity.

Dr. Asakura: In erythrocytes we observed the accumulation of triose-phosphate in the presence of xylitol, probably due to the increased NADH/NAD ratio. However, as you have presented triose-phosphate levels in liver rather decreased in the presence of xylitol. Do you have any idea to explain this difference?

Dr. Hosoya: No, we have no idea at present.

Adaptive Processes Concerned with Absorption and Metabolism of Xylitol

K. H. Bässler

Physiologisch-Chemisches Institut, University of Mainz, West Germany

The turnover rate of a monosaccharide or polyalcohol in most cases depends on the mode of application. With the exception of glucose or galactose carbohydrates are metabolized faster after intravenous application than after oral application. This shows, that absorption from the gastro-intestinal tract is a rate limiting factor. Dealing with adaptive processes we therefore have to be concerned with absorptive mechanisms on one hand and enzymatic reactions in the cellular metabolism on the other hand.

Adaptive mechanisms connected with the process of absorption could only be effective on oral administration. Adaptive changes in cellular metabolism would effect the turnover rate of intravenously given compounds and eventually the rate of absorption, as we will see later.

One of the early observations during oral application of xylitol was the incidence of osmotic diarrhoea after high doses, as it is well known from other slowly absorbed sugars or polyols too, like sorbitol, xylose, and other pentoses. In contrast to the latter compounds, however, xylitol-induced diarrhoea in experiments with rats disappeared after 8 to 10 days of feeding. It was our working hypothesis that by adaptive mechanisms the rate of intestinal xylitol absorption was increased, therefore leaving less xylitol to exert osmotic effects in the lower parts of the intestinal tract.

This hypothesis was tested by the CORI-technique on adapted and non-adapted rats, whereby adaptation was achieved by feeding a standard diet with 10% xylitol for 10—14 days [1]. Adapted as well as non-adapted rats were divided into 10 groups of 12 animals each. 200 mg of xylitol were given by stomach tube and then the groups were killed in 1 hr intervalls from $1^1/_2$ to $10^1/_2$ hours after xylitol application. The gastro-intestinal tract was quickly removed, divided into 3 parts in which the remaining xylitol was determined.

The results of these experiments are shown on the first slide (Table I).

In the first two lines the non absorbed amount of xylitol in percent of the dose in the total gastro-intestinal tract of non-adapted rats is

Table I. *Absorption of xylitol in adapted and non-adapted rats*

		Time after xylitol application (hrs)							
		3.5	4.5	5.5	6.5	7.5	8.5	9.5	10.5
Total gastro-intestinal tract	non-adapted	71	54	52	48	47	44	34	24
	adapted	62	40	32	24	10	9	3	0
Stomach and small intestine	non-adapted	29	10	5	5	3	3	3	1
	adapted	35	9	7	3	1	1	<1	0
Cecum	non-adapted	28	40	33	35	35	32	25	20
	adapted	28	25	20	18	7	6	2	0
Colon	non-adapted	14	9	8	7	8	8	6	3
	adapted	5	5	5	2	1	<1	<1	0

200 mg of xylitol were given by stomach tube. The figures indicate the remaining xylitol in the gastro-intestinal tract in percent of the dose.

compared with adapted rats. The differences are significant at $4^1/_2$ hrs and later. $10^1/_2$ hrs after xylitol administration still 24% of the dose are found in the intestine of non-adapted rats, whereas in adapted rats the dose is completely absorbed. The half life time of absorption is $6^1/_2$ hrs in non-adapted and $4^1/_2$ hrs in adapted rats. Comparison of the three parts of the gastro-intestinal tract shows that passage of the stomach and small intestine is relatively fast and considerable amounts remain in the cecum for a longer time. Since less xylitol arrives the cecum in adapted rats, it seems that absorption in the small intestine is faster in adapted than in non-adapted rats. A considerable amount of ingested xylitol is absorbed in the cecum and evidently quite faster in adapted than in non-adapted animals. The amounts arriving in the colon are small and smaller in adapted rats.

The large portions of xylitol in the lower parts of the intestine and the long time it stays there and exerts osmotic effects explain the observed diarrhoea. The different results in adapted animals would count for the observation that diarrhoea disappears after prolonged xylitol feeding.

The question is now, what is the mechanism of this adaptation.

First we considered active transport mechanisms as the point of action. However, all efforts to demonstrate active transport had been in vain. Experiments in vitro showed no difference in the permeation rate of xylitol through the intestinal wall of adapted and non-adapted rats. There is no influence of 2,4-dinitrophenol or phlorizin. There is no transport against a concentration gradient. Finally, the temperature coefficient of xylitol permeation through the intestinal wall is 1.2. This is in the order of physical processes rather than of enzymatic reactions.

For this reason we proposed that xylitol crosses the intestinal wall by passive diffusion and the rate of this diffusion is governed by the xylitol level in the blood. If this is correct, adaptation mechanisms should be suspected at the level of cellular metabolism.

The blood level of xylitol as a regulatory factor for the absorption rate is favored by two observations:

1. Adapted rats on a diet with 10% xylitol get diarrhoea again as soon as they are treated with subcutaneous injections of xylitol in order to keep the xylitol level in the blood high.

2. Adaptation can also be achieved by parenteral xylitol application. Animals prepared this way get no diarrhoea after change to oral xylitol application.

For these reasons we studied the rate of disappearance of xylitol from the blood in adapted and non-adapted rats. Direct measurement of the elimination rate of xylitol from the blood after an intravenous load showed a significant increase in adapted animals:

Table II. *Rate constants*[a] *of xylitol disappearance and half-life of xylitol in the blood of adapted and non-adapted rats*

	Control group	Adapted group
Number of animals	13	20
Rate constant	0.022 min^{-1}	0.037 min^{-1}
Standard deviation	0.004	0.008
Half-life	31 min	19 min
Significance of difference	$P < 0.001$	

[a] The rate constants of xylitol disappearance were calculated as the slopes of the regression lines of the logarithm of the blood concentration on time.

The half life of xylitol in adapted rats is 19 minutes compared to 31 minutes in non-adapted rats.

Further we were concerned with the question, what enzyme or enzymes are responsible for this increase in turnover rate.

Comparison of different subcellular fractions in the liver showed that the enzymes for the metabolism of exogenous xylitol are located in the cytoplasm. Among these, the cytoplasmatic polyol dehydrogenase (E.C. No. 1.1.1.14), catalyzing the first step of xylitol turnover, is rate limiting:

Table III. *Activities of enzymes of xylitol metabolism in rat liver*

	Polyol dehydrogenase (xylitol as substrate)	Xylulo-kinase	Trans-ketolase
Mean[a]	0.0224	0.0461	0.0483
Standard deviation	0.0032	0.013	0.010
Mean[b]	2.46	5.03	4.47
Standard deviation	0.35	1.31	1.00

[a] μmoles per minute and mg protein.
[b] μmoles per minute and gram fresh weight.

The mitochondrial xylitol dehydrogenases are not of quantitative importance in the metabolism of exogenous xylitol.

The liver accounts for 80 to 90% of the xylitol turnover of the total organism. This figure is in good agreement with data of F. Müller *et al.* [2] obtained with different methods on rabbits.

Table IV. *Activities of xylitol dehydrogenation*

	Nanomols per minute and	
	mg protein	total organ
Liver	24.7	27,000
Kidneys	6.5	2,138
Testes	1.5	238
Small intestine	0.26	76
Heart	0.40	24
Lungs	0.13	20
Epididymal fat pad	0.56	16
Brain	0.10	10

On the basis of these data it seems reasonable to assume that the cytoplasmatic polyol dehydrogenase of the liver limits in first line the turnover rate of exogenous xylitol. Therefore we compared the activities of this enzyme in adapted and non-adapted rats. A significant increase in activity due to adaptation can be demonstrated:

Table V. *Activities of xylitol dehydrogenation in adapted and non-adapted rats (μmoles xylitol dehydrogenated at 30° per minute)*

	Per mg protein		Total liver	
	control	adapted	control	adapted
Mean	0.022	0.032	28	39
Standard deviation	0.003	0.006	5	8
Significance of difference	$P<0.001$		$P<0.001$	

From these results we conclude that prolonged xylitol feeding causes an increase in activity of polyol dehydrogenase in the liver cytoplasma. This in turn causes an increased rate of xylitol elimination from the blood and the result of the latter is an increased absorption rate of xylitol from the intestine by passive diffusion followed by disappearance of osmotic diarrhoea.

I should like to compare a few figures for total turnover capacity of carbohydrates in man:

Table VI. *Turnover of carbohydrates in human subjects*

		k (% per min)	$t/2$ (min)
Xylitol [a]	Babies	3.5 ± 1.1	19.8
	Children (4—8 years)	3.0 ± 0.5	23.1
	Adults (30—40 years)	3.0 ± 1.0	23.1
Glucose [b]	Healthy adults	3.0—4.8	14.5—23
	Diabetic adults	0.3—2.46	28.2—300
Fructose [c]	Healthy adults	3.8 ± 0.3	18.4
	Diabetic adults	2.9 ± 0.17	23.6

[a] From data of [3].
[b] From data of [4] and [5].
[c] From data of [6].

From measurements of the rate of disappearance of xylitol from the blood after an intravenous load we calculated an average half life of 23 minutes in adult men. With the same method other investigators found a half-life of 15—23 minutes for glucose and of 18—20 minutes for fructose in healthy men. That means that in healthy individuums the total capacity of xylitol turnover approaches about 80% of that of glucose or fructose. In diabetic persons the turnover rate of glucose and fructose is reduced and the turnover of xylitol is unimpaired. Thus in diabetics the capacity for xylitol turnover is considerably higher than that for glucose and equals about that for fructose. These comparisons have been made without consideration of adaptation, because we have no knowledge of adaptation in man yet.

Another point of interest is the comparison of the activities of liver polyol dehydrogenase and glucose 6-phosphate dehydrogenase:

Table VII. *Maximal dehydrogenase activities (μmoles per g liver and minute at 30° C)*

Glucose-6-phosphate dehydrogenase[a]	0.78—1.77
Xylitol dehydrogenase (cytoplasmatic)	2.5 —3.5

[a] Calculated from data of [7].

These two figures combined with the fact that the total xylitol turnover passes through the pentose phosphate cycle whereas only a fraction of glucose does so, allow the prediction that xylitol is a very effective precursor of pentoses, especially under metabolic conditions when the glucose flow through the pentose cycle is drastically impaired by low activity of glucose 6-phosphate dehydrogenase. One might speculate that xylitol by this means favors protein synthesis. There is already experimental evidence in favor of this idea. Other investigators [8] have reported close correlation between activity of glucose 6-phosphate dehydrogenase, ribose 5-phosphate formation, RNA synthesis and protein synthesis. On inhibition of glucose 6-phosphate dehydrogenase all these processes are impaired including cell division and immunological reactions.

McKerns [9] observed as one of the primary actions of ACTH an activation of adrenal glucose 6-phosphate dehydrogenase. He concludes that a secondary increase in RNA for increased protein synthesis could occur due to the hormone regulated increase in phosphorylated ribose.

It is reasonable to assume that all these effects could be produced by xylitol independently of glucose 6-phosphate dehydrogenase. Therefore pentose formation could be a factor in the positive action of xylitol on amino acid assimilation even under conditions when other carbohydrates or polyols are not so effective. I think these aspects are worthwhile to be studied in more detail.

One would expect that by xylitol-induced increase in polyol dehydrogenase activity the turnover rate of sorbitol should also be increased. However, the turnover rate of sorbitol remains unchanged despite of an increased activity of polyol dehydrogenase and despite of an increased turnover rate of xylitol. The rate of sorbitol turnover seems to be limited by the further metabolism of fructose. This is reflected by the fact that high intravenous dosage of sorbitol results in fructosemia and fructosuria whereas after high intravenous dosage of xylitol no comparable amounts of intermediates can be detected in the urine [10].

References

1. Bässler, K. H., G. Stein u. W. Belzer: Biochem. Z. **346**, 171 (1966).
2. Müller, F., E. Strack, E. Kuhfahl u. D. Dettmer: Z. ges. exp. Med. **142**, 338 (1967).
3. Bässler, K. H., W. Toussaint u. G. Stein: Klin. Wschr. **44**, 212 (1966).
4. Amatuzio, D. S., F. L. Stutzman, M. J. Vanderbilt, and S. Nesbitt: J. clin Invest. **32**, 428 (1953).
5. Duncan, L. J. P.: Quart. J. exp. Physiol. (Lond.) **41**, 85 (1956).
6. Smith, L. H., R. H. Ettinger, and D. Seligson: J. clin. Invest. **32**, 273 (1953).
7. Glock, G. E., and P. McLean: Biochem. J. **56**, 171 (1954).
8. Beaconsfield, P., J. Ginsburg, and Z. Kosinski: Nature (Lond.) **205**, 50 (1965).
9. McKerns, K. W.: Biochim. biophys. Acta (Amst.) **121**, 207 (1966).
10. Bässler, K. H., H. Holzmann, W. Prellwitz u. J. Stechert: Klin. Wschr. **43**, 27 (1965).

Comment

Inducing Effect of Xylitol in Rats

NORIMASA HOSOYA and NORIKO IITOYO

Nutritional Laboratory, School of Health Sciences, Faculty of Medicine, University of Tokyo, Tokyo, Japan

Male rats (about 70 g) were fed on a laboratory chow (Oriental Yeast Company Ltd., Tokyo, Japan), and adapted to xylitol in the following two ways. In group A, xylitol was provided in successive increases of 5 per cent per week, and a final 20 per cent xylitol diet was provided for 8 or 10 weeks. In group B, xylitol was provided in successive increases of 2.5 per cent per 3.5 days to obtain a 20 per cent xylitol diet (Fig. 1).

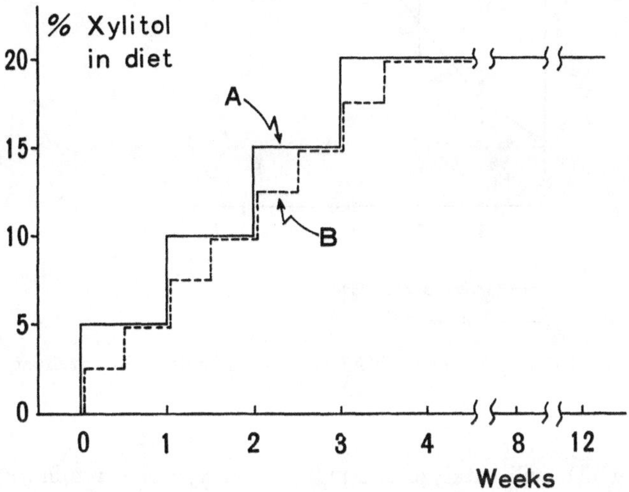

Fig. 1. The feeding of xylitol to rats

Osmotic diarrhoea was observed in the second week of group B, but not in group A. Increasing curves of body weight were not so differet between rats of group A and control normal rats. However, in group B almost 10 per cent body weight loss was observed after diarrhoea. More than 4 days were needed for adaptation to xylitol.

In xylitol metabolism the initial dehydrogenation step was observed in the xylitol adapted rat. A 10 per cent liver homogenate was prepared with 0.25 M sucrose, and cell fractionation was carried out by the method of Littlefield *et al.* [1]. Enzyme activities were measured by the rate of reduction of pyridine nucleotides in the modified Hitachi Fluorophotometer. A unit of enzyme activity is defined as the amount of

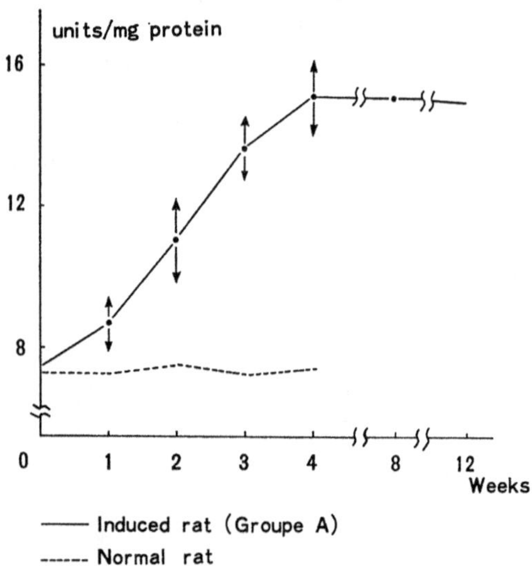

Fig. 2. Xylitol dehydrogenase (NAD linked) activities in rat liver induced with xylitol

enzyme which will catalyze one mμmole on xylitol per minute per mg protein. Touster's method [2], and Hollmann's method [3] were modified, and 2.0 ml of reaction mixture contained 45 mM Tris-buffer (pH 8.1), 2 mM NAD or NADP, 50 mM xylitol, and about 5 units of enzymatic sample.

Mitochondrial dehydrogenation of xylitol is not of quantitative importance in rat liver. The activity of cytoplasmic polyol dehydrogenase (NAD linked) in rat liver increased with xylitol administration. Rat liver adapted with a 20 per cent xylitol diet had 15.08 units/mg protein of NAD-linked xylitol dehydrogenase, and the same value after 8 weeks (Fig. 2). This activity was almost twice that of normal rats. Cytoplasmic NADP-linked xylitol dehydrogenase in rat liver was not affected by xylitol adaptation [normal; 12.26 ± 1.02 (6), induced; 12.04 ± 1.13 (6)].

In xylitol metabolism the initial dehydrogenation step is rate-limiting, and inducing NAD-linked xylitol dehydrogenase causes elevation of the Glucuronic Acid Cycle. Thus xylitol is a very effective precursor of pentoses.

References

1. Littlefield, J. W., and E. B. Keller: J. biol. Chem. **224**, 13 (1957).
2. Touster, O., and G. Montesi: Methods in enzymology, ed. by S. P. Colowick and N. O. Kaplan, vol. 5, p. 317. 1962.
3. Hollmann, S.: Hoppe-Seylers Z. physiol. Chem. **317**, 193 (1959).

Discussion

Dr. Asakura: I would like to comment on the rate-limiting step of xylitol metabolism in erythrocytes. In the cells, initial dehydrogenation is not rate-limiting but the step of xylulokinase is.

Dr. Bässler: It is quite possible that in erythrocytes other steps are rate-limiting.

Dr. Asakura: Did you study xylitol adaptation by administration of sorbitol?

Dr. Bässler: We found no increase in polyol dehydrogenase on sorbitol administration. However Dr. Hollmann told me that he tried sorbitol administration for 4 weeks. We fed only for 10 days.

Dr. Opitz: In toxicological long time feeding experiments with glycols we observed the initial diarrhoea of the rats to disappear. This would indicate that adaptive mechanisms are involved. Am I right?

Dr. Bässler: This is possible indeed.

Dr. Macdonald: As absorption of xylitol is passive, how do you account for the fact that xylitol seems to be absorbed in adapted rats at the lower intestine and cecum?

Dr. Bässler: We did not measure the absorbed amount of xylitol, but determined the nonabsorbed amount remaining in the intestine. Therefore it is difficult to estimate the absorption at the upper small intestine, because transport along the intestine lowers the xylitol content. The fraction of xylitol which is not absorbed in the upper parts of the intestine reaches the cecum and is absorbed there. We have no evidence for a preferential location of xylitol absorption in the intestine. It seems to be of the same magnitude in all parts except the colon.

Dr. Touster: Could the apparent increase in xylitol dehydrogenase be due to cofactor or allosteric effects which cause increased activity, and not actually a consequence of increased enzyme itself? Did you ever use dialyzed liver supernatant?

Dr. BÄSSLER: I am not able to distinguish with certainty between enzyme activation and increase in enzyme concentration. Considering the long time necessary for adaptation, I believe that enzyme synthesis is involved rather than an allosteric effect. I have never used dialyzed liver supernatant. The supernatant, however, was rather diluted, so that eventual cofactors are not very likely to be effective.

Dr. VAN HEYNINGEN: Do you know if it is possible to adapt baby rats by feeding xylitol to the mother?

Dr. BÄSSLER: This is a very interesting suggestion. I have never performed experiments of this kind.

Dr. T. KUZUYA: Is there any change of intestinal bacterial flora which may be responsible for the increased rate of disappearance of xylitol from the intestine in adapted animals?

Dr. BÄSSLER: We did not care for intestinal bacteria because adaptation to xylitol could be obtained by parenteral administration even better than by oral administration. Under parenteral application the influence of the intestinal flora is completely excluded. Furthermore I don't think bacteria are important for reasons Prof. Lang has just pointed out.

Dr. LANG: There are not so many bacteria which are able to metabolize xylitol. Xylitol solution can be stored for long time without any infection. Therefore I suppose that intestinal bacteria do not play an important role in xylitol adaptation phenomena.

Effect of Xylitol on Pyridine Nucleotide Levels in the Liver of Rats

TOSHIO KANEKO, HIROSHI OKA, TOSHITSUGU ODA, and YAWARA YOSHITOSHI

Ist Department of Internal Medicine, Faculty of Medicine, University of Tokyo, Tokyo, Japan

Recently there have been several reports that xylitol stimulates steroid synthesis (*in vitro* [1]) as well as reduction of methemoglobin in erythrocytes [2]. It has been assumed that these stimulations might be induced, at least in part, by altered pyridine nucleotide levels in each of the organ tissues. Actually, the conversion of xylitol to L- or D-xylulose within the cell is accompanied by the reduction of pyridine nucleotides. Therefore the effects of xylitol and other related compounds on the pyridine nucleotide levels have been studied in rat liver slices, and the results are presented in this paper.

Methods and Materials

Approximately 200 mg of rat liver slices were incubated for 15 or 30 min, with 2.5 ml of Krebs-Ringer bicarbonate buffer, pH 7.4, containing 10 mM xylitol, D-xylulose, L-xylulose, ribose or sorbitol. Control slices were incubated with 10 mM glucose. After incubation, slices were taken out and pyridine nucleotides were extracted with 2 ml of 0.25 M glycylglycine buffer, pH 2.0 or pH 13.0, for 2 min at 100° C, and determined by using 6-phosphogluconate-1-^{14}C as described by Epstein *et al.* [3], with minor modifications (Fig. 1).

Results and Discussion

1. Pyridine nucleotide levels in normal rat liver slices were altered during an incubation as shown in Fig. 2. Levels of NAD, NADH as well as NADPH decreased during the first 15 min incubation period, whereas increase of NADP was observed. Levels of pyridine nucleotides during the second 15 min incubation were relatively stable. Fig. 3 shows the effects of xylitol on pyridine nucleotide levels in liver slices from alloxan diabetic rats. Levels of NADH were markedly increased by xylitol with a concomitant decrease of NAD. Although the increased amount of NADH on this occasion was not exactly equivalent to the

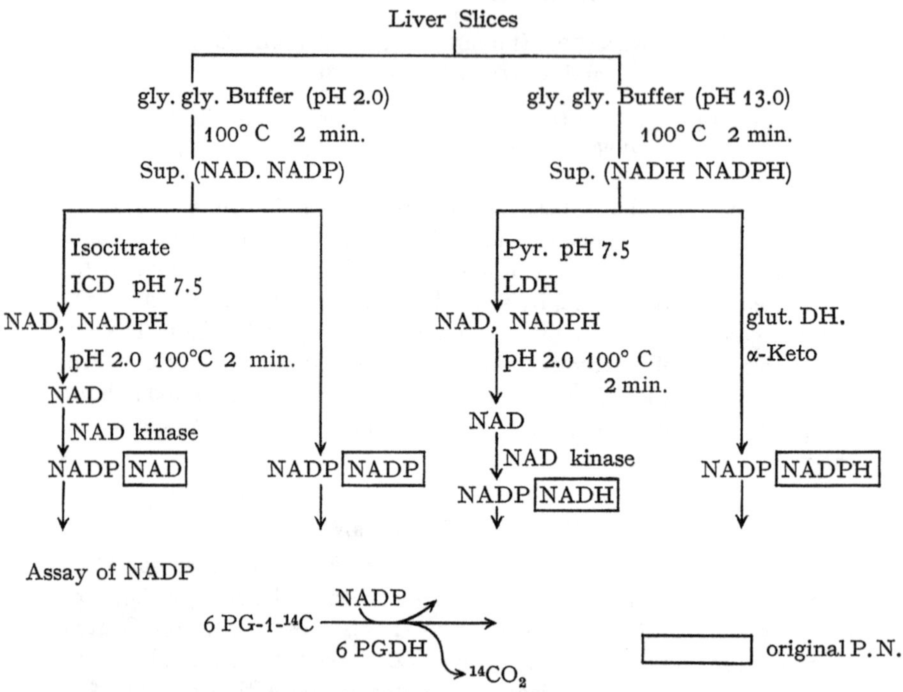

Fig. 1. Assay procedures of pyridine nucleotides

NAD decreased, it is probable that xylitol reduced NAD to NADH with NAD-specific xylitol dehydrogenase. On the other hand, concerning with triphosphopyridine nucleotides the effect was not so obvious; only a slight decrease of NADP was observed in the presence of xylitol. These results suggest that xylitol metabolism to L-xylulose by NADP-specific xylitol dehydrogenase is not a main pathway in liver.

2. As shown in Table I, D-xylulose exerted only slight (or no) effects on NADH, NADPH and NADP, while NAD levels were lower than those in control slices. The effect of L-xylulose on diphosphopyridine nucleotide levels was the same as that of xylitol; increased NADH and decreased NAD were observed. L-Xylulose also caused a rise in NADP levels without affecting NADPH. All these findings with D- and L-xylulose seem to indicate that the effect of xylitol is to generate NADH from NAD with NAD-specific xylitol dehydrogenase, and that L-xylulose is converted to D-xylulose. However, the facts that D-xylulose decreased NAD while L-xylulose did not lower NADPH are not fully explained at present. More complicated mechanisms in the cells must be involved in the regulation of pyridine nucleotide levels.

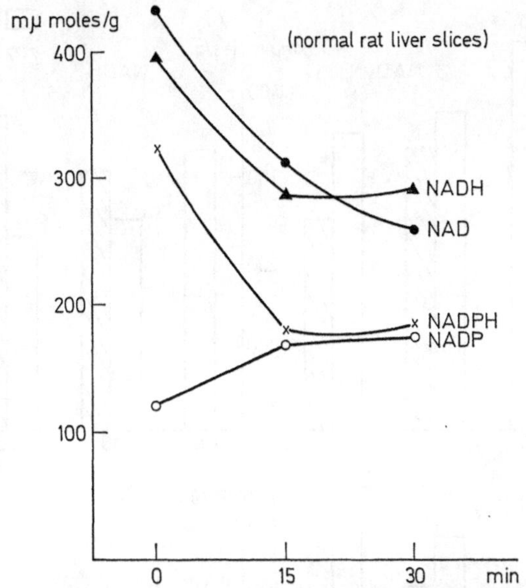

Fig. 2. Changes of pyridine nucleotide levels during incubation. Normal rat liver slices were incubated with 2.5 ml of Krebs-Ringer bicarbonate buffer, pH 7.4, containing 10 mM glucose for 15 or 30 min at 37°C. Pyridine nucleotide levels in slices were assayed as described in Fig. 1

3. Table I presents the effects of ribose and sorbitol on pyridine nucleotide levels in alloxan diabetic rat liver slices. Ribose as well as sorbitol markedly elevated NADH levels and slightly decreased NAD. Sorbitol had no effect on triphosphopyridine nucleotide levels, while ribose resulted in elevated NADPH level. Since sorbitol is easily converted to fructose by sorbitol dehydrogenase (coupled with the reduction of NAD), such an effect might be explained by this mechanism. That of ribose, however, is hard to explain.

4. Since all of the results mentioned above were obtained with alloxan diabetic liver slices of rats, experiments with normal liver slices were performed along the same line. As shown in Fig. 4, the effects of xylitol on pyridine nucleotide levels in normal rat liver slices were quite different from those in diabetic rat liver slices.

With normal rat liver slices no significant effect of xylitol was seen on NAD or NADH levels, and only a slight, but significant decrease in NADPH was brought about by xylitol. As shown in Table II, the effects of D- and L-xylulose on normal liver slices also were different from those observed with diabetic liver slices; the elevations of NADH and NADP induced by L-xylulose in the diabetic liver were not observed in the normal.

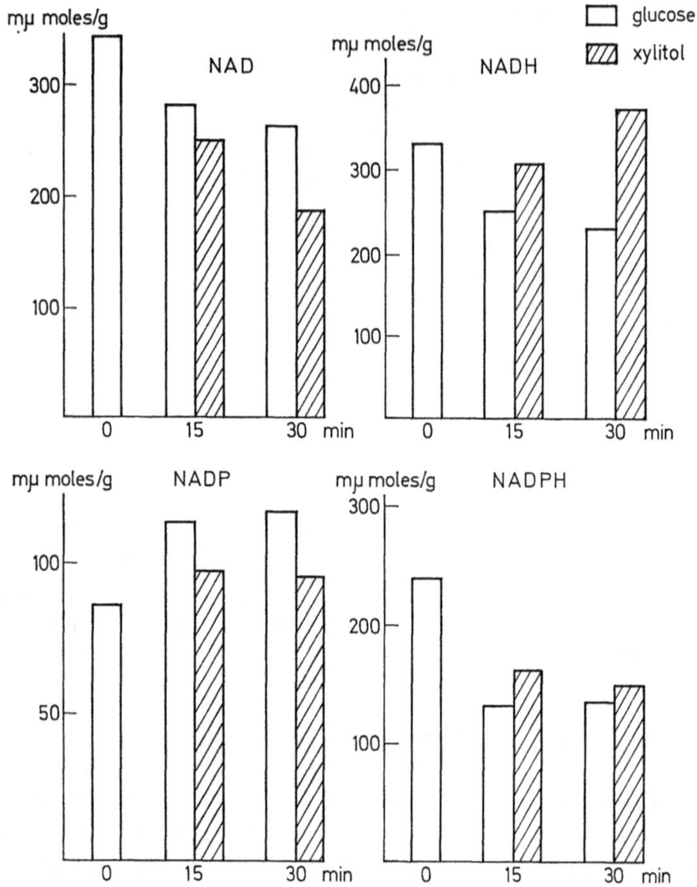

Fig. 3. Effect of xylitol on pyridine nucleotide levels in alloxan diabetic rat liver slices

Table I. *Effects of D-xylulose, L-xylulose, D-ribose and sorbitol on pyridine nucleotide levels in diabetic rat liver slices*

Substrates	Pyridine nucleotides (% of control)			
	NAD	NADH	NADP	NADPH
D-xylulose	64	120	90	104
L-xylulose	65	145	166	99
D-ribose	79	178	111	139
Sorbitol	80	169	97	100

Values are expressed as a percentage of those observed in control slices which were incubated with 10 mM glucose. In each experiment liver slices obtained from the same rat were used. Incubation time was 30 min.

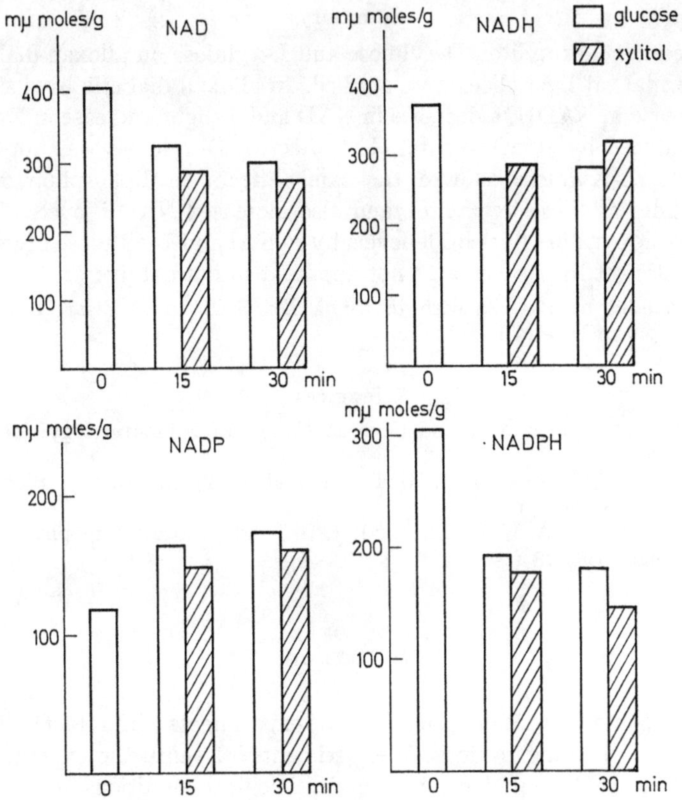

Fig. 4. Effect of xylitol on pyridine nucleotide levels in normal rat liver slices

The reasons for these differences in normal and diabetic liver slices have not been elucidated at present. One possible explanation is that some metabolic disorders present in diabetic liver produce an accumulation of NADH or NADP catalyzed by xylitol dehydrogenase, or on the other hand, some regulating mechanisms may counteract the accumulation of one type of pyridine nucleotide in normal liver.

Table II. *Effects of D-xylulose, L-xylulose, D-ribose and sorbitol on pyridine nucleotide levels in normal rat liver slices*

	NAD	NADH	NADP	NADPH
D-xylulose	111	77	106	90
L-xylulose	102	84	118	76
D-ribose	89	74	114	79
Sorbitol	105	92	96	77

The experimental procedures were as described in the legend to Table I.

Summary

The effects of xylitol, D-xylulose and L-xylulose on alloxan diabetic and normal rat liver slices were studied. In alloxan diabetic liver slices, an increase in NADH, a decrease in NAD and a slight decrease in NADP were induced by 10 mM xylitol. The effect of D-xylulose was not conspicuous. L-Xylulose showed the same effects on diphosphopyridine nucleotides as those by xylitol, and also increased NADP levels. These changes in pyridine nucleotide levels by xylitol and L-xylulose observed in diabetic rat liver slices were not apparent in normal liver.

Acknowledgements. We wish to thank Dr. S. Kono of Eisai Co., Ltd., for a supply of D- and L-xylulose.

References

1. Ohnuki, M.: Third Asia and Oceania Congr. of Endocrinology, Manila, 1967.
2. Asakura, T., K. Adachi, S. Minakami, and H. Yoshikawa: J. Biochem. **62**, 2, 184 (1967).
3. Epstein, S. M., A. K. Remer, and J. B. Field: Biochim. biophys. Acta (Amst.) **100**, 13 (1965).

Discussion

Dr. Yoshikawa: Have you ever done experiments with intact animals or are you planning to do such experiments? I think the experiments with slices are done under quite nonphysiological conditions and, therefore, it might be hard to get decisive conclusions.

Dr. Oka: No, I have never done *in vivo* experiments.

Dr. Bässler: Did you measure the total pyridine nucleotide contents of your liver slices or did you use fractionated mitochondrion and cytoplasma? With the slice you should hardly be able to detect changes in NADPH, because most of it is located in the cytoplasm. Only mitochondrial NADPH can be changed due to the presence of xylitol or L-xylulose.

Dr. Oka: I have only measured the total pyridine nucleotides in liver slices without the fractionation of cell components. The changes in triphosphopyridine nucleotides by xylitol were very slight. However, the finding that L-xylulose caused the marked increase in NADP levels suggests that the changes in triphosphopyridine nucleotides due to L-xylulose reductase can be detected by our experimental procedures.

Characteristics of Sugar Transport in the Rabbit Adrenal

Studies by Non-Utilizable Pentoses and Xylitol

SEIICHIRO TARUI and YOSHIKAZU SAITO

Second Department of Internal Medicine, Osaka University Medical School, Osaka, Japan

Passage of sugars and polyols across the cell membrane is the first rate-limiting step in metabolism of these substances and, in particular, glucose transport in some tissues is a site of hormonal regulation. Some tissues show active transport for glucose; that is, a process which must be maintained by energy derived from metabolism in order to move sugars from a lower into a higher concentration. Generally speaking, the tissues having a specialized system of active transport for glucose are situated on the boundary between internal and external environments. Intestinal absorption or tubular reabsorption of glucose is known to be an active transport process. However, up to the present, these systems of active transport for glucose have not been shown to be influenced directly by any hormones. The hormone-sensitive sugar transfer is, so far, limited to a process of downhill transport and the stereospecificity of the hormone effect provides evidence for the existence of a transport carrier system.

The ability of insulin to regulate sugar permeability was first demonstrated in eviscerated animals using D-galactose by Goldstein, Levine and their coworkers [1]. They showed that various sugars were divided into two groups, "responsive" and "non-responsive" to insulin, and proposed that the insulin effect is confined to sugars with the same configuration as D-glucose about carbons 1, 2 and 3. This transfer theory of insulin action received substantial support from subsequent confirmatory observations, in which pentoses such as D-xylose and L-arabinose were often used as non-utilizable model sugars.

Thus the stereospecifically stimulated sugar transfer has been assumed to occur characteristically in a target organ of insulin in the presence of this hormone. However, the author [5] has shown that *in vitro* addition of thyroid stimulating hormone (TSH) enhances the entry of D-xylose and L-arabinose into bovine thyroid slices; the principle of stereospecific regulation of sugar transport is also applicable to peptide hormones other than insulin.

In the course of studying the mode of action of adrenocorticotrophic hormone (ACTH), Eichhorn and his coworkers [3] have revealed that *in vivo* ACTH administration causes an increased penetration of D-xylose into the adrenal in hypophysectomized rats. However, further *in vitro* studies on the permeability characteristics of bisected rat adrenals failed to demonstrate any significant influence of ACTH on the distribution of D-xylose [4]. They reported that the extracellular space in sectioned

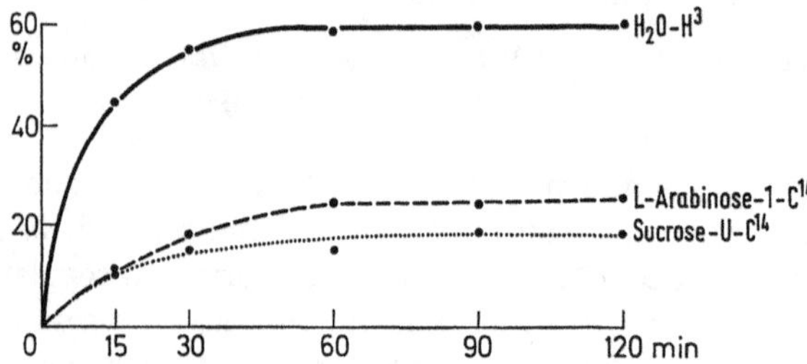

Fig. 1. Permeability characteristics of bisected rabbit adrenals in the absence of ACTH

rat adrenals was expanded as incubation time increased. In our laboratory the extracellular space in *in vitro* experiments has been found to be stable when bisected rabbit adrenals are employed instead of rat adrenals. This communication deals with the characteristics of sugar transport in bisected rabbit adrenals in the presence and absence of ACTH.

Adrenal preparations were obtained from exsanguinated white male rabbits; after being stripped of the thin fibrous capsule, adrenals weighing 120 to 200 mg were bisected for incubation. The tissues were incubated at 37° C with shaking under 100 per cent O_2 in Krebs-Ringer phosphate containing L-arabinose-1-C^{14} and H_2O-H^3. Sucrose space (extracellular space) was determined with sucrose-U-C^{14} in a separate incubation. The concentration of L-arabinose or sucrose was adjusted to 1 mg/ml with non-labelled carrier.

The distribution of a substrate in whole tissue (space) is expressed by the following equation:

$$\text{Space (ml/100 g)} = \frac{\text{content in wet tissues (mg/g)}}{\text{medium concentration (mg/ml)}} \times 100.$$

Fig. 1 illustrates the variation in the spaces of L-arabinose-1-C^{14}, sucrose-U-C^{14} and water-H^3 in bisected rabbit adrenals at varying incubation times. L-arabinose was chosen because it has the same stereochemical property as D-glucose. Even when the incubation time is prolonged to 120 minutes the difference between water-H^3 space and L-arabinose space remains constant and L-arabinose space is always in close proximity to sucrose space. These findings indicate that the permeability barrier is operative in adrenal tissue.

Since the barrier restricts the transfer of sugars across the cell membrane of the target of ACTH, transport of glucose analogs must be facilitated by the trophic hormone *per se.*

Table I. *Effect of β^{1-24} ACTH on the distribution of various sugars in bisected rabbit adrenals*

Space	Control	With ACTH (IU/ml)	Net effect ± S.D.	*P*-value
D-xylose	24.2	28.7	4.5 ± 1.9	< 0.001
L-arabinose	22.5	26.9	4.4 ± 1.6	< 0.02
L-xylose	28.8	27.4	−1.4 ± 1.5	
D-arabinose	28.3	27.5	−0.8 ± 2.2	
^{14}C-sucrose	23.1	24.4	1.3 ± 2.5	

The conditions of incubation (Tables I and II) were essentially the same as in Fig. 1 except that Krebs-Ringer phosphate containing non-labelled pentoses (4 mg/ml) was employed and that sucrose-U-C^{14} space was determined in the presence of pentoses.

As shown in Table I, addition of β^{1-24} ACTH produces a significant increase in the D-xylose and L-arabinose spaces in bisected rabbit adrenals. Calculated from these data, the intracellular distribution of D-xylose is increased several fold by β^{1-24} ACTH, taking sucrose for a measure of the extracellular space. The spaces of L-xylose and D-arabinose are either not affected or rather inclined to be reduced by β^{1-24} ACTH addition. Thus the proposal by Levine *et al.* is compatible with the permeability characteristics of adrenal gland.

The *in vitro* effect on sugar transport depends on the type of ACTH preparations. Synthetic β^{1-24} ACTH peptide and crude ACTH preparations invariably produced a promoting effect on sugar entry. When purified natural ACTH consisting of 39 amino acids was employed instead of synthetic short-chain peptide, three hours or more incubation time was required to demonstrate an excess transfer of "responsive" sugars. Therefore, a considerable discrepancy exists in the time course of hormone effect on sugar transport between 39 and 24 ACTH peptides. However, as for the time course of an effect on corticoid production, the short-chain peptide was found to be quite the same as natural ACTH. It is believed that the immediate effect of ACTH on corticoid production is independent of the change in sugar permeability. The latter is, rather, related to the sustained functioning of the cells. The effect of ACTH on sugar transport presumably constitutes a major factor responsible for the trophic action, which brings about hypertrophy in its excess and atrophy in its absence.

The precise determination of "xylitol space" is not feasible if a fraction of penetrated xylitol is metabolized in adrenal tissue. However, when the distribution of xylitol-U-C^{14} was calculated, depending on the assumption of non-utilization during the incubation time, "xylitol space" was found to be slightly but significantly larger than sucrose space.

Table II. *Effect of cyclic AMP on the distribution of various sugars in bisected rabbit adrenals*

Space	Control	With C-AMP 1.5×10^{-3}M	Net effect $\pm$S.D.	*P*-value
D-xylose	24.9	29.8	4.9 ± 1.3	< 0.005
L-arabinose	25.4	30.9	5.5 ± 2.4	< 0.01
L-xylose	27.1	25.8	-1.3 ± 1.7	
D-arabinose	26.3	24.7	-1.6 ± 0.9	
^{14}C-sucrose	21.6	21.5	-0.1 ± 0.5	

In intact rat hemidiaphragms "xylitol space" was also found to be constantly larger than sucrose space and not to be influenced by insulin addition. Xylitol penetration probably is not affected by the "keeping-out" mechanism of cell membrane.

To date it has been reported that the activity of adenyl cyclase is influenced by several hormones and that cell membrane has been shown to be a principal candidate for the location of this enzyme [6, 7]. The possibility that adenosine 3',5'-monophosphate (cyclic AMP) might stimulate the sugar transport system was studied. As shown in Table II, the strictly stereospecific effect on sugar transport in bisected rabbit adrenals is reproduced when cyclic AMP is added to the medium instead of β^{1-24} ACTH. No effect of 5'-AMP is observed. ACTH has been shown by Haynes [2] to cause an accumulation of cyclic AMP in slices from adrenal cortex. It is suggested that ACTH alters the sugar transport system in adrenal cortex by stimulating the formation of cyclic AMP.

Summary

Characteristics of sugar transport in the rabbit adrenal are described. In adrenal tissue, as in diaphragmatic muscle, the presence of a hormone-sensitive membrane restricts the entry of glucose analogs into cells which, without its presence, would be freely permeable. Addition of synthetic β^{1-24} ACTH peptide to the incubation medium promotes the penetration of D-xylose and L-arabinose into bisected rabbit adrenals without causing any increase in the spaces of the corresponding optical isomers. These responses are reproduced when cyclic AMP is added to the medium instead of ACTH. Xylitol penetration probably is not influenced by the hormone-sensitive permeability barrier in cell membrane.

References

1. Goldstein, M. S., W. L. Henry, B. Huddlestun, and R. Levine: Amer. J. Physiol. **173**, 207 (1953).
2. Haynes, R. C., Jr.: J. biol. Chem. **233**, 1220 (1958).
3. Eichhorn, J., I. D. K. Halkerston, M. Feinstein, and O. Hechter: Proc. Soc. exp. Biol. (N.Y.) **103**, 515 (1960).
4. Golden, M., E. Scully, J. Eichhorn, and O. Hechter: Proc. Soc. exp. Biol. (N.Y.) **106**, 354 (1961).
5. Tarui, S., K. Nonaka, Y. Ikura, and K. Shima: Biochem. biophys. Res. Commun. **13**, 329 (1963).
6. Davoren, P. R., and E. W. Sutherland: J. biol. Chem. **238**, 3016 (1963).
7. Rabinowitz, M., L. Desalles, J. Meisler, and L. Lorand: Biochim. biophys Acta (Amst.) **97**, 29 (1965).

Discussion

Dr. Kuzuya: Have you ever studied the effect of cyclic AMP on the diaphragm? Is there any resemblance between the action of cyclic AMP and of insulin on muscle?

Dr. Tarui: Yes, I have done such experiment. When cyclic AMP was added *in vitro*, L-arabinose space of the diaphragmatic muscle was found to be *reduced* significantly. In muscle tissues cyclic AMP does not mimic the action of insulin, so long as the sugar permeability is concerned with. I think, however, it is still possible for insulin to exert its effect through formation of some chemical transmitter. It must be other substance than cyclic AMP.

Stimulation of Insulin Secretion by Xylitol Administration

KINORI KOSAKA

Department of Internal Medicine, Tokyo Women's Medical College, Tokyo, Japan

There have been increasing reports of the clinical use of xylitol solution as a parenteral infusion fluid mainly in Germany and in Japan. Particularly as an infusion fluid for diabetic patients, it is presumed to have several advantages; it does not raise the blood sugar so much [1]; it has an antiketogenic action [2, 3]; and it probably enters the cell independently of insulin action [4].

For more than ten years, I have been studying the mechanism of secretion of insulin with several coworkers. Using dogs as experimental animals, we employed an *in vivo* assay of blood insulin as well as a radioimmunoassay. Here, I would like to present studies concerning stimulation of insulin secretion by xylitol; not only those studies carried out by us, but also other studies performed in various laboratories in Japan.

Soon after xylitol became available for parenteral use, we undertook a series of experiments in which we studied the effects of several different sugars upon insulin secretion. Glucose, fructose, mannitol, and xylitol were chosen in the first series. In these experiments, dogs were anesthetized with sodium pentobarbitol (25—30 mg/kg i.v.), and after taking two control blood samples from the femoral vein, they were given intravenous administration of 0.4 g/kg of each of the above substances in 20% solution. Serial blood samples were withdrawn up to 3 hours after injection, plasma glucose was determined by the glucose oxidase method, and plasma insulin was assayed by a double antibody radioimmunoassay as previously reported [5] (Fig. 1).

After injection of glucose, there were expected rises of both plasma glucose and insulin. After fructose injection, plasma glucose remained unchanged and plasma insulin increased only slightly. Both plasma glucose and insulin remained constant after mannitol injection. To our surprise, following administration of xylitol there was a remarkable increase in plasma insulin. Plasma glucose increased initially, then decreased below the fasting level. Although several sugars other than glucose have been reported to provoke insulin secretion, none of them were more potent than glucose in its capacity to increase plasma insulin.

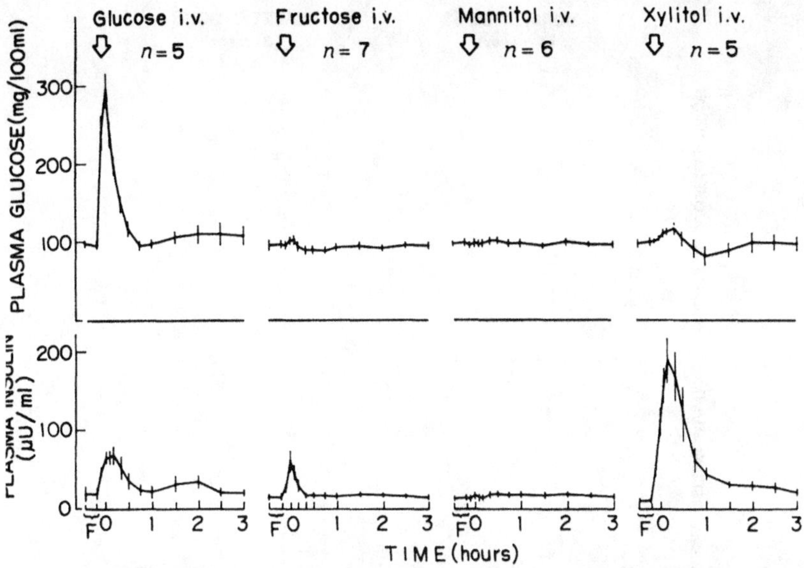

Fig. 1. Effects of intravenous injection of various sugars or sugar alcohols (0.4 g/kg) upon plasma insulin and glucose concentrations in dogs (Mean ± SD)

The effect of 0.4 g/kg xylitol in our experiment much exceeded that of glucose; the insulin concentration was about $2^1/_2$ times as high as that following glucose administration. These results were reported independently in a preliminary form at nearly the same time from us [6] and from Hirata and coworkers in Kyushu University [7]. Furthermore, Hirata and his coworkers performed the same experiment on alloxan-diabetic dogs, finding out that xylitol injection causes a rise in blood sugar without increasing plasma insulin (Fig. 2). The latter finding suggests that the excessive increase in plasma insulin caused by xylitol is a result of augmented secretion of insulin from the pancreas.

That administration of xylitol increases the output of insulin from the pancreas of dogs was confirmed directly by the experiment of Kuzuya and others [8]. Fig. 3 shows the scheme of their experiment. After laparotomy, a catheter was inserted into the superior pancreaticoduodenal vein and the other end of the catheter was inserted into the portal vein in end-to-side fashion. In order to prevent blood coagulation within this catheter, dilute heparin was slowly infused upstream through another thin catheter. By detaching the connections of the catheter, pancreatic vein blood was collected directly into glass tubes. By this method, both insulin concentration and blood flow in the pancreaticoduodenal vein can be measured. From these data and hematocrit, output of insulin through this vein was calculated.

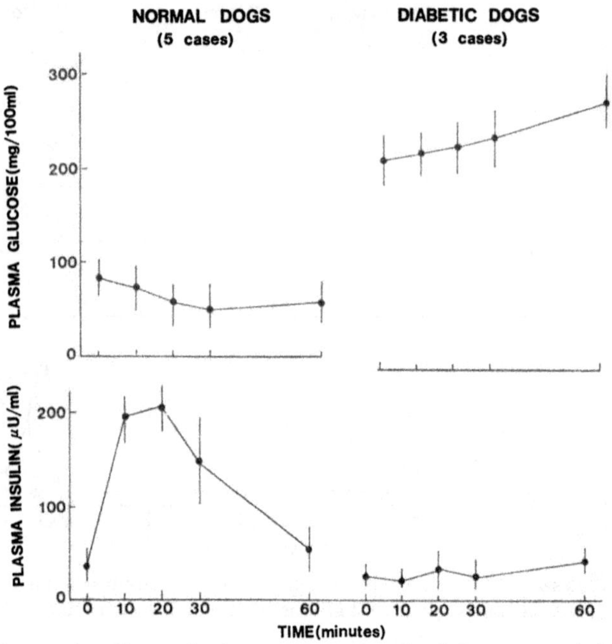

Fig. 2. Plasma insulin and glucose response to intravenous injection of 0.4 g/kg xylitol in normal and diabetic dogs (Y. Hirata *et al.*)

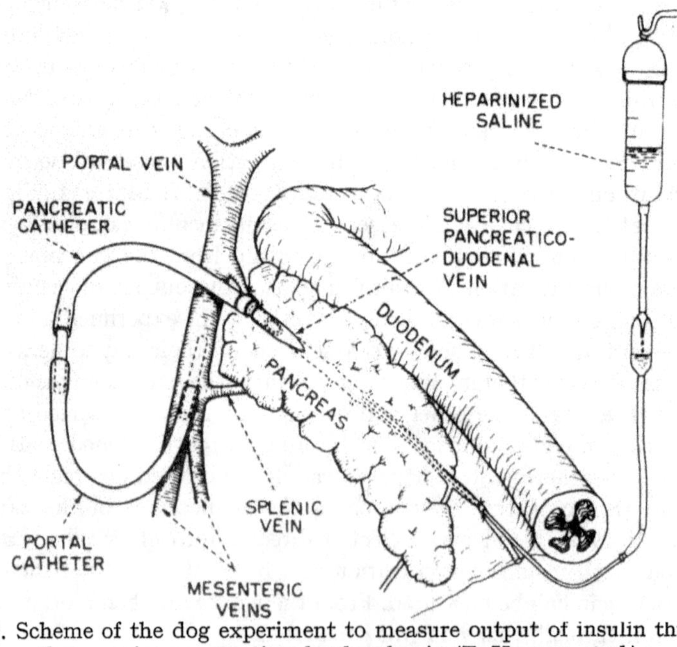

Fig. 3. Scheme of the dog experiment to measure output of insulin through the superior pancreatico-duodenal vein (T. Kuzuya *et al.*)

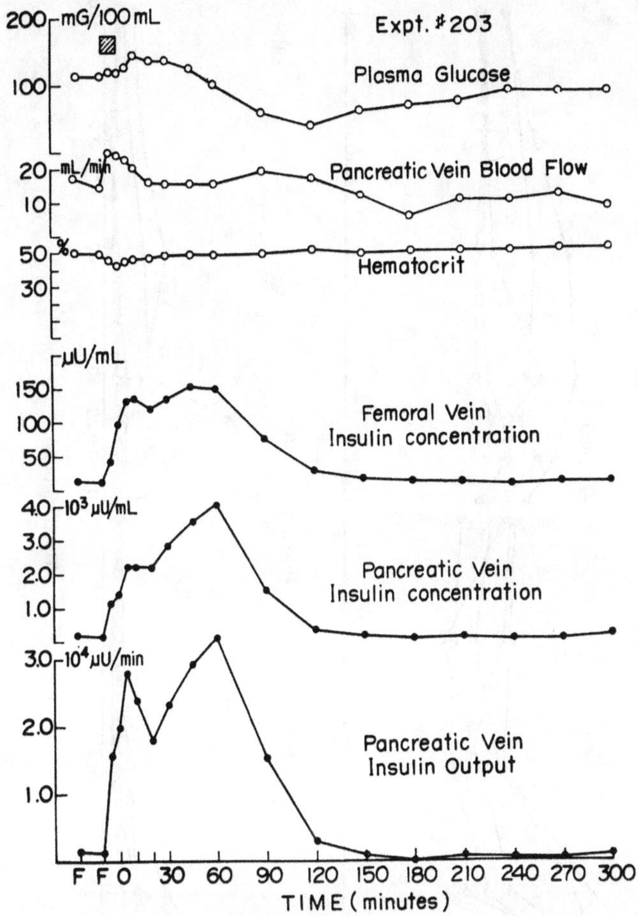

Fig. 4. Effect of intravenous injection of 1.0 g/kg xylitol upon insulin output via the superior pancreatico-duodenal vein in a dog (T. Kuzuya *et al.*)

One example of such an experiment is shown in Fig. 4. After intra venous injection of 1.0 g/kg xylitol, both plasma insulin concentration and blood flow in the pancreatic vein increased quickly, and the hematocrit decreased slightly for a short time. Therefore, output of insulin increased quickly, indicating that there was an actual increase of insulin release after xylitol administration. It may be concluded that plasma insulin response to xylitol in dogs is ascribed, at least in part to augmented output of insulin.

So far, I have been talking about the effect of xylitol in dogs. It is important to determine whether this strong insulinogenic effect of xylitol

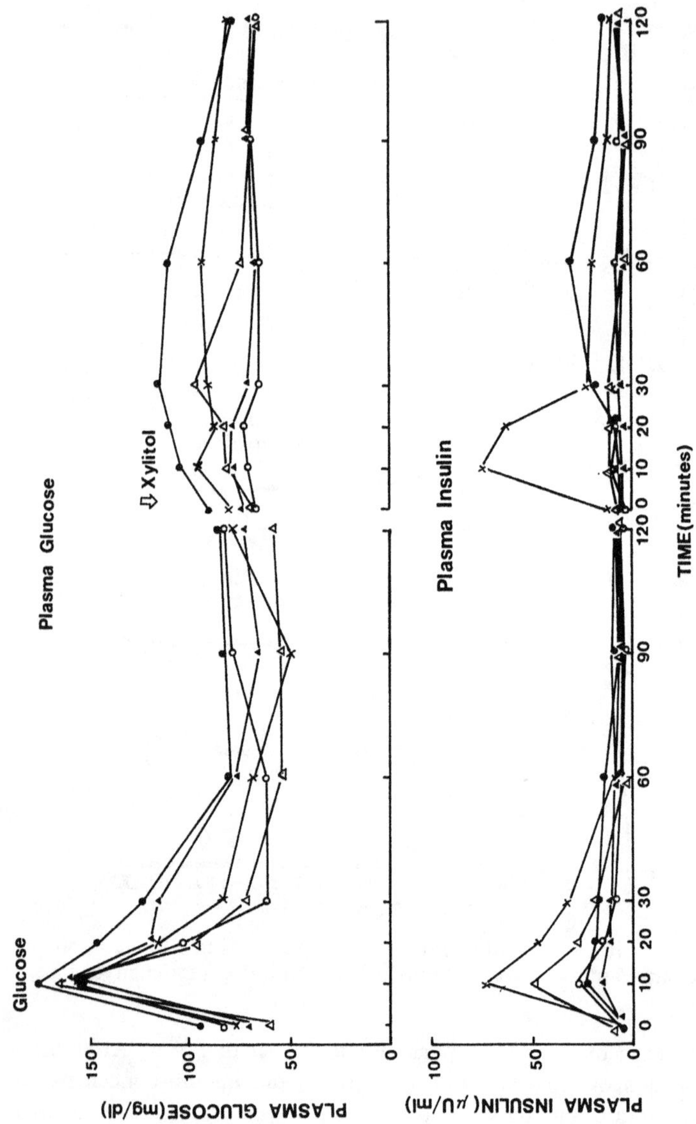

Fig. 5. Effect of intravenous administration of 12 g glucose or xylitol upon plasma glucose and insulin in 5 healthy subjects. The same symbol represents the same subject

is reproduced in man, because this substance is now increasingly used in clinical medicine. Therefore, we have studies the effect of administration of xylitol upon plasma insulin in man [9].

Twelve grams of glucose or xylitol (about 0.2—0.25 g/kg) was injected intravenously to 5 healthy volunteers and 5 diabetic patients. To compare the effect of both substances in the same subject, each

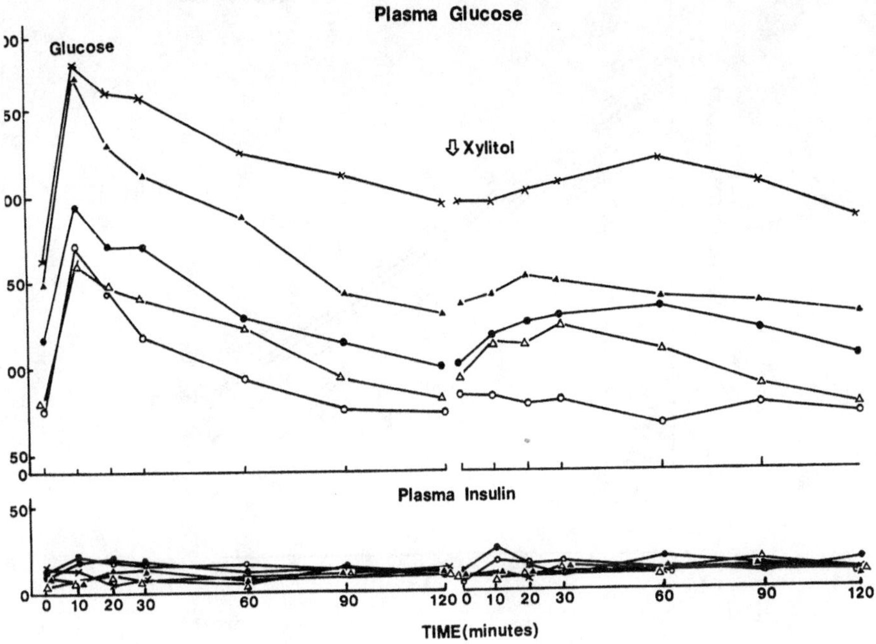

Fig. 6. Effect of intravenous administration of 12 g glucose or xylitol upon plasma glucose and insulin in 5 diabetic subjects. The same symbol represents the same subject

subject received both tests within an interval of a few days. Venous blood was withdrawn several times, and plasma glucose and insulin were determined. As shown in Fig. 5, in contrast to dogs, there was a smaller rise of plasma insulin after xylitol administration than after that of glucose in healthy people. However, it is interesting that there was one person who responded to xylitol as well as to glucose. Plasma glucose showed a slight inconsistent increase after xylitol injection.

Fig. 6 shows similar experiments carried out on several diabetic patients. Plasma insulin response to xylitol was even less conspicuous in these people. The rates of disappearance of administered glucose or xylitol in plasma also were studied (Fig. 7). While the rate of disappearance of glucose was much slower in diabetic subjects than in normal people, that of xylitol was about the same in both diabetic and normal groups. The diabetic state apparently does not affect the disappearance of xylitol.

The poor response of plasma insulin in man also was reported by Ogawa and coworkers in Wakayama University [10] and by Ohnishi and coworkers in Okayama University [11].

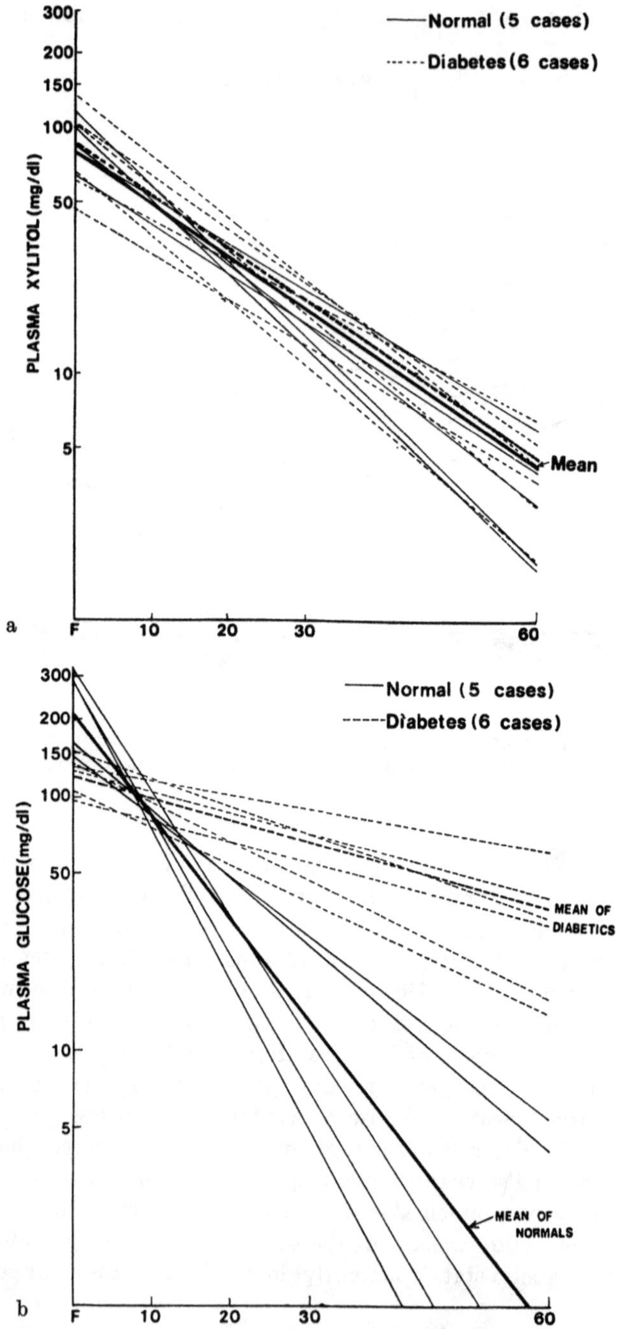

Fig. 7. Changes of plasma xylitol (a) or glucose (b) concentration after injection of 12 g of xylitol or glucose in normal and diabetic subjects. Solid lines represent normal subjects while dotted lines represent diabetics

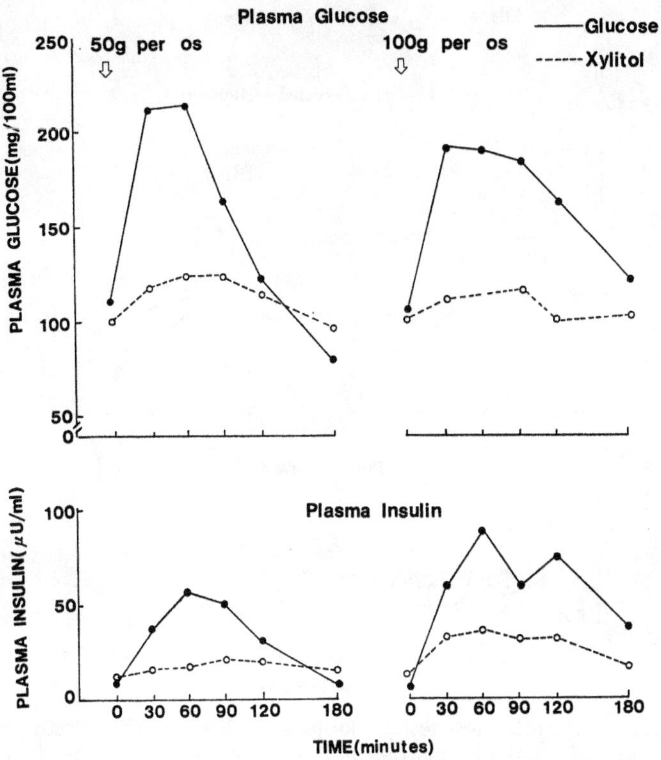

Fig. 8. Effect of oral administration of glucose and xylitol upon plasma glucose and insulin in normal and mildly diabetic subjects

We have also studied the effect of oral administration of xylitol in normal or mildly diabetic subjects (Fig. 8). When 50 to 100 g of xylitol was given orally, plasma glucose did not increase at all or increased slightly. Plasma insulin increased only slightly, usually much less than that after glucose administration. Occasionally there also was a moderate increase in plasma insulin comparable to that caused by glucose.

Because the absorption rate of xylitol from the intestine is fairly slow and diarrhea occurs often after ingestion of large amount of xylitol, several volunteers received daily oral administration of 30 g of xylitol for 14 days before oral xylitol tolerance tests were performed (Fig. 9). Bässler and others [12] demonstrated that in rats the intestinal absorption rate of xylitol is much improved by adaptation to daily administration of xylitol. However, in our human experiments, even after this pretreatment, plasma insulin response after oral xylitol was poor, and much less than that after oral glucose.

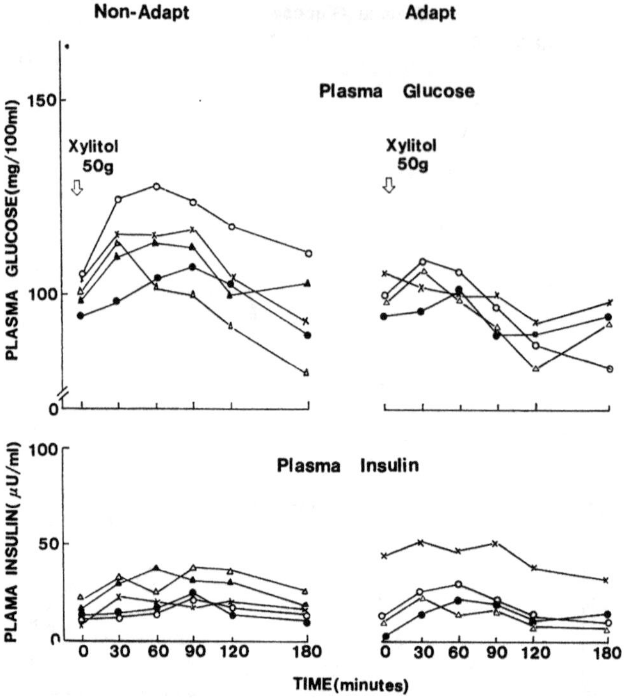

Fig. 9. Oral xylitol tolerance test in subjects without or with pretreatment of daily oral administration of 30 g xylitol for 2 weeks

The difference between plasma insulin responses in man and dog seems to be due to a species difference of inherent character, because the recent experiments of Kuzuya and coworkers [13] indicated that dogs respond well not only to intravenous but aslo to oral administration of xylitol by an excessive increase in plasma insulin, accompanied by severe hypoglycemia (Fig. 10). Plasma insulin after oral xylitol administration reached a far higher peak than that after oral administration of glucose.

Whatever the resaon for this species difference in plasma insulin response is, it seems interesting to study further the strong insulinemic effect of xylitol in dogs. Xylitol is known to enter the pentose-phosphate pathway via D-xylulose 5-phosphate. Therefore, it would seem reasonable to assume that some common metabolic process or intermediate shared by glucose and xylitol may be directly connected with the control mechanism of insulin secretion. In this respect, investigation of insulin secretion by xylitol may give some clue also to the mechanism of glucose-induced secretion of insulin.

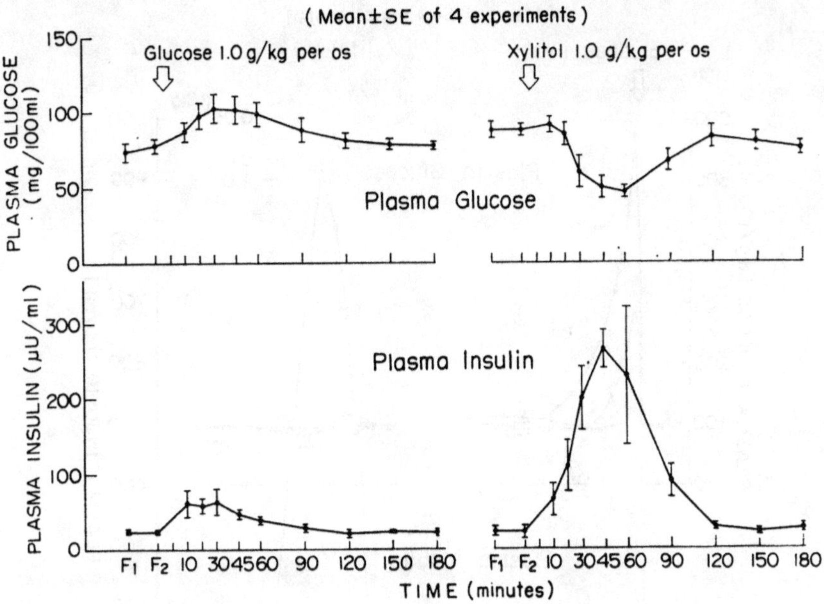

Fig. 10. Effect of oral administration of 1.0 g/kg glucose or xylitol in dogs (Mean ± SE of 4 experiments) (T. Kuzuya *et al.*)

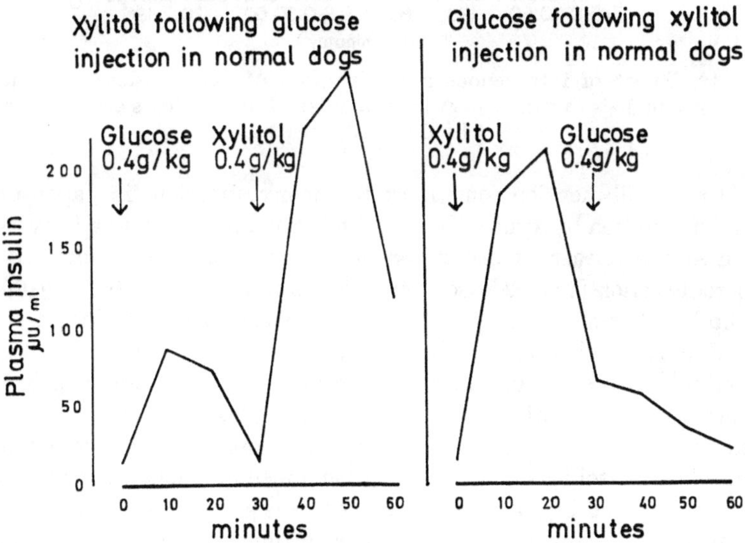

Fig. 11. Plasma insulin response to successive injections of glucose and xylitol or vice versa (Y. Hirata *et al.*)

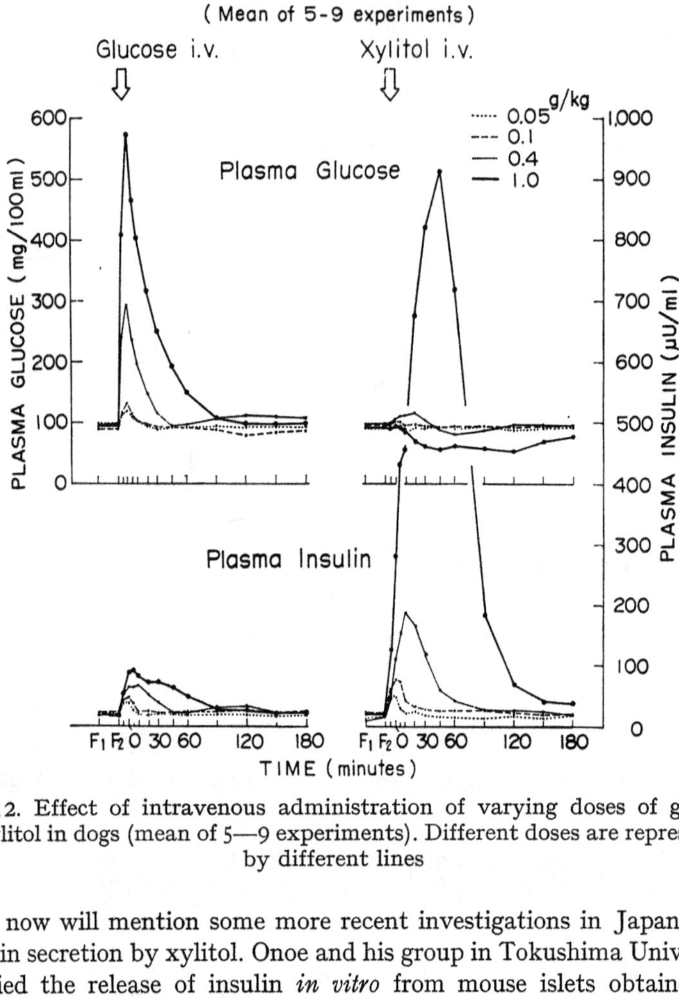

Fig. 12. Effect of intravenous administration of varying doses of glucose or xylitol in dogs (mean of 5—9 experiments). Different doses are represented by different lines

I now will mention some more recent investigations in Japan upon insulin secretion by xylitol. Onoe and his group in Tokushima University studied the release of insulin *in vitro* from mouse islets obtained by microdissection [14]. Release of insulin was stimulated by 300 mg/100 ml xylitol, but less markedly by the same concentration of glucose.

Fujisawa and coworkers [15] in Kyushu University injected 0.4 g/kg of either glucose or xylitol twice at an interval of 30 minutes (Fig. 11). When glucose was injected twice, the rises of plasma glucose and insulin after the second injection were smaller than those after the first injection, but when xylitol was given twice, plasma insulin reached a higher peak after the second injection. When glucose was given first and xylitol was given later, plasma insulin increased each time after injection. Contrary to this, when xylitol was given first and glucose was given secondly, no increase in plasma insulin was noted after the second injection.

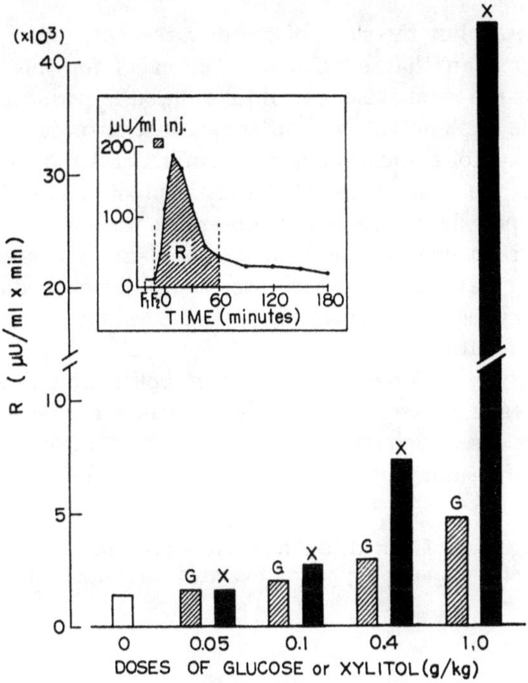

Fig. 13. Dose-response relationship of plasma insulin response during 60 minutes after injection of glucose or xylitol

Recently, Kosaka and coworkers [13] compared the insulinemic effects of glucose and xylitol in anesthetized dogs by means of intravenous administration of varying doses of these substances (Fig. 12). When the dose of glucose was increased, the peak values of plasma glucose and insulin reached higher levels. With smaller doses of xylitol, plasma insulin reached slightly higher peaks than those after administration of equivalent doses of glucose. The stronger insulinemic effect of xylitol relative to glucose became much more pronounced at higher doses. Particularly after administration of 1.0 g/kg of xylitol, plasma insulin reached surprising high levels, near 1,000 μU/ml. The peak insulin concentration tended to appear progressively delayed after injection of higher doses of xylitol. Plasma glucose remained unchanged after small doses of xylitol. At a moderate dose it increased initially, then decreased to a hypoglycemic level. After a large dose, the dogs became overtly hypoglycemic without preceding hyperglycemia.

If the degree of plasma insulin response is expressed by the area encircled by the insulin curve and the abscissa from the beginning to 60 minutes after injection, we can draw a dose-response chart as shown in Fig. 13. Insulin response to glucose and xylitol both increases with increasing doses, but the slope of the dose-response curve by xylitol is much steeper than that by glucose. The cause for this difference is obscure, as is the mechanism of xylitol-induced hyperinsulinemia itself. One plausible explanation is a difference in transport mechanisms of glucose and xylitol through beta-cell membrane. Although no data are available for the sugar-transport mechanism of beta-cell at present, glucose transport across the cell membrane in various cells may involve a specific carrier, which would be saturated at high glucose concentration. If the transport of xylitol into the cell is not mediated by such a carrier, it is likely that xylitol can enter the cell more freely, especially at high concentrations.

Further investigation of the *in vivo* metabolism of xylitol in different animal species is necessary to clarify the cause of species differences in insulin response. Such studies may also cast a light on the mechanism of secretion of insulin, not only by xylitol but also by glucose and other stimuli.

Acknowledgement. I would like to express my thanks to Drs. Y. Hirata and T. Kuzuya for kindly supplying me with their data and figures before publication.

References

1. Bässler, K. H., W. Prellwitz, V. Unbehaun u. K. Lang: Xylitstoffwechsel beim Menschen. Klin. Wschr. **40**, 791 (1962).
2. Haydon, R. K.: The antiketogenic effects of polyhadric alcohols in rat-liver slices. Biochim. biophys. Acta (Amst.) **46**, 598 (1961).
3. Bässler, K. H., u. G. Dreiss: Antiketogene Wirkung von Xylit bei alloxandiabetischen Ratten. Klin. Wschr. **41**, 593 (1963).
4. —, u. W. Prellwitz: Insulin und der Verteilungsraum von Xylit bei eviscerierten Ratten. Klin. Wschr. **42**, 94 (1964).
5. Kanazawa, Y., T. Kuzuya, T. Ide, and K. Kosaka: Plasma insulin responses to glucose in the femoral, hepatic and pancreatic veins in dogs. Amer. J. Physiol. **211**, 442 (1966).
6. Kuzuya, T., Y. Kanazawa, and K. Kosaka: Plasma insulin response to intravenously administered xylitol in dogs. Metabolism **15**, 1149 (1966).
7. Hirata, Y., M. Fujisawa, H. Sato, T. Asano, and S. Katsuki: Blood glucose and plasma insulin responses to xylitol administrated intravenously in dogs. Biochem. biophys. Res. Commun. **24**, 471 (1966).
8. Kuzuya, T., Y. Kanazawa, and H. Sando: Effect of xylitol upon insulin secretion in dogs (Abstract in Japanese). Folia endocr. Jap. **43**, 281 (1967).
9. Kosaka, K., Y. Kanazawa, T. Kuzuya, M. Shimoshige, and Y. Mizuno: Plasma insulin responses to xylitol in man (Abstract in Japanese). J. Jap. Diabetic Soc. **10**, 142 (1967).
10. Ogawa, Y., K. Ino, T. Miyamura, and T. Mizoguchi: Radioimmunoassay of blood insulin (Abstract in Japanese). Folia endocr. Jap. **42**, 1373 (1967).

11. Ohnishi, Y., T. Kusumoto, T. Takano, T. Arimichi, A. Itani, S. Haraoka, T. Yamabuki, and K. Kosaka: Blood insulin after xylitol administration (Abstract in Japanese). J. Jap. Diabetic Soc. **10**, 65 (1967).
12. Bässler, K. H., G. Stein u. W. Belzer: Xylitstoffwechsel und Xylitresorption. Biochem. Z. **346**, 171 (1966).
13. Kuzuya, T., Y. Kanazawa, M. Shimoshige, and K. Kosaka: Stimulation of insulin secretion by xylitol in dog and in man. Excerpta Medica, Internat. Congr. Series No. **140**, 77 (1967).
14. Onoe, K., K. Kobashi, T. Yamamoto, S. Fujinaka, and T. Aburaya: *In vitro* studies on secretion and release of insulin by the intact islets of Langerhans microdissected from mice pancreas (Abstract in Japanese). Folia endocr. Jap. **43**, 277 (1967).
15. Fujisawa, M., H. Sato, Y. Hirata, and S. Katsuki: Effects of glucose, fructose, sorbitol, xylitol and ribose on plasma insulin in dogs (Abstract in Japanese). Folia endocr. Jap. **42**, 1370 (1967).

Comment

Effect of Intravenous Injection of Xylitol on Plasma Insulin

YUKIMASA HIRATA, MASAJI FUJISAWA, and TOSHIO OGUSHI

Second Department of Internal Medicine,
Faculty of Medicine, Kyushu University, Fukuoka, Japan

It already has been reported that marked hypoglycemia with a significant increase of plasma insulin level was observed in normal dogs following intravenous administration of xylitol. These characteristic responses of blood sugar and plasma insulin to xylitol were not found in diabetic dogs in our previous experiment [1].

In the present study, the influence of epinephrine infusion on the hyperinsulinemic effect of xylitol in normal dogs has been investigated. The intravenous administration of a large dosage of xylitol in man has also been tested.

Methods

After one week feeding with a standard diet, five normal dogs were anesthetized with sodium pentbarbital (30 mg per kg body weight) following 16 to 20 h starvation. Epinephrine infusion was carried out in the anesthetized dogs for 40 minutes at a rate of 2.5 μg per kg body weight per minute, 20 minutes after the starting point of the epinephrine infusion, 1.0 g per kg body weight of xylitol was administered intravenously.

In the human experiment a 50% xylitol solution was administered intravenously in 3 normal male volunteers with no family history of diabetes. All of them were within 10% of ideal weight and their ages ranged from 21 to 22 years. The amount of xylitol used in this human experiment was 0.5 g per kg body weight, which already had been established as sufficient to produce the remarkable hyperinsulinism in normal dogs described in the previous report.

Blood sugar was determined with the glucose oxidase method. Plasma insulin in men and in one dog was determined with the double antibody immunoassay reported by Hales and Randle [2], and in the other 4 dogs with a single antibody immunoassay using an ethanol precipitation method (Fujisawa and Hirata [3]).

Results

Effect of Epinephrine Infusion on the Hyperinsulinemic Effect of Xylitol in Normal Dogs. A decrease of plasma insulin level was produced by epinephrine infusion; the concentration of plasma insulin approached zero following the xylitol injection performed during the epinephrine infusion as shown in Fig. 1. In Fig. 1 are shown changes of plasma

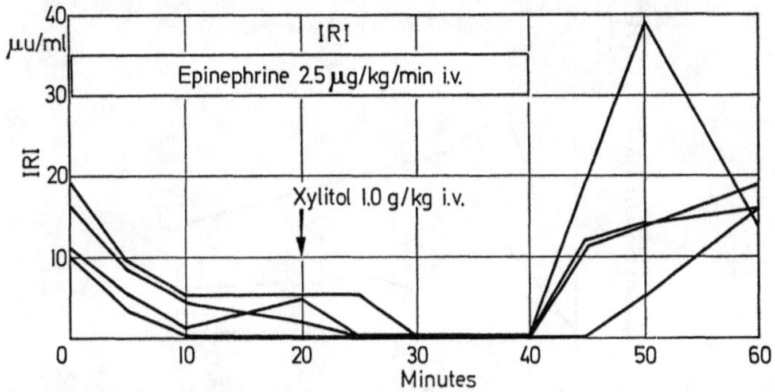

Fig. 1. The effect of epinephrine infusion on the hyperinsulinemic effect of xylitol in normal dogs

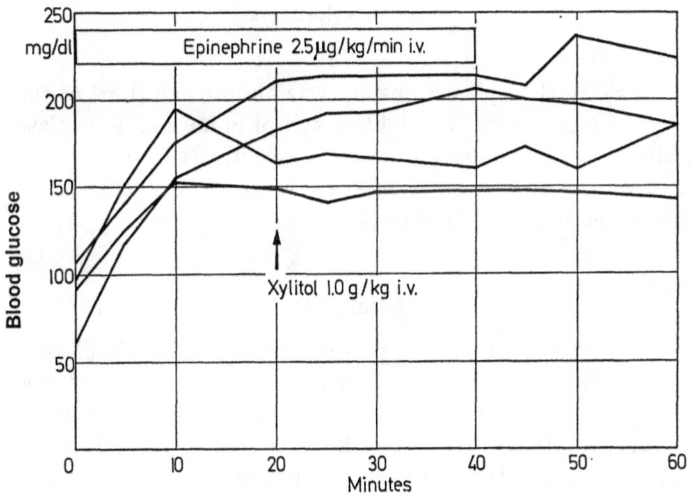

Fig. 2. The effect of epinephrine infusion on the hypoglycemic effect of xylitol in normal dogs

insulin concentration of 4 dogs, determined with the ethanol precipitation method. Plasma insulin of the other one dog, determined with the double antibody method, also showed the same result.

Hyperglycemia produced by epinephrine infusion was not affected by xylitol administration, as shown in Fig. 2.

Effect of Xylitol Injection on Plasma Insulin Level in Normal Persons. In three normal healthy subjects the response of plasma insulin to xylitol injection was variable as shown in Fig. 3. Case No. 1 showed a good response in plasma insulin concentration to xylitol injection,

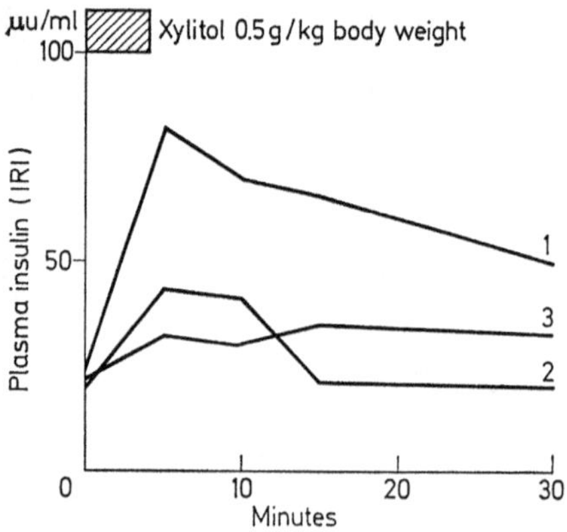

Fig. 3. The effect of xylitol injection on plasma insulin concentration in healthy volunteers

and the peak level of plasma insulin was almost one third of that produced by 0.4 g per kg body weight of xylitol in normal dogs. Case No. 2 and 3 showed fair and poor response, respectively.

There was no change in blood sugar level following the intravenous administration of xylitol in normal persons.

Discussion

In our previous experiment it was shown that a stimulating effect of xylitol on insulin secretion in normal dogs was much stronger than that of glucose. whereas in the present study it was demonstrated that epinephrine completely suppressed the hyperinsulinemic effect of xylitol. Although the mechanism of increase in plasma insulin following xylitol administration is not clear, this phenomenon is inhibited by epinephrine infusion.

The inhibitory effect of epinephrine upon insulin secretion induced by glucose *in vivo* was reported by Coore and Randle [4]. Porte *et al.* [5] showed that epinephrine inhibites the common stimuli of insulin secretion in normal subjects. Porte [6] also reported that isoproterenol, beta-adrenergic receptor stimulator, causes an increase of insulin in plasma, and suggested that this result would help to explain the observation that insulin levels were lowered by the beta-receptor blocking drug, propranolol, administered during epinephrine infusion [7].

In the present study it was observed that plasma insulin concentrations are almost zero following xylitol injection performed during epinephrine infusion. However xylitol is not a beta-receptor blocking compound, and a mechanism of collaboration of xylitol with epinephrine to decrease plasma insulin level is obscure.

Xylitol injected intravenously in normal dogs produced a significant hyperinsulinemia and hypoglycemia, whereas there were less marked changes of plasma insulin and blood sugar in normal subjects. There also might be an apparent individuality in the response of plasma insulin to xylitol in the normal healthy persons. In another experiment with 2 rabbits, an increase of plasma insulin with no change in blood sugar level was observed following xylitol injection. The peak of increased plasma insulin produced by xylitol in rabbits was almost one half of that produced in normal dogs.

It seems that individual or species differences of the plasma insulin response to xylitol are based on the difference of enzymatic activities in pancreatic beta-cells of each person or species.

Summary

It was shown in the previous report that plasma insulin increased significantly in normal dogs following intravenous administration of xylitol.

The results in the present study dealing with the hyperinsulinemic action of xylitol were summarized as follows:

1. The hyperinsulinemic effect of xylitol in normal dogs was suppressed completely by epinephrine infusion.

2. In human the effect of xylitol injection on plasma insulin was much less than that in dog, and an apparent individual difference was observed.

References

1. Hirata, Y., M. Fujisawa, H. Sato, T. Asano, and S. Katsuki: Biochem. biophys. Res. Commun. **24**, 471 (1966).
2. Hales, C., and P. J. Randle: Biochem. J. **88**, 137 (1963).
3. Fujisawa, M., and Y. Hirata: Unpublished.
4. Coore, H. G., and P. J. Randle: Biochem. J. **93**, 66 (1964).
5. Porte, D., Jr., A. Graber, T. Kuzuya, and R. H. Williams: J. clin. Invest. **44**, 1087 (1965).
6. — Diabetes **16**, 150 (1967).
7. — J. clin. Invest. **46**, 86 (1967).

Comment

Some Recent Observations on Xylitol-Induced Insulin Secretion

TAKESHI KUZUYA

Third Department of Internal Medicine,
University of Tokyo, Faculty of Medicine, Tokyo, Japan

In order to study whether xylitol itself has a direct effect upon islets or whether this effect is mediated by some of its metabolites, direct infusion of xylitol into the pancreatic artery was performed. Blood samples were taken from the corresponding pancreatic vein and from the femoral artery.

Fig. 1 shows the result of such experiment. When xylitol was infused into the femoral vein at a rate of 0.5 mg/kg/min for 30 minutes, the concentration of insulin did not increase in the femoral artery nor in

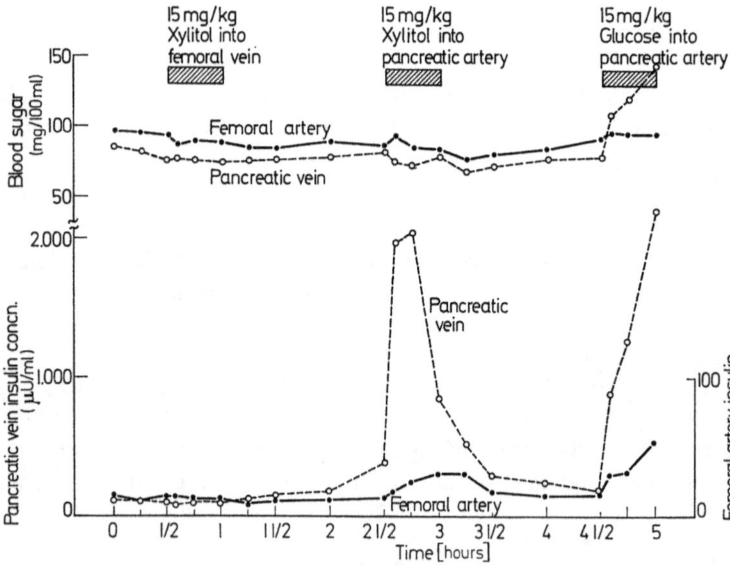

Fig. 1. Effect of a small amount of xylitol injected into the femoral vein and the pancreatic artery upon pancreatic vein insulin level

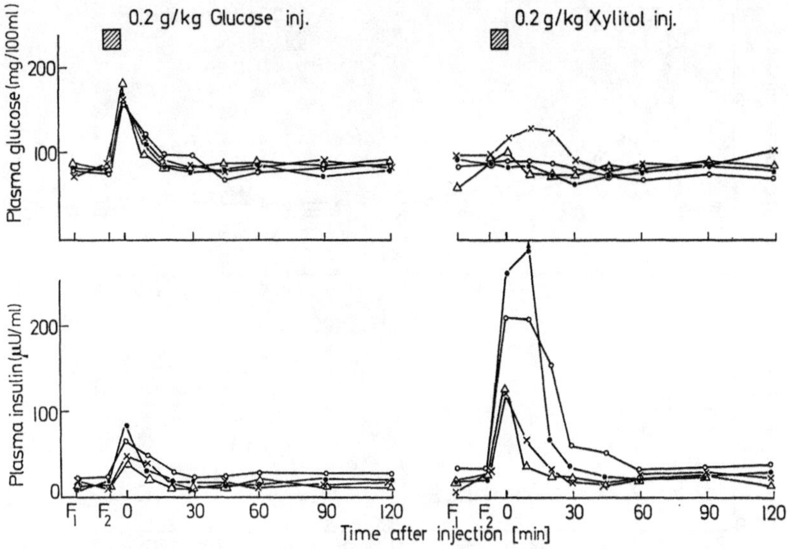

Fig. 2. Insulin response following intravenous injection of 0.2 g/kg glucose or xylitol in dogs

the pancreatic vein. In contrast, when the same dose of xylitol was infused into the pancreatic artery, there was a significant rise of insulin in the corresponding pancreatic vein. This experiment seems to suggest that xylitol does not need to be converted into some other compounds by the liver or other organs in order to be effective upon islets to stimulate insulin release.

It has been reported that there is a marked difference between man and dog in the plasma insulin response to xylitol administration. Therefore, in the next experiment, the effect of intravenous injection of 0.2 g/kg of glucose or xylitol was compared with various mammals, as shown in Figs. 2—5. At this concentration xylitol is enough to produce a definitely higher insulin response in dogs than glucose. In cows, 0.2 g/kg glucose and xylitol produces nearly the same order of insulin response. In horses, plasma insulin response is poor by both glucose and xylitol. The response in goats is similar to that in cows; both glucose and xylitol increases plasma insulin moderately. From these data, it appears that stimulation of insulin secretion by xylitol occurs differently in a variety of mammals. At present, however, we are entirely ignorant of the reason that xylitol stimulates insulin secretion so markedly in dogs but not so strongly in other animal species investigated.

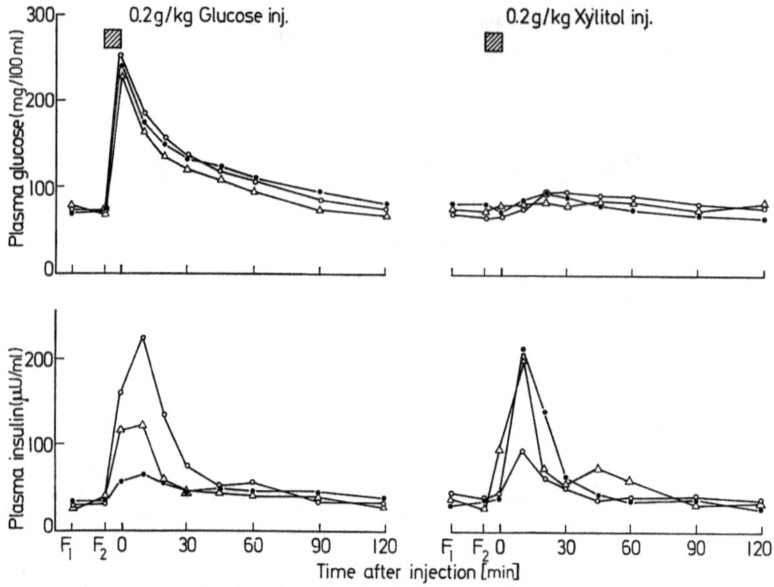

Fig. 3. Insulin response following intravenous injection of 0.2 g/kg glucose or xylitol in cows

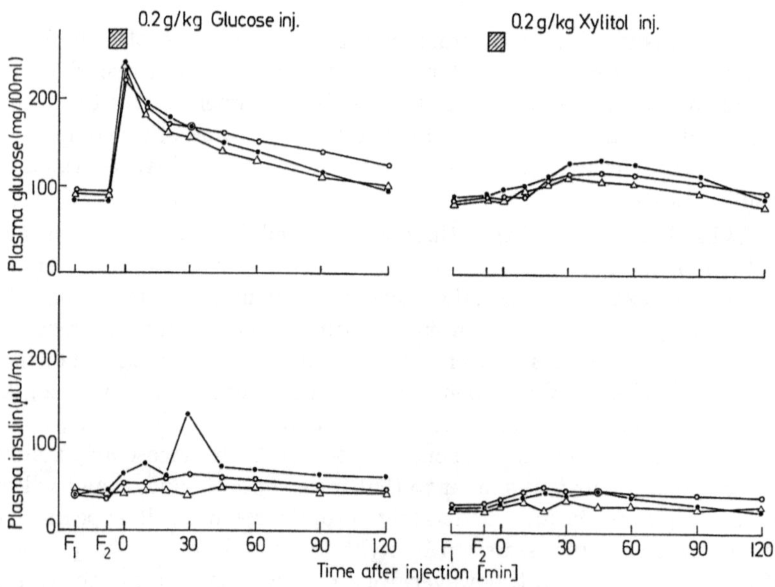

Fig. 4. Insulin response following intravenous injection of 0.2 g/kg glucose or xylitol in horses

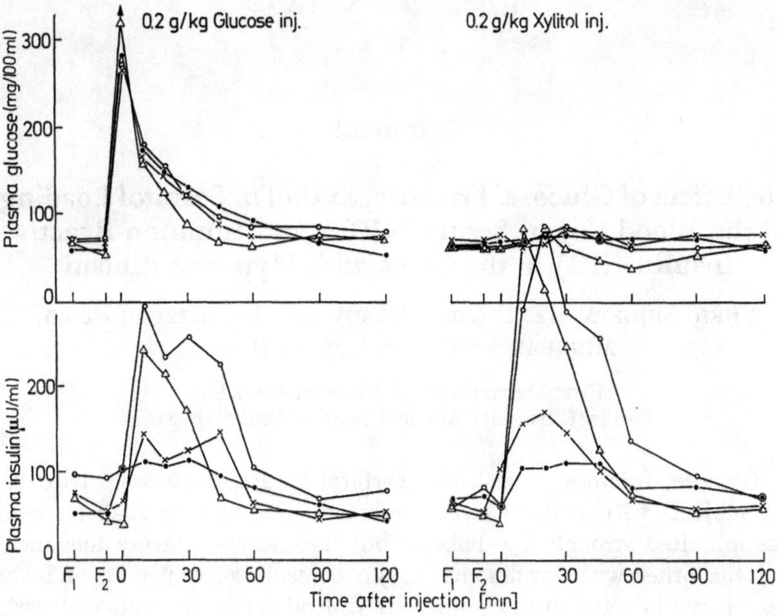

Fig. 5. Insulin response following intravenous injection of 0.2 g/kg glucose or xylitol in goats

Comment

The Effect of Glucose, Fructose, Xylitol or Sorbitol Loading on the Blood Sugar, Serum NEFA and Immuno-Reactive Insulin (IRI) in the Cases with Hyperinsulinism

YUKIO SHIGETA, KANJI IZUMI, NOBUYOSHI OJI, MITSURU HOSHI, MINHUAN KANG, and YUTAKA HARANO

First Department of Internal Medicine,
Osaka University Medical School, Osaka, Japan

Glucose, fructose, xylitol, and sorbitol loading tests were tried on the subjects with marked hyperinsulinism; one with prediabetes who was inherited strongly by diabetes but had normal glucose tolerance, and the other with insulinoma. 0.33 g of each sugar per kg of body weight was injected intravenously, and blood sugar (by Autoanalyzer), serum NEFA (by Novak's method) and IRI (by Hales and Randle's method) were measured intermittently for one hour after injection.

In the subject with prediabetes, the elevation of blood sugar was highest after glucose or fructose, moderately after sorbitol, or slightly after xylitol loading. The removal of serum NEFA was stimulated

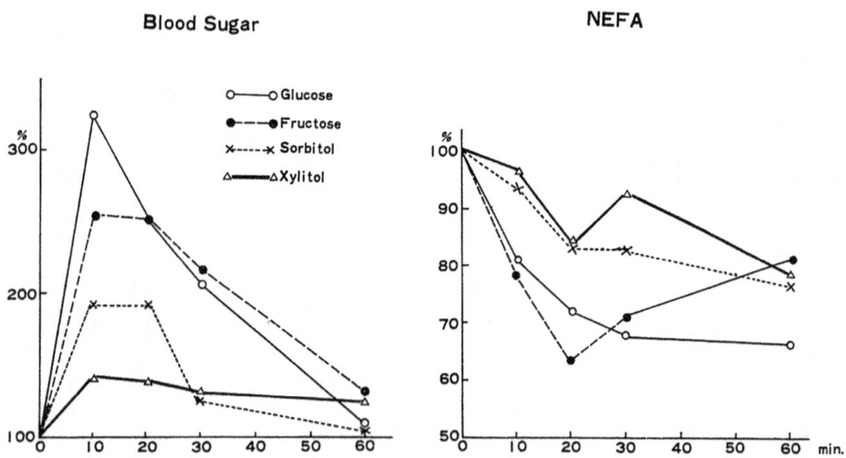

Fig. 1. Blood sugar and serum NEFA after intravenous injection of glucose, fructose, sorbitol or xylitol in a prediabetic subject

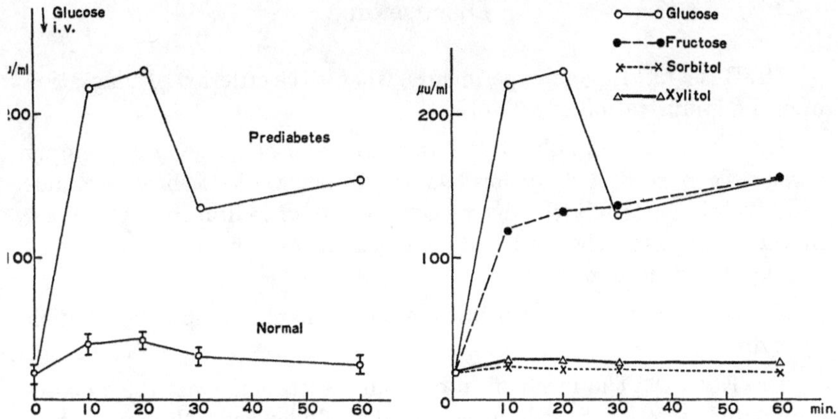

Fig. 2. Serum immuno-reactive insulin after intravenous injection of glucose, fructose, sorbitol or xylitol in a prediabetic subject

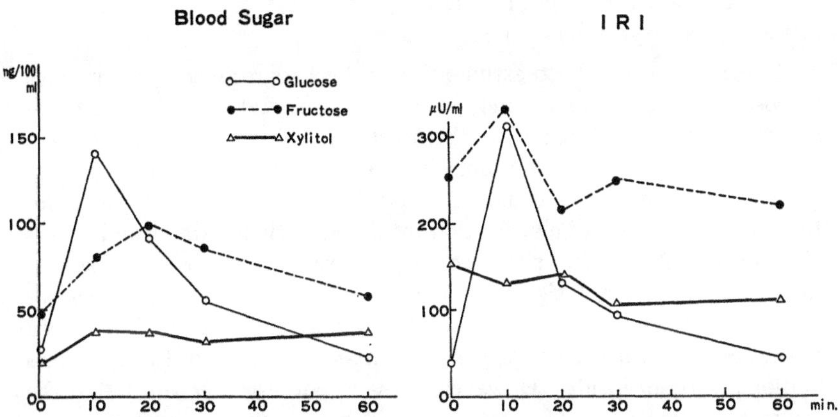

Fig. 3. Blood sugar and serum immuno-reactive insulin after intravenous injection of glucose, fructose or xylitol in a patient with insulinoma

markedly after glucose or fructose, or slightly after sorbitol or xylitol loading (Fig. 1). The serum IRI increased strikingly only after glucose or fructose loading (Fig. 2).

In the patient with insulinoma, blood sugar and IRI were elevated very much after glucose or fructose, or very little after xylitol loading (Fig. 3).

From these results, it is suggested that xylitol is converted to glucose only in small amounts, and that it inhibits slightly the release of NEFA from adipose tissue and stimulates very slightly insulin secretion from the beta cells of pancreas.

Discussion

Dr. Bässler: In your experiments, what is the criterion of adaptation of oral administration of xylitol?

Dr. Kosaka: Actually, there was no criterion. They were simply given 30 g of xylitol daily for two weeks before the xylitol tolerance test. We did not study the disappearance rate of xylitol from the blood in these subjects. They all had diarrhea in 2—3 hours after 50 g of oral xylitol given in one dose.

Dr. Hosoya: What do you think is the mechanism of insulin secretion by xylitol?

Dr. Kosaka: The mechanism of insulin secretion, even the glucose-stimulated secretion, is still unclear at the cellular level. Various factors will come into mind as a possible direct trigger of insulin release. It may be the extracellular concentration of glucose, intracellular concentration of free glucose, some specific metabolites or the changes of the ratio of specific coenzymes and so on.

It seems reasonable to assume that xylitol shares some process with glucose in its effect to produce insulin release. Xylitol is known to be incorporated into glycogen in the liver. If we assume that glycogen synthesis from xylitol also occurs in the beta-cells, any member of the glycolytic intermediates could be a possible trigger. Recently, in order to explain the effect of glucagen on insulin secretion, Samols and others [Diabetes **15**, 855 (1966)] proposed a hypothesis that glucose 6-phosphate formed by glycogenolysis in the beta-cells may be a direct trigger for the insulin release. Another possibility is that the increase in reduced pyridine nucleotides may play some important role for the synthesis and/or release of insulin. However, these hypothesis are still within the range of speculation. The data are still too scanty to construct any definite idea.

Dr. Touster: I wonder whether anyone here knows whether xylitol is utilized by the pancreas?

Dr. Kuzuya: I don't think there is any information of xylitol dehydrogenases in the islet cells. As the percentage of islet cells in the pancreas is very low, the direct determination of enzyme activity of the whole pancreas would give only very poor information about the presence of the enzyme in beta-cells. However recently I noticed a literature [Stiller, D., and J. Gorski, Histochemie **5**, 407 (1965)] that xylitol dehydrogenases, both NAD- and NADP-dependent, are detectable by histochemical technics. So I hope this technique would bring some information about the presence of these enzymes in pancreas.

Dr. ASAKURA: Did you study the effect of other sugar alcohols which were used as substrates of polyol dehydrogenase on insulin secretion?

Dr. KUZUYA: I have not tested the effect of sorbitol, but I am now very much interested in that.

Dr. ASAKURA: Do you think methylene blue or Vitamin K, which are known to accelerate the pentose phosphate pathway with formation of NADPH, stimulate insulin secretion?

Dr. KUZUYA: I do not think there is any report concerning the effects of these substances upon insulin secretion. It will be an interesting theme for future investigation.

Dr. ISHII: I would like to ask Dr. Kuzuya whether you could tell me about the insulin response to xylitol in other experimental animals, for example, rat, mouse or guinea pig.

Dr. KUZUYA: I have not studied the small animals yet, mainly because of the difficulty in obtaining sufficient blood samples to measure insulin.

The Influence of Xylitol and other Polyols and Sugars on Fat Mobilization

KLAUS OPITZ

Institute of Pharmacology, University of Munster, West Germany
(Director: Prof. Dr. med. Dr. phil. A. Loeser)

The so-called non-esterified or free fatty acids (NEFA, FFA) constitute a small but very dynamic plasma lipid fraction. These FFA are important sources of energy. Their physiological significance is easily appreciated if one realizes that in the fasting state more than 50 per cent of the energy consumption is covered by oxidation of fatty acids [2] and their clinical importance is evident from the fact that very high FFA levels are found in the blood with several disorders, e.g. severe burns, pheochromocytoma, hyperthyroidism, and diabetes mellitus. In this latter condition the excessive fat mobilization and its consequences (fatty liver, hyperlipemia, ketonemia) are no longer considered secondary to the disturbances of carbohydrate metabolism. Fatty acids have been shown, for instance, to inhibit the action of insulin on glucose transport [16].

The FFA in blood plasma originate from the triglycerides deposited in the vast fat stores of the organism. Adipose tissue is permanently mobilizing fatty acids which are then taken up by the blood stream, attached to plasma albumins, and transported to various organs, e.g. skeletal muscle, heart, and liver. Those fatty acids which are taken up by the peripheral organs are oxidized and (in the liver) also re-esterified. Since fatty acid utilization is normally proceeding at a rather constant rate the level of circulating FFA is generally agreed to indicate the rate of fat mobilization.

The mobilization of fatty acids from adipose tissue is an extremely rapid process; its rate can be doubled or halved within a few minutes. The FFA output is determined by the amount of FFA present within the fat cells. This intracellular FFA-pool is controlled by two antagonistic processes, namely lipolysis and re-esterification.

The intracellular FFA-pool is replenished by lipolysis, which means the formation of FFA and glycerol by hydrolysis of triglycerides. This is effected by two different enzymes, the hormone-sensitive or triglyceride lipase and the diglyceride lipase. Lipolysis is stimulated by various hormones, e.g. epinephrine, norepinephrine, corticotropin, growth hormone, vasopressin, lipotropin, and glucagon and by some polypeptides, e.g. the fat-mobilizing substance (FMS) of Chalmers [3]. These agents are supposed to activate the triglyceride lipase. In the case of epinephrine this is, however, denied by Okuda *et al.* [9].

Re-esterification tends to diminish the intracellular FFA-pool by re-synthesis of triglycerides from fatty acids and glycerol which is derived from the breakdown of glucose.

Thus fat mobilization can be increased either by stimulation of lipolysis or by inhibition of re-esterification. Conversely, fat mobilization can be decreased by inhibition of lipolysis or by increased re-esterification.

This somewhat intricate mechanism is perhaps best understood if one considers the response of the organism to glucagon. Glucagon is one of the so-called adipokinetic hormones, and it is well established that glucagon strongly stimulates lipolysis. Therefore one would expect the FFA level to rise after the injection of glucagon. But this hormone, as is well known, also stimulates the breakdown of liver glycogen, and the resulting hyperglycemia favours the re-esterifiation of FFA and moreover triggers the release of insulin. Thus falling plasma FFA levels occasionally have been observed following injections of glucagon [4, 7].

Apart from hormonal influences and sympathetic nervous activity it is the carbohydrate metabolism which controls fat mobilization. An increased glucose metabolism (a higher blood-sugar level and/or insulin activity) decreases the mobilization of fatty acids, and their concentration in the plasma falls rapidly. This reduction of fat mobilization is commonly related to increased re-esterification caused by the availability of greater amounts of alpha-glycerophosphate in the fat cells. In the fasting state when the glycogen stores of the organism are nearly exhausted and the blood-glucose concentration is low fat mobilization increases and plasma FFA levels are high. They drop instantly when glucose or insulin is injected.

It was not very surprising that glucose could be replaced by sugars which are rapidly converted into glucose in the organism, e.g. fructose. But how about glycols and polyols? In earlier investigations of the glycogenetic activity of polyhydric alcohols it has been found that 1,2-propanediol, 1,2-butanediol, and 1,2,4-butanetriol like glycerol are converted into glycogen in the amphibian and mammalian liver [10—12]. In further experiments we have been able to show that in fasted rats the injection of 1,2-propanediol, glycerol, and sorbitol is followed by a significant decline of plasma FFA concentration as a result of decreasing fat mobilization. In vitro the FFA output of adipose tissue samples excised from animals treated with the said polyols was significantly reduced as compared with that of adipose tissue from untreated rats or from animals which had been treated with meso-inositol [13].

In other experiments the above mentioned polyols have been added to the incubation medium of adipose tissue taken from untreated animals, and again fat mobilization was distinctly reduced. This result, however, is not easy to explain since many authors are convinced that adipose tissue being devoid of the enzyme alpha-glycerokinase is not able to esterify free glycerol. On the other hand there seems to exist some kind of feedback control of fat mobilization: When in long-term experiments a certain amount of glycerol is formed from triglyceride by lipolysis and this free glycerol reaches a critical concentration in the incubation medium fat mobilization slows down and finally ceases [17].

Recently we have examined the effect of several sugars, sugar alcohols, and related substances on plasma FFA and blood-glucose levels in fasted rats [14]. There was a significant decrease of plasma FFA levels following the intraperitoneal administration of 25 millimoles per kg body weight of glucose, fructose, mannose, xylose, sorbitol, glucosamine, or ascorbic acid. These substances as well as galactose caused slight to moderate hyperglycemic responses (blood-glucose concentrations were determined enzymatically, of course). Only xylitol caused a significant drop of plasma FFA level without any concomitant change in blood-glucose concentration (Table I).

This unique property of xylitol in connection with its increasing significance as an adjuvant in the treatment of diabetes mellitus has induced us to study the influence of this substance on fat mobilization in detail.

As to the mechanism by which xylitol normalizes elevated plasma FFA levels it may be different from that of glucose.

Firstly, injections of glucose cause an additional release of insulin which is known to be an antilipolytic agent, and it has been shown that xylitol also stimulates the secretion of insulin in some species but that does not explain the direct action of xylitol on fat mobilization.

Secondly, according to Goodner and Tustison [5], any change in the blood-glucose concentration or utilization is monitored by specific glucose receptors in the hypothalamus. Thus injections of glucose, even in minute amounts, are registered and promptly answered by a decrease of central sympathetic activity which means reduced fat mobilization in adipose tissue. Certainly this hypothesis, however plausible it may seem, does not explain the effect of xylitol on plasma FFA concentration as described here.

Thirdly, it is not only conceivable but highly probable that xylitol like glucose or fructose inhibits fat mobilization by a direct action on adipose tissue.

Table I. *Influence of sugars, polyols, and related substances on plasma non-esterified fatty acid (FFA) and blood-glucose levels in fasted rats. Means ± S.E. (n = 10). Dosage: 25 millimoles per kg BW intraperitoneally (sucrose 12.5 millimoles per kg BW). Interval: 1 hour*

Substance	Non-esterified fatty acids (FFA) (μval/l)				Blood-glucose (mg/100 ml)			
	before	after	difference		before	after	difference	
Glucose	503 ± 26	202 ± 16	−60%	$P < 0.001$	63 ± 2.6	289 ± 17.9	+359%	$P < 0.001$
Fructose	507 ± 25	155 ± 18	−70%	$P < 0.001$	65 ± 3.5	90 ± 7.1	+ 38%	$P < 0.001$
Mannose	580 ± 21	339 ± 34	−42%	$P < 0.001$	63 ± 1.9	90 ± 6.2	+ 43%	$P < 0.005$
Xylose	574 ± 25	428 ± 38	−25%	$P < 0.01$	58 ± 1.8	83 ± 4.5	+ 43%	$P < 0.001$
Galactose	677 ± 58	662 ± 47	− 2%	n.s.[a]	60 ± 2.1	95 ± 4.7	+ 35%	$P < 0.001$
Sucrose	562 ± 34	554 ± 17	− 1%	n.s.	65 ± 1.8	63 ± 2.8	− 3%	n.s.
Sorbitol	638 ± 42	298 ± 18	−53%	$P < 0.001$	66 ± 1.9	103 ± 7.8	+ 56%	$P < 0.001$
Xylitol	561 ± 26	374 ± 14	−33%	$P < 0.001$	66 ± 2.1	66 ± 1.8	± 0%	n.s.
Mannitol	683 ± 46	689 ± 45	+ 1%	n.s.	65 ± 3.2	64 ± 2.1	− 2%	n.s.
Dulcitol	545 ± 35	593 ± 32	+ 9%	n.s.	67 ± 1.9	61 ± 2.1	− 9%	n.s.
Glucosamine	467 ± 33	156 ± 28	−67%	$P < 0.001$	58 ± 2.0	79 ± 4.4	+ 36%	$P < 0.001$
Ascorbic acid	441 ± 31	224 ± 35	−49%	$P < 0.001$	77 ± 1.3	90 ± 4.9	+ 17%	$P < 0.01$

[a] n.s. = not significant.

This possibility has been tested by in vitro experiments in which xylitol was added to the incubation medium of isolated epididymal adipose tissue taken from normal fasted rats.

When investigating fat mobilization in vitro one has to distinguish between media which contain albumin as a fatty acid acceptor and albumin-free incubation fluids. In the presence of albumin FFA formed from triglycerides by lipolysis partly leave the tissue and are taken up by the medium. At the end of the incubation period this FFA output is determined by measuring the FFA concentration in an aliquot of the medium. The amount of fatty acids which are mobilized and do not enter the medium but remain in the tissue is usually neglected in this kind of experiments.

In the absence of albumin virtually no FFA enter the medium; in other words: there is no FFA output. Fat mobilization, however, proceeds at the same rate as in experiments where albumin is present. When incubation is finished the adipose tissue sample is homogenized in Dole's extraction mixture and its FFA content is measured. Thus the amount of mobilized fatty acids is determined as a whole which means that fat mobilization is estimated with greater accuracy in experiments using albumin-free media. In this type of experiments the FFA-pool present within the fat cells at the beginning of the incubation period is often neglected.

In a pilot experiment with fresh blood plasma from fasted donor rats as an incubation medium a high concentration of xylitol (33 micromoles per millilitre) caused a 27 per cent decrease of fat mobilization. This was undoubtedly effected by direct influence of xylitol on the adipose tissue.

Inhibitory effects as a rule are the more pronounced the higher is the rate of the process to be inhibited. Lipolysis partly depends upon the ional composition of the surrounding fluid, and its rate increases in potassium-rich media [1]. Using the potassium chloride — TRIS buffer solution described by Bleicher *et al.* [1] we have tested the inhibitory influence of several concentrations of xylitol and sorbitol on spontaneous fat mobilization in isolated epididymal adipose tissue taken from normal fasted young rats (Table II and Fig. 1). There were significant inhibitory effects in the concentration range from 2 to 30 micromoles per millilitre, and xylitol was clearly more active than sorbitol but a maximal inhibition of 35 per cent was never surmounted with xylitol even when very high concentrations were used. (With glucose a decrease of 70 to 80 per cent is easily achieved.)

The practical significance of this inhibitory action of xylitol on fat mobilization must be considered in view of the use of xylitol in the treatment of diabetes mellitus.

The excessive fat mobilization which is typical for this condition

Table II. *Inhibitory effects of sorbitol and xylitol on fat mobilization in vitro. Each experiment included 7 pairs of adipose tissue samples taken from the two epididymal fat pads of 7 fasted young rats, one sample from each animal serving as a control. Incubation: 2 hours at 37° C in albumin-free 0.025 M TRIS buffer containing 0.15 M KCl. Free fatty acid release expressed in microequivalents FFA per g fresh tissue per hour incubation (μval/g/h), means (± S.E.)*

Substance	Concentration μM/ml	FFA release (μval/g/h)		Difference		
		control	experiment			
Sorbitol	100	3.13	1.69	-1.44 ± 0.07	−46%	$P < 0.001$
Sorbitol	30	3.86	2.72	-1.14 ± 0.12	−30%	$P < 0.001$
Sorbitol	5	3.07	2.71	-0.36 ± 0.11	−12%	$P < 0.02$
Xylitol	100	3.86	2.82	-1.04 ± 0.12	−27%	$P < 0.001$
Xylitol	30	4.35	2.92	-1.43 ± 0.45	−33%	$P \approx 0.02$
Xylitol	5	3.76	2.57	-1.19 ± 0.21	−32%	$P < 0.002$
Xylitol	2	3.72	3.11	-0.61 ± 0.21	−16%	$P < 0.05$
Xylitol	1	3.81	3.46	-0.35 ± 0.18	− 9%	n.s.[a]
Xylitol	0.5	3.32	3.13	-0.19 ± 0.16	− 6%	n.s.

[a] n.s. = not significant.

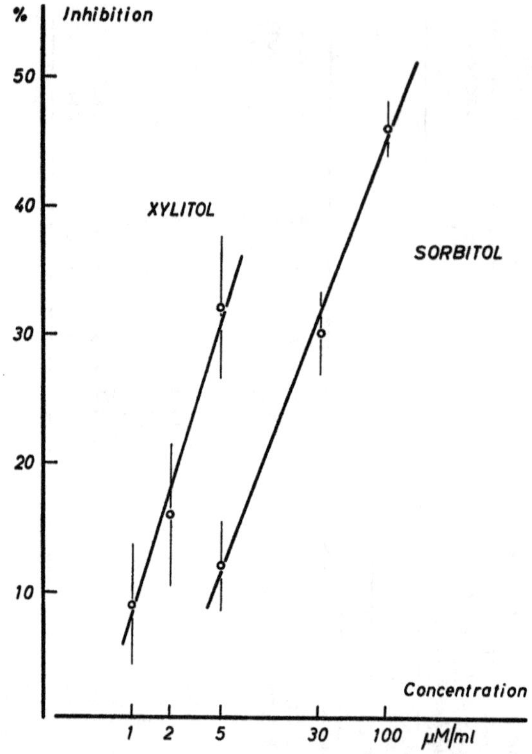

Fig. 1. Influence of xylitol and sorbitol on fat mobilization in vitro (means ± S.E.). Abscissa: concentration in micromoles per millilitre, logarithmical scale. Ordinate: inhibition of fat mobilization in per cent from control

is partly due to the impaired glucose utilization and the resulting deficiency of alpha-glycerol phosphate which is indispensable for the re-esterification of fatty acids. In alloxan diabetes as well as in the juvenile or insulin-deficiency diabetes the lack of insulin is an additional cause of hyperlipacidemia, for insulin (apart from its indirect influence on lipid metabolism) is now known to have direct antilipolytic properties [6]. Insulin is supposed to inhibit the lipolytic effects of various hormones, and in its absence these hormones freely promote fat mobilization [8]. Therefore it appeared essential to investigate if xylitol is capable to counteract the effects of these adipokinetic hormones.

The influence of xylitol on fat mobilization stimulated by the hormones norepinephrine, glucagon, corticotropin, or growth hormone respectively has been investigated in vivo, and it was quite impressive to see that xylitol greatly reduced the effect of norepinephrine and even abolished the adipokinetic actions of glucagon, corticotropin, and growth hormone (Table III).

Table III. *Influence of xylitol on the lipolytic effects of several adipokinetic hormones in fasted rats (Reference: sucrose). Means ± S.E. (n = 5)*

Treatment		Plasma non-esterified fatty acids (FFA) μval/l				
		before	(Interval)	after	Difference	
Norepinephrine[a]	Sucrose[e]	518 ± 31	(1 hr)	1,000 ± 91	+93%	$P < 0.02$
Norepinephrine	Xylitol[f]	573 ± 25	(1 hr)	686 ± 82	+20%	
Glucagon[b]	Sucrose	537 ± 27	(1 hr)	738 ± 56	+37%	$P < 0{,}005$
Glucagon	Xylitol	523 ± 6	(1 hr)	448 ± 31	−14%	
Corticotropin[c]	Sucrose	632 ± 58	(1 hr)	933 ± 52	+48%	$P < 0{,}001$
Corticotropin	Xylitol	626 ± 48	(1 hr)	603 ± 70	− 4%	
Growth hormone[d]	Sucrose	559 ± 44	(2 hrs)	774 ± 92	+38%	$P < 0{,}05$
Grwoth hormone	Xylitol	695 ± 65	(2 hrs)	566 ± 62	−19%	

[a] 1 mg/kg intramuscularly.
[b] 1 mg/kg intramuscularly.
[c] 5 iU/kg (3rd int. standard) intramuscularly.
[d] 15 iU/kg intramuscularly.
[e] 12.5 mMol/kg (21,395% g/v) intraperitoneally.
[f] 25 mMol/kg (19,025% g/v) intraperitoneally.

While it is obvious that in the diabetic state (apart from the impairment of fatty acid re-esterification) lipolysis must be stimulated we do not know exactly the individual hormone which is responsible for this effect.

The catecholamines do not seem to play a major role in this context since it is not possible to correct the diabetic hyperlipacidemia with beta-adrenergic blocking agents. Growth hormone may be involved. But there is another possibility: Recently, Peterlik *et al.* [15] have reported that the excretion of the so-called fat-mobilizing substance (FMS) is materially increased in the diabetic state. Is this polypeptide which regularly appears in the urine of fasting humans and animals (but not in the urine of hypophysectomized individuals) responsible for the increased fat mobilization in diabetes? Further work on the influence of xylitol on the adipokinetic effect of FMS is in progress.

References

1. Bleicher, S. J., A. Lewis, L. Farber, and M. G. Goldner: Effect of mono- and divalent cations on in vitro lipolysis. Proc. Soc. exp. Biol. (N.Y.) **121**, 980 (1966).
2. Carlson, L. A.: Regulation and consequences of the mobilization of lipids from adipose tissue. In: Die Pathogenese des Diabetes mellitus/ Die endokrine Regulation des Fettstoffwechsels, S. 182—187. Berlin-Heidelberg-New York: Springer 1967.
3. Chalmers, T. M., G. L. S. Pawan, and A. Keckwick: Fat-mobilising and ketogenic activity of urine extracts: Relation to corticotrophin and growth hormone. Lancet **1960 II**, 6.
4. De Plaen, J., and G. Galansino: Effects of glucagon and epinephrine in fasted dogs. Proc. Soc. exp. Biol. (N.Y.) **121**, 501 (1966).
5. Goodner, C. J., and W. A. Tustison: Autonomic mediation of the effect of raised arterial glucose upon free fatty acids. Science **146**, 770 (1964).
6. Jeanrenaud, B., u. W. Stauffacher: Die Wirkung des Insulins am Fettgewebe. In: Die Pathogenese des Diabetes mellitus / Die endokrine Regulation des Fettstoffwechsels, S. 188—199. Berlin-Heidelberg-New York: Springer 1967.
7. Lefebvre, P., et A. Luyckx: Glucagon et métabolisme du tissu adipeux. Journées Ann. Diabètol. Hôtel-Dieu, p. 67. Paris: Ed. Méd. Flammarion 1967.
8. Mahler, R., W. S. Stafford, M. E. Tarrant, and J. Ashmore: The effect of insulin on lipolysis. Diabetes **13**, 297 (1964).
9. Okuda, H., I. Yanagi, and S. Fujii: The mechanism of in vitro stimulation of lipolysis by adrenaline. J. Biochem. (Tokyo) **59**, 438 (1966).
10. Opitz, K.: Über die Bildung von Glucose und Glykogen aus 1,2-Propandiol in vitro. Naunyn-Schmiedebergs Arch. exp. Path. Pharmak. **233**, 355 (1958).
11. — Über die glykogenbildende Wirkung von Glykolen. Naunyn-Schmiedebergs Arch. exp. Path. Pharmak. **234**, 448 (1958).
12. —, u. A. Loeser: Zur Wirkung von 1,2-Propandiol auf den Glykogengehalt von Kaulquappen. Z. ges. exp. Med. **129**, 526 (1958).

13. Opitz, K., u. F. Weintz: Über den Einfluß mehrwertiger Alkohole auf die Freisetzung von Fettsäuren aus dem Fettgewebe. Naunyn-Schmiedebergs Arch. exp. Path. Pharmak. **245**, 390 (1963).
14. — Über den Einfluß von Zuckern, Zuckeralkoholen und verwandten Substanzen auf die Fettmobilisierung. Naunyn-Schmiedebergs Arch. Pharmak. exp. Path. **255**, 192 (1966).
15. Peterlik, M., K. Reissert u. B. Rudas: Eine fettmobilisierende Substanz (FMS) im Harn alloxandiabetischer Ratten. Med. Pharmacol. exp. **16**, 75 (1967).
16. Randle, P. J., P. B. Garland, C. N. Hales, and E. A. Newsholme: The glucose fatty-acid cycle. Its role in insulin sensitivity and the metabolic disturbances of diabetes mellitus. Lancet **1963 I**, 785.
17. Weinges, K. F.: Fettsäure- und Glucosestoffwechsel am Fettgewebe unter den Bedingungen der hormonsensitiven Lipolyse. In: Die Pathogenese des Diabetes mellitus / Die endokrine Regulation des Fettstoffwechsels, S. 174—181. Berlin-Heidelberg-New York: Springer 1967.

Comment

Xylitol Effect on Ketogenesis in Alloxan Diabetic Rat Liver Slice

NORIMASA HOSOYA, and TADAHIKO MACHIYA

Nutritional Laboratory, School of Health Sciences,
University of Tokyo, Faculty of Medicine, Tokyo, Japan

Haydon [1] has observed that xylitol is antiketogenic in liver slices of starved rats, and Bässler and Dreiss [2] also found that xylitol exerts antiketogenic action in alloxan diabetic rats. In this experiment xylitol effect on ketogenesis was studied in alloxan diabetic rat liver slices with palmitate-1-C^{14} used as substrate.

Male rats (80—100 g) were injected i.p. with alloxan (150 mg/kg), and after two weeks rats having over 300 mg/dl blood sugar level were used as the diabetes. 300 mg of liver slices were suspended in 2.5 ml of Krebs-Ringer phosphate buffer (pH 6.9) containing palmitate-1-C^{14} (0.25 μc/μmole), 1 mg/ml bovine serum albumin and 1 mM xylitol, under oxygen in conventional Warburg vessels. Produced labeled carbon dioxide was trapped as barium carbonate and counted by the method of Villee *et al.* [3]. Extraction of total lipids from slices was carried out by the method of Folch *et al.* [4]. Acetoacetate in the medium was β-decarboxylated with 4-aminoantipyrine by the method of Tawara [5], and produced acetone was trapped as phenylhydrazone [3], and carbon dioxide as barium carbonate [3].

Metabolism of palmitate-1-C^{14} is presented in Table I. Conversion of palmitate-1-C^{14} to products is expressed in mμmoles per g of wet tissue per hour $\pm$ s.e. Carbon dioxide production from palmitate-1-C^{14} was almost at the same level either with or without xylitol in both normal and diabetics. Xylitol did not affect the complete oxidation of palmitate.

Labeled carbon of palmitate will appear in the carboxyl carbon and the carbonyl carbon in acetoacetate moiety, and the amount will be identical theoretically. In our experiment C^{14}-carbon converted to the carbonyl carbon in acetoacetate was higher than that to the carboxyl carbon. Another carbonyl compound metabolized from palmitate-1-C^{14} would be trapped in our analysis.

However acetoacetate production from palmitate-1-C^{14} was decreased in the presence of xylitol in both normal and diabetics. While palmitate

Table I. *Palmitate-1-C^{14} metabolism in rat liver slices (mμmoles/g/h)*

Rats \ Incorporation into		Carbon dioxide	Total lipid	Acetoacetate trapped as phenylhydrazone	Acetoacetate trapped as barium carbonate
Normal	without	72.5 ±6.3	54.1 ±5.6	31.8 ±3.7	24.2 ±1.3
	with xylitol	54.5 ±4.7	57.0 ±6.1	21.5 ±2.3	11.6 ±1.3
Diabetes	without	48.5 ±4.5	29.1 ±3.2	65.4 ±6.2	26.9 ±2.4
	with xylitol	50.0 ±4.7	61.1 ±5.4	31.5 ±3.0	8.5 ±1.1

Rat liver slice (300 mg) was suspended in 2.5 ml of Krebs-Ringer phosphate buffer (pH 6.9) containing palmitate-1-C^{14} (0.25 μc/μmoles), 1 mg/ml bovine serum albumin and 1 mM xylitol. 60 min at 37° C in oxygen.

incorporation into tissue lipid fractions in the diabetics was increased with xylitol, contrary to decrease of acetoacetate production.

From these results, it seems that diabetic liver with xylitol is not necessary to obtain energy for β-oxidation of palmitate from other system. Xylitol would provide this energy in its metabolic utilization process. An amount of palmitate which produced acetoacetate would combine with glycerol derivatives. These glycerol derivatives would also be produced from the metabolism of xylitol. The antiketogenic action of xylitol involves a diminution in the ketone bodies formed and not merely reduction of the conversion of acetoacetate to β-hydroxybutyrate [6]. Although this point was not tested by determination of β-hydroxybutyrate in the experiment, it is likely that xylitol will decrease the accumulation of both ketone bodies.

References

1. Haydon, R. K.: Biochim. biophys. Acta (Amst.) **46**, 598 (1961).
2. Bässler, K. H., u. G. Dreiss: Klin. Wschr. **41**, 593 (1963).
3. Villee, C. A., and A. B. Hastings: J. biol. Chem. **179**, 675 (1949).
4. Folch, J., M. Lee, and G. H. Sloane-Stanley: J. biol. Chem. **226**, 497 (1957).
5. Tawara, Y.: Nisshin-igaku **44**, 554 (1957).
6. Blakley, R. L.: Biochem. J. **49**, 257 (1951).

Comment

Effect of Xylitol on Individual Free Fatty Acids in the Plasma

YUICHIRO GOTO, SHIRO NAGANO, and HARUO NAKAMURA

Department of Medicine, Keio University Hospital
Tokyo, Japan

In 1956, during the course of a study on pentosuria in guinea pig, a metabolic conversion of L-xylulose to D-xylulose via xylitol was discovered by Touster and his coworkers [1—3], and named the Glucuronic Acid Pathway which is metabolically connected with the Pentose Phosphate Pathway by phosphorylation. Evidence has accumulated concerning the metabolism of xylitol since that time in both basic and clinical fields. Lang [4] and his group [5] in Mainz administered xylitol preparation to diabetic patients and observed that xylitol is utilized well in the body and exerts little influence on the blood glucose level. Plasma free fatty acids (FFA) were found to be reduced by administration of xylitol [6, 7]. The present communication concerns with the further studies on the effect of xylitol on plasma fatty acids in hyperlipemic and diabetic patients.

Seven cases of cerebrovascular accidents and 1 diabetic patient hospitalized in Keio University Hospital were subjected to this study. 7 cases of cerebral stroke showed hyperlipemic sera, namely over 100 mg/dl in triglyceride and over 200 mg/dl in cholesterol. One diabetic patient revealed apparent glucose intolerance.

Xylitol (100 mg/kg) was injected intravenously into the cubital vein in the fasting state. Blood samples were obtained prior to injection, and 15, 30, 45, 60 and 120 min after injection.

Blood glucose levels were assayed by the method of Somogy-Nelson, and lactic acid was measured enzymatically. FFA were analyzed by the method of Dole and Meinertz, and their fatty acid composition was determined by gas liquid chromatography.

Table I represents the change of FFA during the experiment. Case No. 4 and 5 were the same subject on which the study was performed twice. FFA values in the fasting state of hyperlipemic subjects were generally high. FFA decrease was elicited by injection of xylitol in

15 minutes except in case 8, and the reached maximum in 30 minutes. They tended to increase in 60 minutes, and exceeded the initial values.

Tables II and III show the change of fatty acid composition in the individual case during the experiment. Palmitic acid ($C_{16:0}$) showed an

Table I. *Effect of xylitol on FFA concentrations (uEq/l)*

	0	15′	30′	45′	60′	120′
1. K.Y.	786	760	700	710	762	900
2. S.K.	820	756	760	700	804	890
3. S.T.	1,254	1,000	922	896	1,054	1,100
4. H.W.	796	774	654	686	722	826
5. H.W.	826	650	656	640	1,146	1,200
6. K.N.	1,280	884	488	550	666	890
7. R.H.	786	745	667	684	826	965
8. K.U.	868	907	480		1,272	
9. K.S.	394	306	300		354	

Table II. *Effect of xylitol on plasma individual FFA (%)*

		14:0	16:0	16:1	18:0	18:1	18:2	20:4
1. K.Y.	0	3.2	26.4	8.3	11.1	30.9	19.2	1.0
	30′	3.6	29.0	7.4	12.7	28.7	17.5	1.1
	60′	3.0	26.3	8.4	11.0	31.1	19.3	0.9
2. S.K.	0	2.5	26.1	8.7	10.9	32.1	18.3	1.4
	30′	2.6	27.4	8.5	11.9	30.8	17.6	1.2
	60′	2.4	26.0	8.9	10.5	32.5	18.6	1.1
3. S.T.	0	1.2	25.0	7.8	8.4	39.5	17.0	1.1
	30′	1.6	26.2	7.0	10.3	37.1	17.0	0.8
	60′	1.1	25.4	7.6	8.9	39.0	17.2	0.8
4. H.W.	0	3.0	26.2	8.2	12.9	31.0	18.7	
	30′	3.5	28.0	7.5	14.0	28.3	17.5	1.2
	60′	3.2	26.0	8.4	13.1	30.9	18.4	

Table III. *Effect of xylitol on plasma individual FFA (%)*

		14:0	16:0	16:1	18:0	18:1	18:2	20:4
5. H.W.	0	0.4	37.3	2.1	15.2	22.7	22.4	
	30′	0.5	41.1	1.9	18.6	20.0	16.3	1.7
	60′	0.6	34.7	2.1	14.9	35.2	12.5	
8. K.U.	0	0.9	26.0	7.4	9.2	36.1	18.9	1.6
	30′	1.2	23.2	8.2	7.6	40.8	18.5	0.6
	60′	0.9	18.0	9.6	5.6	43.6	21.9	0.4
9. K.S.	0	0.8	23.9	8.3	11.3	46.1	9.6	
	30′	1.1	28.0	7.4	10.4	43.5	9.5	
	60′	1.0	31.9	6.8	9.2	40.8	10.4	

Table IV. *Effect of xylitol on plasma individual FFA (uEq/l)*

		14:0	16:0	16:1	18:0	18:1	18:2	20:4
1. K.Y.	0	25.2	207.5	65.2	87.2	242.9	150.9	7.9
	30′	22.4	203.0	51.8	88.9	200.9	122.5	7.7
	60′	22.9	200.4	64.0	83.8	237.0	147.1	6.9
2. S.K.	0	20.5	214.0	71.3	89.4	263.2	150.1	11.5
	30′	19.8	208.2	64.6	90.4	234.0	133.8	9.1
	60′	19.3	209.0	71.6	84.4	261.3	149.5	8.8
3. S.T.	0	15.0	313.5	97.8	105.3	495.3	213.2	13.8
	30′	14.8	241.6	64.5	95.0	342.1	156.7	7.4
	60′	11.6	267.7	80.1	93.8	411.1	181.3	8.4
4. H.W.	0	23.9	208.6	65.3	102.7	246.8	148.9	
	30′	22.9	183.1	49.1	91.6	185.1	114.5	7.8
	60′	23.1	187.7	60.6	94.6	223.1	132.8	

Table V. *Effect of xylitol on plasma individual FFA (uEq/l)*

		14:0	16:0	16:1	18:0	18:1	18:2	20:4
5. H.W.	0	3.3	308.1	17.3	125.6	187.5	185.0	
	30′	3.2	269.6	12.5	122.0	131.2	106.9	11.2
	60′	6.9	397.7	24.1	170.8	403.4	143.3	
8. K.U.	0	7.8	225.7	64.2	79.9	313.3	164.1	13.9
	30′	5.8	111.4	39.4	36.5	195.8	88.8	2.9
	60′	11.4	229.0	122.1	71.2	554.6	278.6	5.1
9. K.S.	0	3.2	94.2	32.7	44.5	181.6	37.8	
	30′	3.3	84.0	22.2	31.2	130.5	28.5	
	60′	3.5	112.9	24.1	32.6	144.4	36.8	

increase in 30 minutes except in case 8, and tended to return to the initial value in 60 minutes. However, in case 9, a diabetic subject, palmitic acid continued to increase. Stearic acid in the cases 1 through 5 changed concomitantly with palmitic acid. Oleic acid decreased in 30 minutes with a subsequent increase in 60 minutes in all except case 8.

However, oleic acid in diabetic case continued to decrease as shown in Table III. In Tables IV and V individual fatty acids are expressed in net concentration. Each fatty acid concentration decreased in 30 minutes, especially oleic acid. In 60 minutes, the increase of oleic, palmitoleic and linoleic acid was observed in every case.

There was no definite change in plasma triglyceride.

Blood glucose levels decreased slightly and lactate levels also decreased (Table VI).

Table VI. *Effect of xylitol on blood glucose (mg/dl) and lactate (mg/dl)*

	0	15′	30′	45′	60′	120′
1. K.Y.	98	78	84	80	86	96
2. S.K.	114	106	98	106	99	120
3. S.T.	99	102	94	96	102	99
4. H.W.	116	124	116	126	104	96
5. H.W.	137	150	152	137	104	81
6. K.N.	114	77	89	41	123	77
7. R.H.	107	98	96	104	96	90
5. H.W.	20.0	15.3	14.1	8.0		6.5
6. K.N.	16.8	15.9	13.5	11.8	10.6	9.4
8. K.U.	8.3	15.7	7.4		7.3	
9. K.S.	27.8	22.4	25.9		26.7	

The metabolic relationship between carbohydrate and lipid in the body is based upon very complicated mechanisms involving the action of insulin, catecholamine and several enzymes.

Xylitol and other intermediates of the Glucuronic Acid Pathway are metabolized primarily in the liver and kidney. According to reports accumulated recently, after xylitol injection blood glucose is unchanged [8] or slightly decreases [9], lactate rises [10], pyruvate falls [11], NEFA decreases [6] and glycogen tends to accumulate in liver [12]. Insulin secretion is stimulated in dogs by xylitol [13]. Thus Glucuronic Acid Pathway appeares to be functioning actively in the fasting state and in diabetes.

It is known that free fatty acid levels in the plasma usually fall after the loading of carbohydrate. The present study confirmed this change. In the change expressed in percentage, an increase of palmitic acid was observed. However, in the change of net concentration, each fatty acid decreased to some extent. Among these changes, that of oleic acid was significant.

The mechanism of decrease in individual free fatty acids after xylitol administration is an important subject. Glucose is converted to G-6-P in the presence of insulin, and is metabolized further through the Embden-Meyerhof, Pentose Phosphate and Glucuronic Acid Pathways. These three pathways are closely connected at F-6-P, F-D-P, and xylulose 5-phosphate.

The Glucuronic Acid Pathway involving xylitol as an intermediary component functions actively in liver and kidney. Therefore administered xylitol could be metabolized in tissue as the source of energy without requiring insulin.

According to the results of Furukawa *et al.*, xylitol administered is converted mostly to D-xylulose, and is joined to the Pentose Phosphate

Pathway via D-xylulose 5-P. The study on metabolism of C^{14}-xylitol in adipose tissue [14] indicated that $C^{14}O_2$ production is almost $^1/_{10}$—$^1/_{20}$ the amount from C^{14}-glucose.

Moreover, the synthesis of glyceril-glycerol from xylitol is enhanced in adipose tissue, according to the investigation of Jeanrenaud *et al.* [15]. This result indicates that the incorporation of xylitol into triglyceride is stimulated by the pathway of xylitol to GA-3-P, and α-glycerophosphate, subsequently, supplying glyceril-glycerol.

The present study clearly demonstrated the marked decrease of FFA, suggesting enhanced esterification of FFA after administration of xylitol. However, antilipolytic action by insulin via decreased cyclic AMP [16] should be considered also.

The fasting animal, according to the results of Rizack [17], is in a state of stimulated sympathetic nervous system and thus is under the control of epinephrine. The action of epinephrine on metabolism weakened by xylitol administration resulted in increased adenyl-cyclase activity and decreased activity of hormone sensitive lipase in adipose tissue.

These actions could be produced by any kind of carbohydrate which is utilized in the body. Therefore, the possible role of xylitol on adenylcyclase should be investigated by further experiments.

The decrease of FFA release from adipose tissue and acceleration of glyceril-glycerol synthesis could be responsible for the decrease of plasma FFA.

The result on insulin behavior in the dog after administration of xylitol indicates the possible participation of glycolysis. This also is to be investigated further.

Among the changes of individual fatty acids in 30 minutes after xylitol administration, the percentage increase of palmitic acid appeared to be reflected by the decrease of oleic acid. This is reasonable on the basis of FFA composition in which oleic acid is the major fraction in both plasma and adipose tissue, reflecting significantly the stage of esterification and antilipolysis.

References

1. Touster, O., V. H. Reynols, and R. M. Hutcheson: J. biol. Chem. **221**, 697 (1956).
2. —: Amer. J. Med. **26**, 724 (1959).
3. —: Fed. Proc. **19**, 977 (1960).
4. Lang, K.: Med. u. Ernähr. **4**, 45 (1963).
5. Bässler, K. H., W. Prellwitz, V. Unbehaun u. K. Lang: Klin. Wschr. **40**, 791 (1962).
6. Yamagata, S., Y. Goto, A. Ohneda, M. Anzai, S. Kawashima, M. Chiba, Y. Maruhama, and Y. Yamauchi: Lancet **1965 II**, 918.
7. Kanaami, T.: Tonyobyo **10**, 108 (1967).

8. Prellwitz, W., u. K. H. Bässler: Klin. Wschr. **41**, 196 (1963).
9. Fujisawa, M., and Y. Hirata: Tonyobyo **9**, 131 (1966).
10. Mehnert, H., J. D. Summa u. H. Förster: Klin. Wschr. **42**, 382 (1964).
11. Fujisawa, T., and Y. Hirata: Tonyobyo **9**, 263 (1966).
12. Bässler, K. H., u. D. Heesen: Klin. Wschr. **41**, 595 (1963).
13. Hirata, Y., M. Fujisawa, H. Sato, T. Asano, and S. Katsuki: Biochem. biophys. Res. Commun. **24**, 471 (1966).
14. Kanaami, T.: Tonyobyo **10**, 97 (1967).
15. Jeanrenaud, B., and A. E. Renold: J. biol. Chem. **234**, 3082 (1959).
16. Sutherland, E. W., and T. W. Rall: Pharmacol. Rev. **12**, 265 (1960).
17. Rizack, M. A.: J. biol. Chem. **236**, 657 (1961).

Comment

The Substitution of Dietary Carbohydrate with Xylitol and its Effect on the Plasma Lipids

YUICHIRO GOTO, SHIRO NAGANO, and HARUO NAKAMURA

Department of Medicine, Keio University Hospital
Tokyo, Japan

There is increasing evidence that various kinds of carbohydrate have a lipogenic effect. The present study was carried out to investigate the effect of xylitol on the plasma lipid and compare the effect with those of sucrose, and starch.

5 maturity onset diabetic patients were administered daily with 20—50 g xylitol for 2 to 6 weeks, substituting sucrose or starch with that on the isocaloric dietary bases. Every case except 4 was put on the diet in which 20—50 g of sucrose was substituted and then put on the second diet in which 20—50 g of starch was substituted with the same amount of xylitol. During each experimental diet plasma lipids were measured. Differences of the effect by each dietary design were compared.

Major plasma lipids decreased more on the sugar-xylitol substitution diet than on the starch-xylitol diet. 50 g of xylitol administration to the 3rd case appeared to increase plasma lipids on starch-xylitol diet.

Table I. *The substitution of dietary carbohydrate with xylitol and its effect on the plasma lipids*[a]

Case	Age	Xylitol (g/day)	Duration (weeks)	Trig. (mg-%)	Chol. (mg-%)	PL. (mg-%)
1. I.Y.	34M	20	6(Sugar-Xyl.)	26/104	20/224	19/214
			6(Starch-Xyl.)	5/78	3/204	2/195
2. S.Y.	58M	20	6(Sugar-Xyl.)	52/196	49/336	32/274
			6(Starch-Xyl.)	17/144	12/287	13/242
3. H.M.	69M	50	4(Sugar-Xyl.)	25/215	18/306	12/285
			4(Starch-Xyl.)	+7/190	+5/288	2/273
4. H.W.	71M	30	4(Starch-Xyl.)	13/143	16/247	11/232
5. M.H.	57M	30	2(Sugar-Xyl.)	96/247	44/192	15/187
			2(Starch-Xyl.)	10/151	+9/148	17/172

[a] Decreased amount/Initial concentration.

These results indicate that xylitol might be less potent in lipogenesis than sucrose and aslo that lipogenic action must be dependent on the amount of administered xylitol.

Discussion

Dr. KUMAGAI: Is there any possibility that xylitol has some effect on the excretion of FMS or that xylitol excretes something to inhibit lipolysis?

Dr. OPITZ: It is probable that FMS-excretion decreases after the administration of xylitol in fasting individuals. The action of xylitol on fat mobilization is probably determined mainly by increased re-esterification.

Dr. BÄSSLER: Do you believe that metabolism of xylitol in adipose tissue accounts for the anti-lipolytic action? Is there activity enough to provide sufficient amounts of glycerol phosphate?

Dr. OPITZ: I think so.

Dr. GOTO: Does glucose also depress the increase in free fatty acids produced by hormones such as glucagon, norepinephrine?

Dr. OPITZ: Certainly, but there is hyperglycemia following the administration of glucose, whereas xylitol does not cause significant changes in blood glucose concentrations.

The Activity of the Glucuronic Acid Pathway in Human and Experimental Diabetes *

ALBERT I. WINEGRAD, REX S. CLEMENTS JR., PAUL J. BEISSWENGER, and CONSTANCE L. BURDEN-RUSSELL

George S. Cox Medical Research Institute Department of Medicine
Hospital of the University of Pennsylvania
Philadelphia, Pennsylvania, U.S.A.

The nature of the genetic defect in diabetes mellitus is unknown, and at present the data do not exclude its inheritance as a multifactorial trait. It is common knowledge that the experimental production of insulin deficiency in many mammalian species leads to gross abnormalities in glucose and lipid metabolism similar to those observed in patients with the juvenile form of diabetes. These patients characteristically have no detectable insulin on radio-immunoassay of their plasma, and require exogenous insulin for survival [1]. However, the relationship between gross insulin deficiency, or a more subtle abnormality in its normal regulatory role, and the development of the vascular complications of diabetes mellitus is obscure. Certainly no simple correlation has been established between the usual criteria of clinical "control" and the progression of disease of small or large blood vessels. Siperstein [2] has recently reported that if one biopsies the skeletal muscle of patients who have normal glucose tolerance but are presumed to be genetically constituted diabetics, because both of their parents are known to have the disease, one finds significant thickening of the capillary basement membrane in 50%. This observation has lead Siperstein [2] to suggest that thickening of the basement membrane reflects a primary inherited defect in the metabolism of small blood vessels which is followed only secondarily by the development of manifestations of insulin deficiency. This provocative conclusion rests upon two assumptions; first, that diabetes mellitus is inherited as an autosomal recessive; second, that a normal glucose tolerance excludes the presence of any derangement in metabolism resulting from an altered insulin regulatory mechanism.

The vascular complications of diabetes mellitus are characterized by the accumulation of polysaccharide containing materials. In the case of the specific intra-mesangial nodules found in the renal glomerulus these have the staining characteristics of a glycoprotein, as does the diffusely

* These studies were supported in part by the following USPHS Grants: AM 04722, 1 TOI AM 05556, 5-K3-GM-6405.

thickened basement membrane that occurs in the capillaries of many organs [3]. Recent studies of the structure of the glomerular basement membrane of man and cattle by Spiro [4] and by Dische [5] suggest certain similarities to vertebrate collagen. Two types of carbohydrate units have been found in this material, one a disaccharide composed of glucose and galactose, and the other a heteropolysaccharide which in human tissue contains sialic acid, fucose and hexosamine. In addition to the characteristic microangiopathy, human diabetics are predisposed to the development of clinically significant atherosclerosis and, as noted by many workers, these lesions contain mucopolysaccharide-rich collagenous tissue [6].

If one could relate the development of these accumulations of glycoprotein and mucopolysaccharide to a quantitative or qualitative alteration in glucose metabolism it would be unnecessary to postulate a separate inherited defect in the metabolism of small or large blood vessels. There is some evidence that such a relationship may exist. Spiro [7] has observed that the utilization of glucose for the synthesis of the glucosamine components of liver and serum glycoprotein is essentially unaltered in alloxan diabetic rats. In addition, studies in our own laboratory suggest that the relative contribution of the glucuronic acid pathway to total CO_2 production from glucose is increased in the adipose tissue of starved or alloxan diabetic rats [8]. These observations have lead to the suggestion that in insulin deficient states glucose may be shunted in increased quantities into non-insulin sensitive pathways that normally account for only a small fraction of total glucose utilization; these pathways provide some of the nucleotide precursors of glycoprotein and mucopolysaccharides, and it is possible that such a derangement might influence the rates of synthesis of these materials, or lead to the synthesis of subtly altered materials that are poorer substrates for the enzymes concerned with their degradation.

There is at present strong prejudice against the use of experimental models in the study of the pathogenesis of vascular lesions in diabetes mellitus. This makes it necessary to demonstrate a possible etiologic factor in humans before seeking an experimental model in which it can be studied in greater detail. The studies to be discussed are concerned with the activity of the glucuronic acid pathway in human diabetics, but let us first consider the studies in experimental animals which prompted them.

Rat epididymal adipose tissue produces $C^{14}O_2$ from glucuronolactone-6-^{14}C at a slow but easily measurable rate, and labeled glucuronic acid can be isolated from the tissue after incubation with uniformly labeled glucose [8]. There is then presumptive evidence for the existence of portions of the glucuronic acid pathway in adipose tissue. When paired samples of adipose tissue are incubated with uniformly labeled glucose-^{14}C the addition of a pool of unlabeled glucuronolactone produces a significant decrease in the recovery of ^{14}C in the CO_2 produced by starved or alloxan diabetic rats but not in that produced by adipose tissue from normal fed rats [8] (Fig. 1). A dilution effect in tissue from starved or alloxan diabetic animals can be demonstrated when glucose-6-^{14}C is substituted for uniformly labeled glucose, but is not observed with

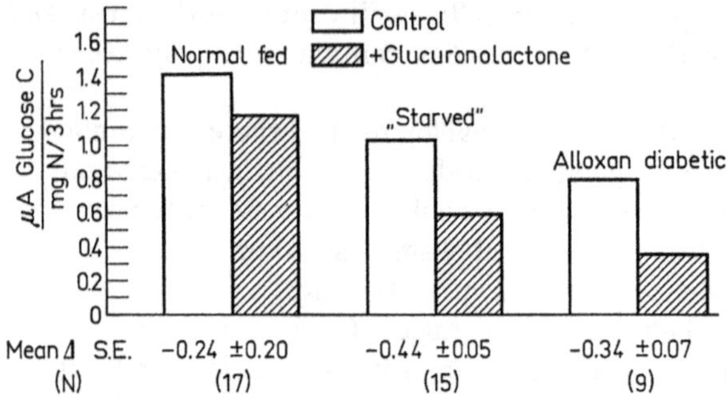

Fig. 1. Effect of unlabeled glucuronolactone on recovery of ^{14}C from glucose-U-^{14}C in CO_2 produced by adipose tissue from rats. Paired epididymal fat pads were incubated in 3.0 ml Krebs bicarbonate buffer, pH 7.4 (gas phase 5% Co_2 — 95% O_2) containing uniformly labeled glucose-C^{14} (5 mM). One member of each pair was incubated in the same buffer containing in addition unlabeled glucuronolactone (20 mM). (Data from [8])

glucose-1-^{14}C. This is consonant with the interpretation that the effect of unlabeled glucuronolactone is largely upon $^{14}CO_2$ produced by the glucuronic acid pathway. The relative contribution of the glucuronic acid cycle to total $^{14}CO_2$ production from glucose is also increased in adipose tissue from animals pretreated with bovine growth hormone or barbital, an agent known to increase the activity of this pathway in rat liver [8]. These observations lead us to speculate that glucose utilization by the glucuronic acid pathway is unimpaired or increased in the diabetic state. This speculation was further stimulated by the observations of Straumfjord and West [9] who found that the urinary excretion of ascorbic acid was four to five times greater in alloxan diabetic rats maintained on an ascorbate deficient diet than in normal rats on the same diet. Fisher and Weaver [10] in our own laboratory have extended this observation and demonstrated that the increased urinary ascorbic acid excretion of alloxan diabetic rats is decreased by insulin therapy and rises to elevated levels when the hormone is withdrawn (Fig. 2).

Our problem was to determine whether a similar alteration in the activity of the glucuronic acid pathway might be demonstrated in human diabetics. Since a number of the intermediates of this cycle are non-phosphorylated it seemed possible that alterations in the activity of the pathway might be reflected in changes in the plasma concentrations of these intermediates. L-xylulose proved to be the most suitable

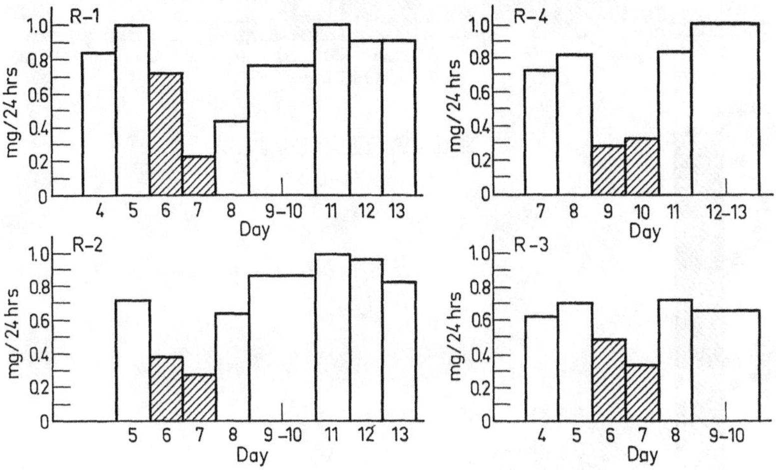

Fig. 2. Effect of insulin on urinary ascorbic acid excretion by alloxan diabetic rats. Animals maintained on an ascorbate deficient diet throughout the experimental period. Subcutaneous injections of 15 units and 10 units of NPH insulin were given respectively on the first and second days indicated by the dark bars. (Fisher and Weaver, unpublished data)

for study since NADP-xylitol dehydrogenase prepared from acetone powders of guinea pig liver provides the basis of an enzymatic assay [11]. At the time of our initial studies it was not recognized that D-erythrulose is also a substrate for this enzyme [12]. For reasons to be discussed we believe that the material in human plasma assayed with NADP-xylitol dehydrogenase primarily represents L-xylulose. The ammonium sulphate fraction used in these studies did not react significantly with D-xylulose, D-glucuronolactone, or D-glucuronic acid in the assay procedure of Hickman and Ashwell [11].

Blood samples were obtained from a heterogeneous group of hospital inpatients after an overnight fast and before any physical activity was permitted, the latter precaution is necessary to minimize fluctuations in plasma growth hormone levels [13]. As shown in Fig. 3, there is a significant difference in the distribution of plasma (or serum the levels being the same) L-xylulose concentrations in diabetic and non-diabetic

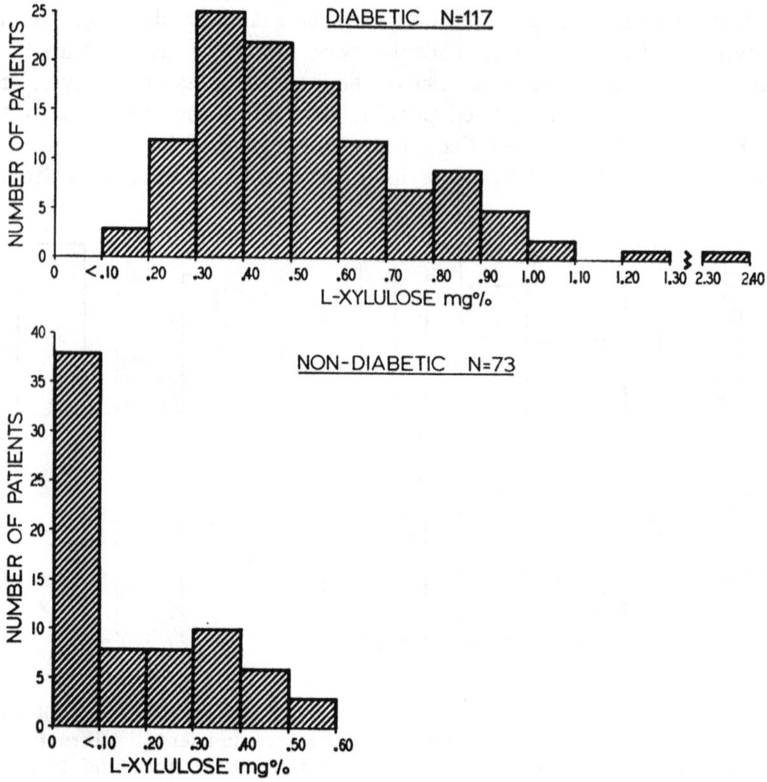

Fig. 3. L-xylulose concentration in plasma (or sera) of diabetic and non-diabetic hospital in-patients

patients. At least half of the non-diabetic subjects had values too low to be determined by this technique. To avoid including patients with hereditary pentosuria, only non-Jewish subjects were included in the study [14, 15]. In addition patients with liver disease and those receiving drugs known to stimulate the glucuronic acid pathway were also excluded [16].

The fasting plasma (or serum) L-xylulose concentration could not be correlated with age or sex, nor as shown in Fig. 4 was there any apparent relationship between fasting serum L-xylulose and blood glucose concentrations [17].

The elevation of fasting plasma L-xylulose concentration observed in diabetic patients could result from impaired utilization, from impaired urinary excretion, or as postulated from increased production either via the glucuronic acid pathway or from myoinositol. These points have been considered in our interpretation of the findings in human diabetics.

First with regard to the possibility of impaired utilization. Due in large part to the studies of Dr. Touster [14], it is known that patients with hereditary pentosuria have a defect in the utilization of L-xylulose. Hiatt and his co-workers have demonstrated the presence of elevated

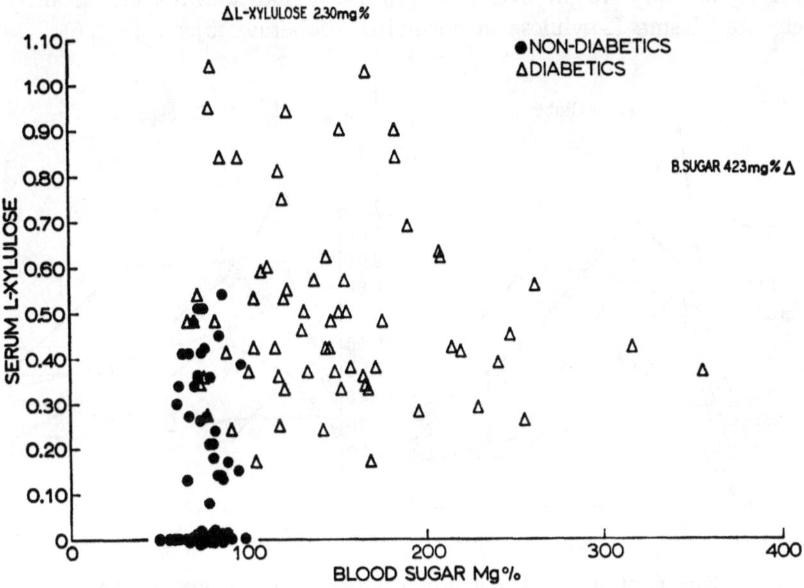

Fig. 4. Comparison of serum L-xylulose and blood glucose in fasting diabetic and non-diabetic subjects. Illustration reproduced through the courtesy of New Engl. J. Med. **274**, 298 (1966) [17]

levels of L-xylulose in the fasting serum of these patients and demonstrated that the administration of oral loads of glucuronolactone produces excessive rises in the serum concentration of L-xylulose in both homozygotes and heterozygotes [15]. This response to the administration of glucuronolactone reflects the presumed deficiency of NADP-xylitol dehydrogenase activity in the livers of these subjects [15]. As shown in Fig. 5 the administration of glucuronolactone loads produced comparable rises in plasma L-xylulose concentration in diabetic and non-diabetic subjects, and it is therefore unlikely that the elevated fasting levels found in the diabetics reflect a defect in the reduction of L-xylulose such as that present in essential pentosuria.

The studies reported from the laboratories of Dr. Lang and of Dr. Yamagata [19] and by Mehnert *et al.* [20] as well as the reports presented elsewhere in this conference indicate that the utilization of xylitol is unimpaired in human or experimental diabetes. There is, therefore, little reason to suspect that the elevated plasma L-xylulose levels in human diabetics reflect an abnormality in the utilization of xylitol. Moreover, the rapid intravenous infusion of 0.1 g of xylitol per kg of body weight over a 10-minute period does not significantly elevate plasma L-xylulose in normal or diabetic subjects (Fig. 6).

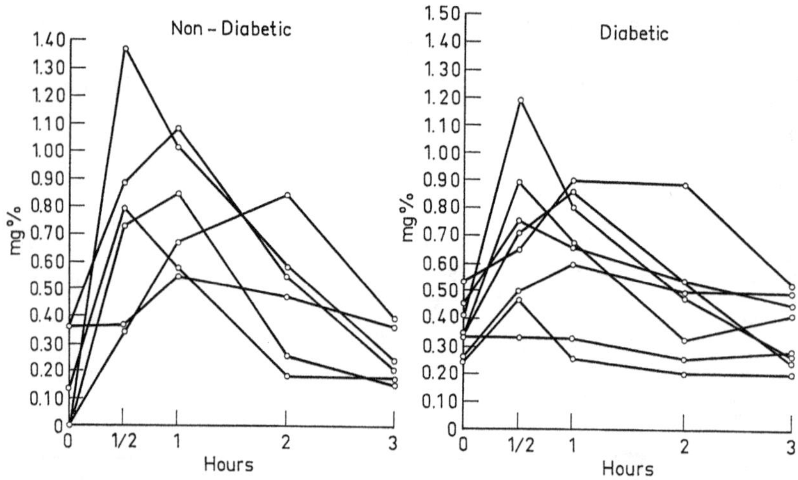

Fig. 5. Effect of an oral glucuronolactone load (25 grams) on plasma L-xylulose concentrations in diabetic and non-diabetic patients. Illustration reproduced through the courtesy of New Engl. J. Med. **274**, 298 (1966) [17]

Impaired urinary excretion might have contributed to the higher fasting L-xylulose levels observed in the diabetic patients. There was, however, no relationship between blood urea nitrogen or even gross evidence of renal disease and the serum level of the pentulose [17]. This was surprising for Kumahara *et al.* [21] in Hiatt's laboratory had compared endogenous creatinine and L-xylulose clearance in a pentosuric patient with a serum level of 1.70 mg-% and in a heterozygote whose plasma level had been elevated to 1.0 mg-% by a glucuronolactone load at the time of study; their results indicated that most, if not all, of the filtered L-xylulose appeared in the urine. We have compared endogenous creatinine and L-xylulose clearance in a few diabetic and non-diabetic patients whose levels remain quite constant in the range of 0.3 to 0.5 mg-% when resting in the fasted state [17]. At the lower filtered loads observed in these patients only 2% of the filtered load is excreted. These observations suggest that there may be a Tm for L-xylulose reabsorption. This would explain the poor correlation between the BUN and plasma L-xylulose concentration. Impaired urinary excretion does

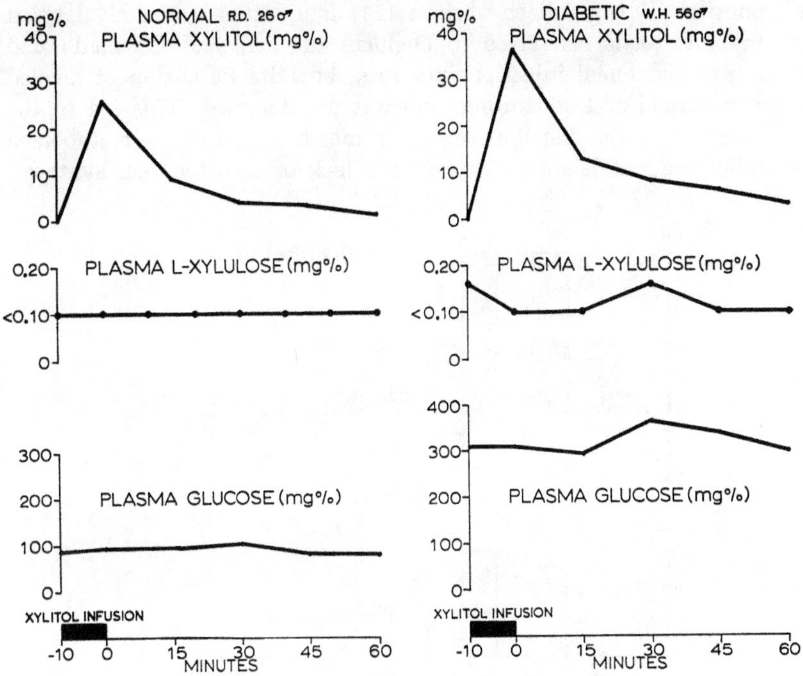

Fig. 6. The effects of an I.V. xylitol infusion (0.1 g/kg) on plasma L-xylulose and glucose concentrations

not appear to be responsible for the elevated plasma L-xylulose concentrations of diabetic patients.

Charalampous [22] has demonstrated that the supernatant fraction of rat kidney contains an enzyme which catalyzes the cleavage of myoinositol to racemic glucuronic acid. This finding was confirmed by Burns *et al.* [23], who demonstrated that the product of the reaction was D-glucuronic acid whose subsequent metabolism would proceed by way of the glucuronic acid cycle. The elevated levels of L-xylulose observed in the plasma of diabetics might therefore reflect increased formation from myoinositol. This possibility seemed particularly intriguing since Eisenberg [24, 25] has recently demonstrated that the synthesis of inositol in mammalian tissues may proceed by a direct cyclization of glucose-6-phosphate leading to the formation of myoinositol-1-phosphate which can be hydrolyzed by a phosphatase highly specific for equatorial cyclitol phosphates and 2′-adenylic acid.

Although the evidence is not complete it would appear that the elevated levels of L-xylulose found in the plasma of human diabetics are not primarily derived from increased production or utilization of myoinositol. Burns and co-workers [23] found that uniformly labeled ^{14}C myoinositol is converted to D-glucuronic acid and L-gulonic acid after intraperitoneal injection into rats, but the formation of labeled L-ascorbic acid or L-glucuronic acid was not detected. This led to the suggestion that the distribution of enzymes involved in the metabolism of these compounds might explain the lack of ascorbic acid synthesis

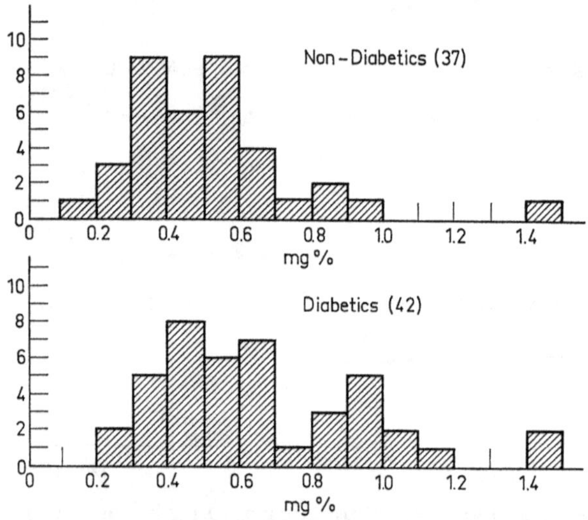

Fig. 7. Fasting plasma free myoinositol concentration. Plasma myoinositol concentrations were determined by bioassay with Kloeckera apiculata

from myoinositol. Inositol is converted to D-glucuronic acid and L-gulonic acid by kidney enzymes but the system required for the formation of L-ascorbic acid from L-gulonic acid is present in the liver. Anderson and co-workers [26, 27] have supported this conclusion by demonstrating that myoinositol-2-^{14}C is not converted to respiratory $^{14}CO_2$ by nephrectomized rats; and by demonstrating that the rates of oxidation of myoinositol-2-^{14}C to $^{14}CO_2$ by rat kidney slices are adequate to account for the degradation of the average dietary intake of the rat. We have found that the levels of free myoinositol are essentially the same in the kidneys of normal and alloxan diabetic rats (normal 7.15 ± 0.13 µM/g, N=6, diabetic 7.04 ± 0.78, N=6). As shown in Fig. 7 the distribution of fasting myoinositol levels is not strikingly different in normal and diabetic patients. There is no apparent relationship between fasting plasma L-xylulose and myoinositol concentrations in normal or diabetic patients (Fig. 8). Moreover, the administration of 3.0 grams of inositol p.o. to normal and diabetic subjects produced no significant rise in plasma L-xylulose over a 4-hour period, although the measurement of plasma myoinositol levels by gas liquid chromatography demonstrated significant absorption. It would appear that increased myoinositol metabolism is unlikely to be the cause of the elevated plasma L-xylulose levels in diabetes mellitus.

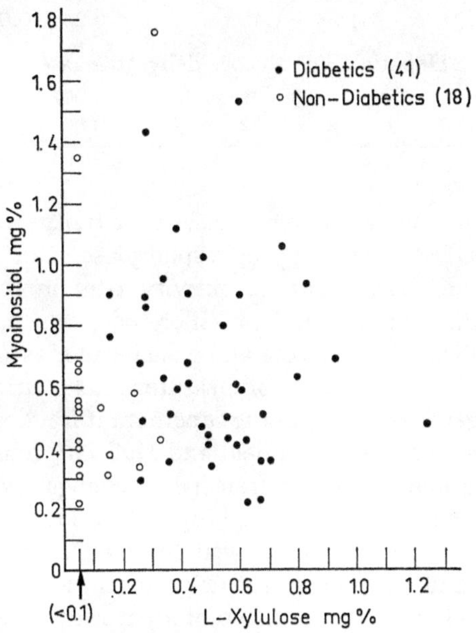

Fig. 8. Comparison of fasting plasma L-xylulose and myoinositol concentration

The data presently available suggest that increased glucose utilization by way of the glucuronic acid pathway is the most likely explanation for the increased plasma levels of L-xylulose observed in diabetic subjects.

The liver appears to be a major source of plasma L-xylulose in the fasting state for in 2 diabetic patients who required cardiac catherization for other reasons we observed higher concentrations of L-xylulose in plasma obtained from the hepatic vein than in samples obtained from the superior vena cava or right ventricle.

Studies of UDPG-pyrophosphorylase and UDPG-dehydrogenase activities in the supernatant fraction of rat liver homogenates suggest that the levels of UDPG dehydrogenase activity are elevated in the diabetic state, although UDPG pyrophosphorylase activity is essentially unaltered (Table I). A number of factors suggest that uridine diphosphoglucose dehydrogenase may be a site of regulation of glucose utilization by way of this cycle. The reaction is essentially irreversible [28]; it is

Table I

	μM/min/g wet wt.	μM/min/mg sup. protein
Hepatic UDP-glucose pyrophosphorylase		
Normal fed [8]	13.0 ± 0.8	0.21 ± 0.02
Alloxan diabetic [7]	18.5 ± 1.4	0.24 ± 0.01
Hepatic UDP-glucose dehydrogenase		
Normal fed [13]	0.13 ± 0.01	0.0021 ± 0.0001
Alloxan diabetic [12]	0.20 ± 0.02	0.0033 ± 0.0003

the first reaction unique to this cycle, its activity is approximately $^{1}/_{100}$th that of UDP glucose pyrophosphorylase, and Hollmann and Touster [29] have shown that its activity rises in response to pretreatment of rats with barbital or chloretone, which are known to stimulate the utilization of glucose and galactose for ascorbate synthesis in rats. However, the significance of measurements of uridine diphosphoglucose dehydrogenase activity in the supernatant fraction of liver homogenates is difficult to assess. Neufield and Hall [30] found that uridine diphosphoglucose dehydrogenase from pea seedlings and calf liver was markedly inhibited by uridine diphospho-D-xylose. In plant tissue UDP-D-xylose is known to be derived from UDP-glucuronic acid by decarboxylation and its effect on the dehydrogenase is considered a case of feedback inhibition of the first step of a pathway by an end product. The kinetics of this inhibition are complex and suggest that UDP-xylose is an allosteric inhibitor of the dehydrogenase. Although the regulatory significance of UDP-xylose in mammalian liver remains to be established, the identification of D-xylose as the residue connecting protein to carbohydrate in a number of polysaccharides [31, 32] suggests that it might play a role in the regulation of mucopolysaccharide synthesis. Neufield and Hall [30] speculated that isoenzymes of UDP-glucose dehydrogenase might exist in mammalian liver since this inhibition by UDP-xylose did not appear metabolically related to the synthesis of glucuronides or ascorbic acid. However, in our laboratory starch-gel electrophoresis of the supernatant fraction of rat liver homogenates demonstrated only a single band of UDP-glucose dehydrogenase activity. Saltis and Oliver [33] have reported that uridine diphosphoglucose dehydrogenase from calf liver is also competitively inhibited by UDP-galactose, and uncompetitively by UDP. It is therefore difficult to assess the significance of UDP-glucose dehydrogenase activity assayed in the supernatant fraction of liver homogenates. Measurements of the steady state concentrations of intermediates of this cycle in isolated perfused liver may clarify this point, and are in progress.

Hormonal effects on the plasma levels of L-xylulose in human subjects have in general followed the responses predicted from studies on the activity of this pathway in isolated adipose tissue. As shown in Fig. 9 the elevated levels of L-xylulose in diabetic subjects fall promptly in response to intravenous insulin. In patients in whom hypoglycemia

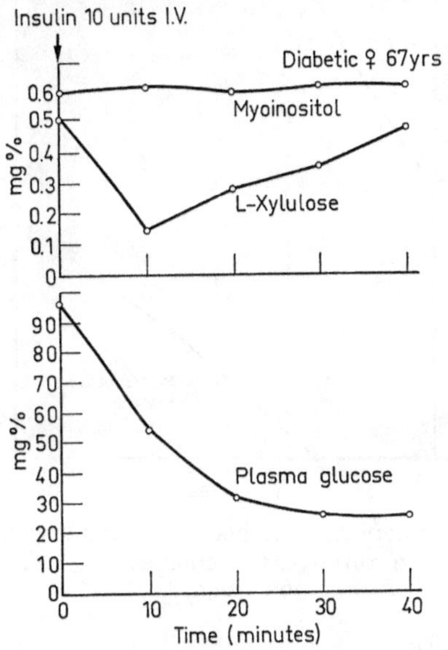

Fig. 9. Effect of insulin on plasma inositol and L-xylulose

develops there may be a secondary rise. Present data do not permit a distinction between decreased production of L-xylulose or an increased volume of distribution as an explanation for this insulin effect. It should be noted that insulin does not produce a significant change in myo-inositol concentration (Fig. 9).

The administration of epinephrine produces a marked increase in fasting X-xylulose concentration in normal subjects. We have some reassurance that the material assayed in the plasma under this circumstances is L-xylulose since we have observed a good correlation between the values for L-xylulose as determined enzymatically and those for total xylulose obtained by gas liquid chromatography of the trimethysilylethers of protein free filtrates of plasma (Fig. 10). Independent study of the same samples by Dr. Marjorie Horning have yielded similar results, and analysis of the derivative by mass spectoscopy in her laboratory have confirmed its identity as (a derivative) of xylulose.

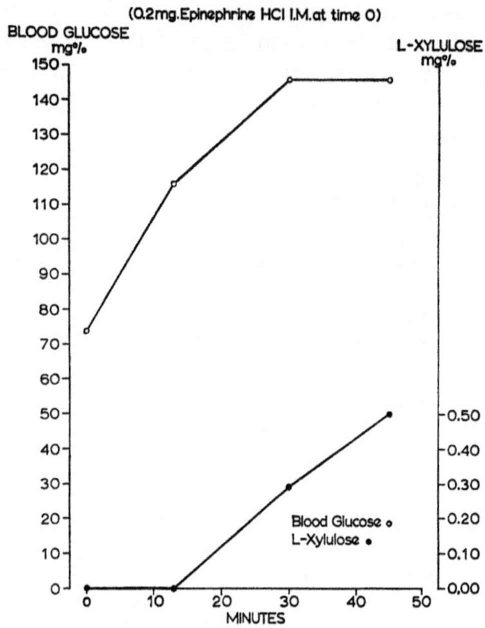

Fig. 10. Effect of epinephrine on plasma L-xylulose in a normal subject. Illustration reproduced through the courtesy of New Engl. J. Med. **274**, 298 (1966) [17]

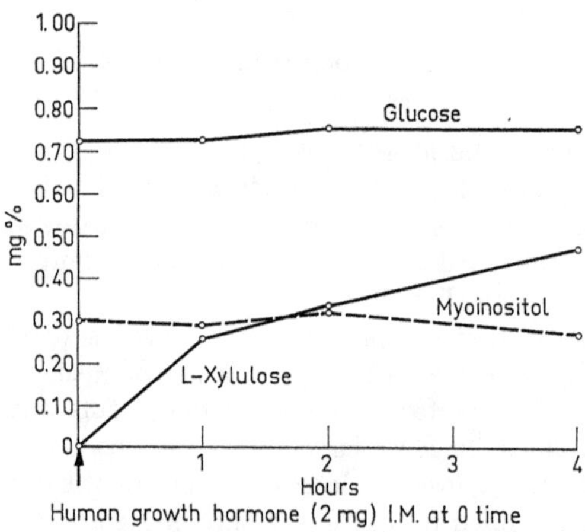

Fig. 11. Effect of human growth hormone on plasma L-xylulose and myoinositol in a hypopituitary dwarf

The administration of human growth hormone to normal subjects or to pituitary dwarfs (Fig. 11) results in a prompt rise in plasma L-

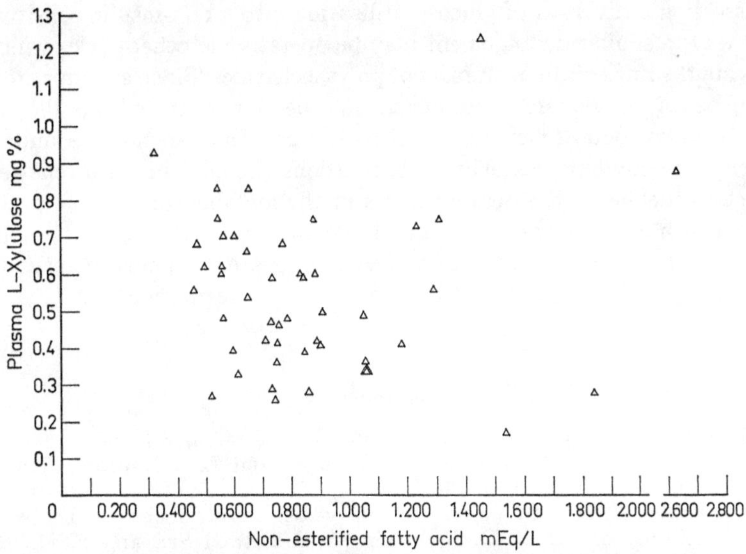

Fig. 12. Comparison of plasma L-xylulose and NEFA in fasting diabetic subjects. Illustration reproduced through the courtesy of New Engl. J. Med. **274**, 298 (1966) [17]

xylulose concentration without a significant change in myoinositol concentration. The mechanism of these hormonal effects is under study in isolated tissues. However, one obvious consideration must be mentioned. The administration of growth hormone [34] or epinephrine [35] to human subjects produces a significant increase in plasma unesterified fatty acid concentration as the result of increase release from adipose tissue. It is known that insulin decreases the rate of free fatty acid release from adipose tissue [36] and that it lowers the plasma FFA concentration. Randle *et al.* [37], and Steinberg [38] have considered the possibility that certain hormonal effects may be mediated by means of changes in plasma FFA concentration since this is a major determinant of their rate of uptake in many tissues. There is, however, no apparent correlation between fasting plasma L-xylulose and FFA concentrations in diabetic subjects (Fig. 12).

The observations reported would suggest that the utilization of glucose by way of the glucuronic acid pathway may be increased in diabetes mellitus. This derangement in carbohydrate metabolism is not assessed by measurements of blood glucose or plasma FFA and would therefore not be reflected in present systems for judging the adequacy of clinical control. The fact that the glucuronic acid pathway has nonphosphorylated intermediates may have fortuitously provided a means of assessing a diverson of glucose utilization into a non-insulin sensitive pathway; a similar derangement may be operative in other cycles which provide the nucleotide precursors of polysaccharides. Since a comparable disturbance in glucose metabolism can be demonstrated in alloxan diabetic rats, detailed examination of the problem can now be undertaken in isolated tissues. These observations should stimulate renewed efforts to delineate the derangements in the metabolism of trace pathways which may occur in human and experimental diabetics in the hope of relating the development of vascular lesions in diabetes mellitus to the metabolic consequences of ineffective insulin regulation.

References

1. Berson, S. A., and R. S. Yalow: Amer. J. Med. **40**, 676 (1966).
2. Siperstein, M. D., W. Norton, R. H. Unger, and L. L. Madison: Trans. Ass. Amer. Phycns **79**, 330 (1966).
3. Lecompte, P. M.: In: Small blood vessel involvement in Diabetes mellitus, ed. by M. D. Siperstein, A. R. Colwell Sr., and K. Meyer. Baltimore: Garamong / Pridemark Press 1964.
4. Spiro, R. G.: J. biol. Chem. **242**, 1915 (1967).
5. Dische, R. M., G. D. Pappas, A. Grauer, and Z. Dische: Biochem. biophys. Res. Commun. **20**, 63 (1965).
6. Curran, R. C.: In: Biological aspects of occlusive vascular disease, ed· by D. G. Chalmers and G. A. Gresham. Cambridge: Cambridge University Press 1964.
7. Spiro, R. G.: Diabetes **12**, 223 (1963).
8. Winegrad, A. I., and W. N. Shaw: Amer. J. Physiol. **206**, 165 (1964).
9. Straumfjord, J. V., Jr., and E. S. West: Proc. Soc. exp. Biol. (N.Y.) **94**, 566 (1957).
10. Fisher, W. B., and J. P. Weaver (Unpublished data).
11. Hickman, J., and G. Ashwell: J. biol. Chem. **234**, 758 (1959).
12. Batt, R. D., F. Dickens, and D. H. Williamson: Biochem. J. **77**, 272 (1960).
13. Glick, S. M., J. Roth, R. S. Yalow, and S. A. Berson: Recent Progr. Hormone Res. **21**, 241 (1965).
14. Touster, O.: Amer. J. Med. **26**, 724 (1959).
15. Hiatt, H. H.: In: The metabolic basis of inherited disease, ed. by J. B. Stanbury, J. B. Wyngaarden, and D. S. Fredrickson. New York: McGraw-Hill Book Co. 1966.
16. Conney, A. H., G. A. Bray, C. Evans, and J. J. Burns: Ann. N.Y. Acad. Sci. **92**, 115 (1961).
17. Winegrad, A. I., and C. L. Burden: New Engl. J. Med. **274**, 298 (1966).
18. Schmidt, B., M. Fingerhut u. K. Lang: Klin. Wschr. **42**, 1073 (1964).
19. Yamagata, S., Y. Goto, M. Anzai, M. Chiba, A. Ohneda, S. Kawashima, Y. Maruhama, and Y. Yamauchi: Lancet **1965II**, 918.
20. Mehnert, M., J. D. Summa u. M. Forster: Klin. Wschr. **42**, 382 (1964).
21. Kumahara, Y., D. S. Feingold, I. M. Freedberg, and H. H. Hiatt: J. clin. Endocr. **21**, 887 (1961).
22. Charalampous, F. C., and C. Lyras: J. biol. Chem. **228**, 1 (1957).

23. Burns, J. J., N. Trousof, C. Evans, N. Papapopoulous, and B. W. Agranoff: Biochim. biophys. Acta (Amst.) **33**, 215 (1959).
24. Eisenberg, F., Jr., and A. M. Bolden: Biochem. biophys. Res. Commun. **14**, 419 (1964).
25. — J. biol. Chem. **242**, 1375 (1967).
26. Anderson, L., and R. M. Coots: Biochim. biophys. Acta (Amst.) **28**, 666 (1958).
27. Howard, C. F., Jr., and C. Anderson: Arch. Biochem. **118**, 332 (1967).
28. Strominger, J. L., E. S. Maxwell, J. Axelrod, and H. M. Kalckar: J. biol. Chem. **224**, 79 (1957).
29. Hollmann, S., and O. Touster: Biochim. biophys. Acta (Amst.) **26**, 338 (1962).
30. Neufeld, E. F., and C. Hall: Biochem. biophys. Res. Commun. **19**, 465 (1965).
31. Roden, L.: Presentation in this symposium.
32. — J. Gregory, and T. Laurent: Biochem. J. **91**, 2p. (1964).
33. Saltis, G., and I. T. Oliver: Biochim. biophys. Acta (Amst.) **81**, 55 (1964).
34. Raben, M. S., and C. H. Hollenberg: In: Ciba Foundation Colloquia on Endocrinology, vol. 13. Boston: Little, Brown & Co. 1959.
35. Gordon, R. S., Jr., and A. Cherkes: J. clin. Invest. **35**, 206 (1956).
36. Dole, V. P.: J. clin. Invest. **35**, 150 (1956).
37. Randle, P. J., P. B. Garland, C. N. Hales, and E. A. Newsholme: Lancet **1963I**, 785.
38. Steinberg, D.: In: The control of lipid metabolism, ed. by J. K. Grant. London: Academic Press 1963.

Discussion

Dr. Touster: Could the accumulation of L-xylulose induced by hormones be attributed to changes in the oxidation state of the pyridine nucleotide coenzymes? An increase in NADP as compared with NADPH would increase L-xylulose.

Dr. Winegrad: This is certainly a reasonable possibility but we have not studied this problem in isolated tissues as yet and we could only speculate at this point.

Dr. Touster: It would be useful to test the conversion of ^{14}C-glucuronolactone to urinary ribose, as previously done by Hiatt, to demonstrate an increased level of the cycle.

Dr. Winegrad: Such studies, i.e. the administration of uniformly labeled glucuronolactone-^{14}C and imidazoleacetic acid would be of interest but are present not feasible because of the lack of uniformly labeled glucuronolactone-^{14}C. These studies would, however, demonstrate only that the pathway from glucuronolactone to D-xylulose was unimpaired. The administration of specifically labeled glucose and the isolation and degradation of the riboside might be of great help.

Dr. LANG: You showed us figures on the myo-inositol blood level. Are the figures concerned with free inositol or with total inositol? It may possible, that the free inositol level is constant during great variation in bound phospholipid. The equilibrium between free and bound is not yet very well established. In several organs the amount of free inositol is very small and the amount of bound inositol is much greater.

Dr. WINEGRAD: These values represent free myoinositol, i.e. that present in protein free filtrates of plasma. The bioassay with Kloeckera apiculata is not significantly affected by inositol in phospholipid, and we have observed good agreement between the values obtained by gas-chromatography analysis and bioassay.

Prevention of Choline-Deficiency Fatty Liver of Rats with Xylitol

YASUYUKI OHTA, TOSHIO TAKANO, and KIYOWO KOSAKA

First Department of Internal Medicine
Okayama University Medical School, Okayama

Fatty liver and cirrhosis produced in rats maintained on a diet low in protein and choline can be modified by the amount and kind of fat included in the diet [1—4]. Recent advances in clinical nutrition, notably carbohydrate [5, 6], prompted us to investigate the effect of changing the carbohydrate in the diet from sucrose to others. In the present communication, it will be reported that the production of fatty liver is modified by the substitution of sucrose for xylitol. In addition a certain unexpected effect of xylitol that might be related to the action of choline will also be described.

Materials and Methods

Male albino rats of the Sprague-Dawley strain, approximately 6 weeks old, weighing about 90 grams, were housed in individual cages. The rats were fed a standard laboratory chow until they weighed about 100 grams. They were then placed on experimental rations (Table I). For protein and fat, 8% casein and 38% lard were used throughout the experiment. For carbohydrate, 48% sucrose was used as the "standard" choline-deficient diet (Group III). Xylitol was substituted for sucrose (10 and 20%), comprising approximately 20 and 40% of the carbohydrate

Table I. *Composition of diet*

I		II	III	IV	V
Fox chow	Xylitol	0%	0%	10%	20%
	Sucrose	48.375	48.375	38.375	28.375
	Lard	37.95	37.95	37.95	37.95
	Casein	8.0	8.0	8.0	8.0
	Cystine	0.625	0.625	0.625	0.625
	Salt mixture	4.0	4.0	4.0	4.0
	Vitamin powder	1.0	1.0	1.0	1.0
	Vitamin A.D.E.	0.05	0.05	0.05	0.05
	Choline chloride	0.4	0	0	0

(Groups IV and V). Two other diets were used as a control; one was a standard laboratory chow (Fox Chow) (Group I), and another was the diet supplemented with choline chloride (0.4%) (Group II). The rats were fed 8 grams of the test diets a day. The calories from daily food intake were 28 Calories for the rats on a standard laboratory chow (Group I), and 45 Calories for the others. The rats were examined daily and weighed twice a week for the entire period. Any rat found dead was immediately removed from its cage and autopsied. The main viscera were fixed in 10% formalin solution for histological examination.

The rats that survived for a period of 180 to 275 days were sacrificed for histological, histochemical and biochemical investigations. Liver, kidney, testis and spleen were weighed immediately after removal. Three pieces of liver and epididymal adipose tissue weighing about 100 mg were incubated with 0.5 μc of 1-C^{14}-glucose or 1-C^{14}-acetate in Krebs-Ringer bicarbonate buffer. The incubation was carried out for 2 hours with constant shaking at 37° C, using a gas phase of 95% O_2 and 5% CO_2. Lipids were extracted according to the method of Folch *et al.* [7]. Washing of the crude lipid extract was repeated twice with 0.73% saline solution. The lipids were then saponified and extracted, using the method of Bjorntorp [8]. The fatty acid fraction was methylated by the method of Metcalfe and Schmitz [9], and separated by gas liquid chromatography using a 15% diethylene glycol succinate column. The fatty acid methyl esters in each peak were trapped on siliconized cellulose acetate filters and the radioactivity in individual fatty acids was determined in a liquid scintillation spectrometer.

Another piece of liver, kidney, lung, spleen, heart muscle or testis was fixed in both alcoholic Bouin's solution and 10% neutral formalin for histological examination. The hematoxylin and eosin stain, Oil red O stain, were used.

Results

Growth Rate and Survival. The animals were well adapted to the amount of xylitol used, and diarrhea was observed only at the beginning. The growth rate was highest for rats fed the choline supplemented diet (Group II), and lowest for rats maintained on the experimental diet with 20% xylitol (Group V). In contrast, the rats on the diet with 10% xylitol (Group IV) showed a growth rate similar to that of rats on the standard hypolipotropic diet (Group III). The rats on Fox Chow (Group I) showed relatively low growth rate, since the daily caloric intake was limited to low level, in comparison with that for the others (Table II).

The rate of survival was also different in each experimental group. A low incidence of death in the rats on the diet with 10% xylitol

Table II. *Growth curves*

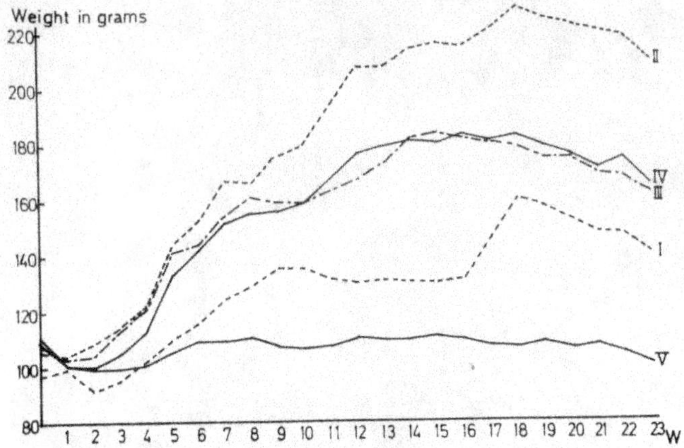

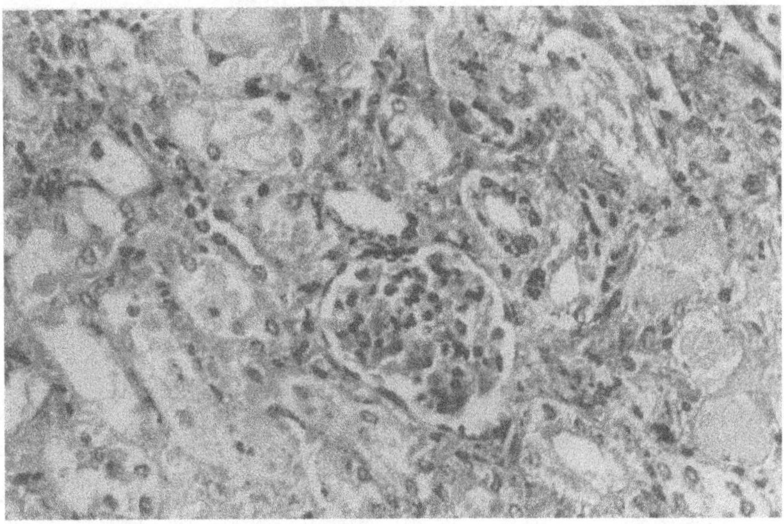

Fig. 1. Kidney of rat fed hypolipotropic diet containing only sucrose as carbohydrate for 43 days, (hematoxylin and eosin stain, ×280)

(Group IV) was noticed, whereas the rats on the diet with 20% xylitol (Group V) showed rather high mortality at the 18th to 31st day, and later, at the 127th to 160th day. The cause of death was obscure, however, and they did not die of hemorrhagic kidney. The rats on the standard choline-deficient diet (Group III) died of typical he orrhagic kidney after 30 to 60 days, as illustrated in the photomicrograph (Fig. 1) which shows necrosis of the proximal convoluted tubules,

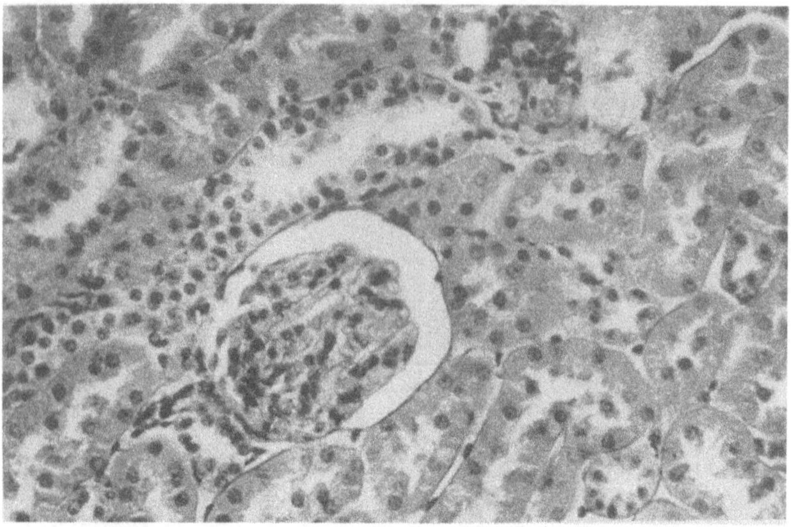

Fig. 2. Kidney of rat fed hypolipotropic diet supplemented with choline for 198 days (hematoxylin and eosin stain, ×280)

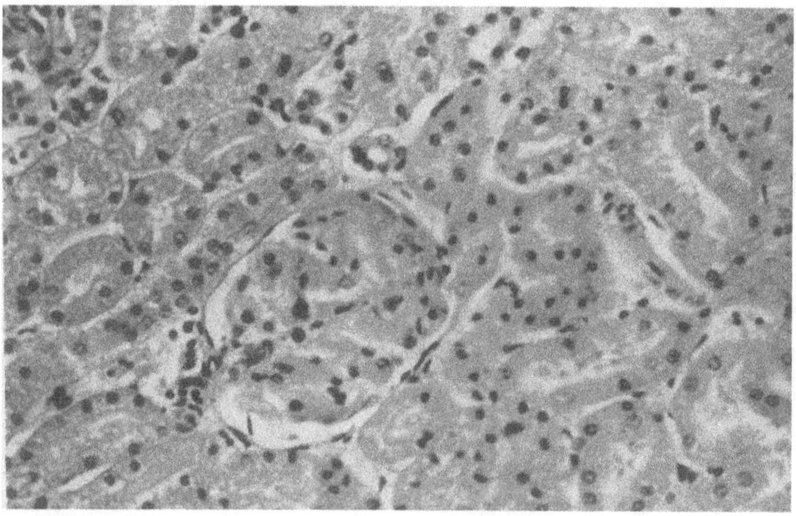

Fig. 3. Kidney of rat fed hypolipotropic diet containing 10% xylitol for 181 days (hematoxylin and eosin stain, ×280)

hyaline casts, and hemorrhagic change. On the other hand the severe hemorrhagic renal necrosis was observed to be protected by either 0.4% choline chloride (Fig. 2) or 10% xylitol (Fig. 3).

Weight of Liver and Testis. The weight of liver and testis was different in each group. Fig. 4 illustrates livers at autopsy performed after 205 days on the experimental diets. The rats on Fox Chow (Group I) and on the diet supplemented with choline (Group II) appeared to be normal, while the animal on the standard choline-deficient diet (Group III) indicated nodular fatty cirrhosis. In contrast, rats on the 10 or 20% xylitol diet (Groups IV and V) were devoid of cirrhosis and their livers

Table III. *Deaths*

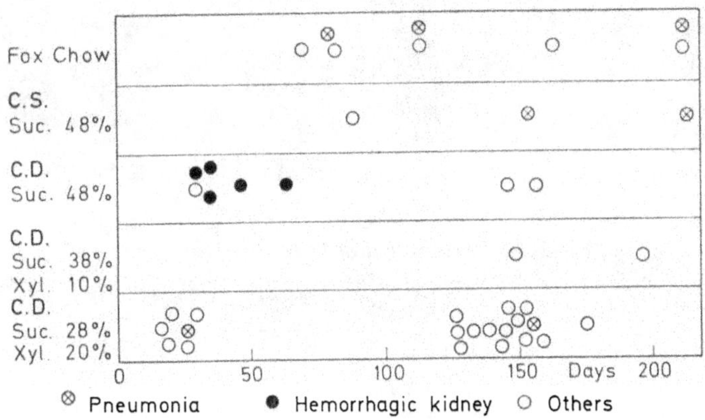

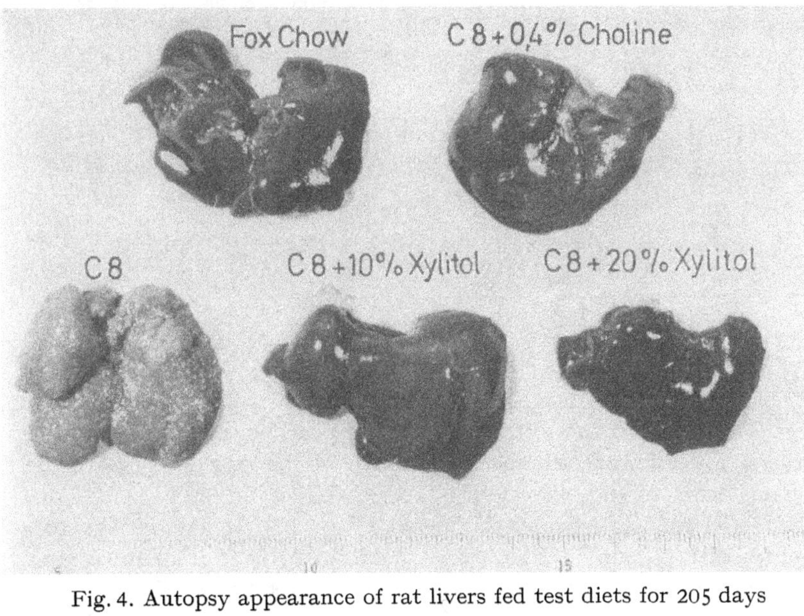

Fig. 4. Autopsy appearance of rat livers fed test diets for 205 days

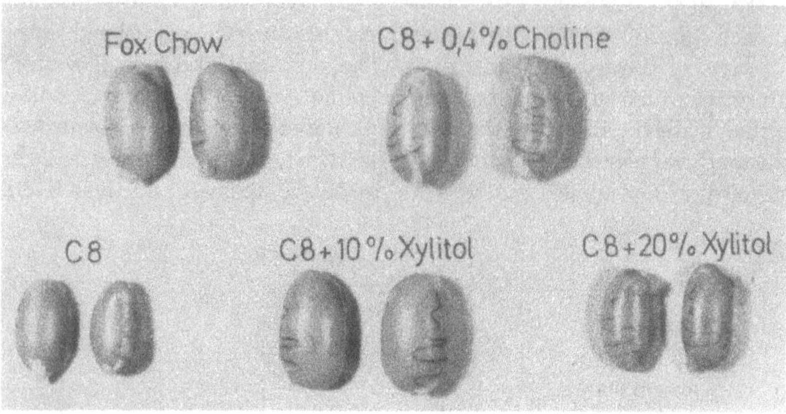

Fig. 5. Autopsy appearance of rat testes fed test diets for 205 days

Table IV. *The weight of body, liver and testis*

Diet	No. of Rats	Body wt.	Liver wt. $\left(\frac{L.W.}{B.W.} \times 100\right)$	Testis wt. R $\left(\frac{T.W.}{B.W.} \times 100\right)$	Testis wt. L $\left(\frac{T.W.}{B.W.} \times 100\right)$
Fox chow	15	186 ± 21	6.11 ± 0.91 (3.31)	1.24 ± 0.24 (0.66)	1.15 ± 0.38 (0.62)
C.S. Suc. 48%	21	227 ± 46	7.91 ± 2.16 (3.48)	1.50 ± 0.22 (0.66)	1.47 ± 0.21 (0.65)
C.D. Suc. 48%	21	169 ± 52	9.97 ± 4.14 (5.89)	0.89 ± 0.33 (0.52)	0.89 ± 0.33 (0.52)
C.D. Suc. 38% Xyl. 10%	24	182 ± 44	7.94 ± 0.59 (4.36)	1.22 ± 0.43 (0.67)	1.17 ± 0.51 (0.64)
C.D. Suc. 28% Xyl. 20%	9	87 ± 14	4.19 ± 0.59 (4.82)	0.54 ± 0.31 (0.62)	0.52 ± 0.31 (0.60)

appeared to be not so fatty, particularly the rats on the 20% xylitol diet (Group V). In the latter animals, the liver looked darker than the others. The difference in the liver weight of the rats on the standard choline-deficient diet (Group III) and those on the 20% xylitol diet (Group V) was statistically significant ($P<0.01$). In rats on the 10% xylitol diet (Group IV), however, the ratio of liver to body weight, was intermediate to the ratio of those on Fox Chow (Group I) and those on the standard choline-deficient diet (Group III). The change of testis was closely related to the state of the liver. A reduction in the size and weight of testis was observed in cirrhotic rats on the standard choline-deficient diet (Group III), while it was prevented by feeding the diet with 10% or 20% xylitol (Fig. 5). The ratio of testis to body weight was not altered in the rats fed the xylitol diets (Table IV).

Histological Studies on the Liver. The livers of 21 rats fed on Fox Chow (Group I) showed normal histology (Fig. 6). The rats on choline-

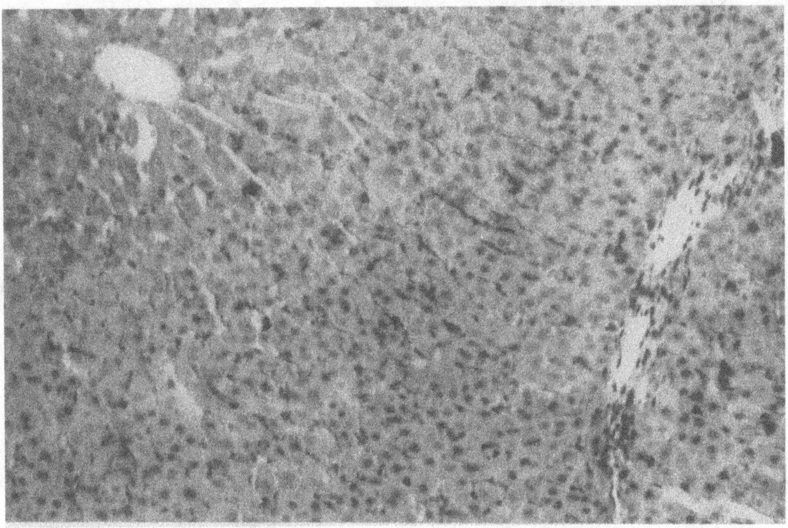

Fig. 6. Liver of rat fed normal laboratory chow for 181 days (Oil red O stain, ×140)

supplemented diet developed minimum periportal fat deposition in 9 of 22 (Fig. 7), and diffuse fatty liver was observed in one case. On other 12 cases the liver remained normal. Rats on standard choline-deficient diet (Group III) for over 6 months developed fatty nodular cirrhosis in 24 out of 28 (Fig. 8). The remaining 4 cases died of hemorrhagic kidney in a relatively short period. In contrast, rats on the 10% xylitol diet (Group IV) showed variable histological changes of the liver. Fatty liver cirrhosis was observed in 8 of 26, 2 cases had fibrous but not so fatty livers. The liver of 3 rats was fatty in diffuse distribution but appeared different from the usual fatty liver observed after such long-term feeding with a choline-deficient diet. Minimum fatty change (centrolobular fat accumulation, but not fibrous change) or almost normal appearance was observed in 13 of 26 rats although the animals

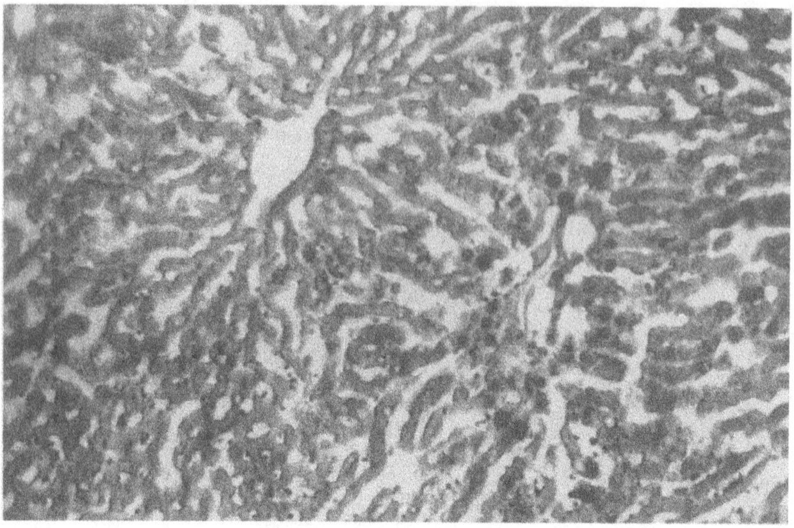

Fig. 7. Liver of rat fed hypolipotropic diet supplemented with choline for 198 days (Oil red O stain, ×140)

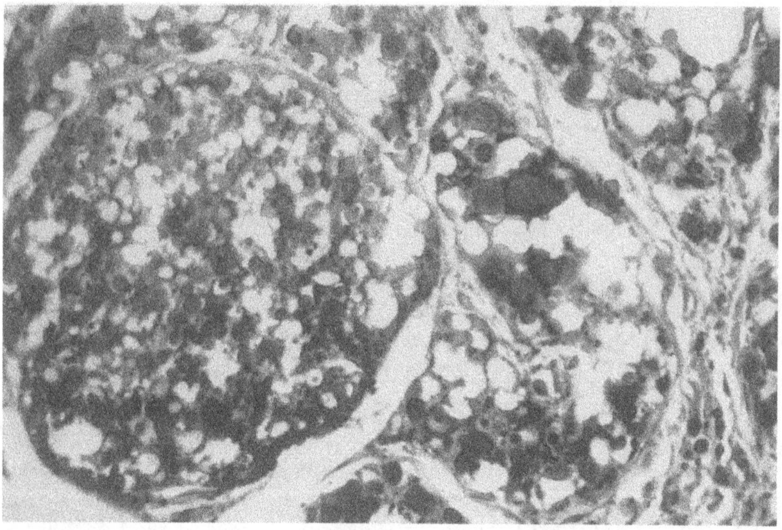

Fig. 8. Liver of rat fed hypolipotropic diet containing only sucrose as carbohydrate for 191 days (Oil red O stain, ×140)

survived and consumed experimental diet for at least 6 months (Fig. 9). The majority of the rats on the 20% xylitol diet (Group V) did not proceed to fatty cirrhosis. In 20 of 29, the rats died of probable mal-

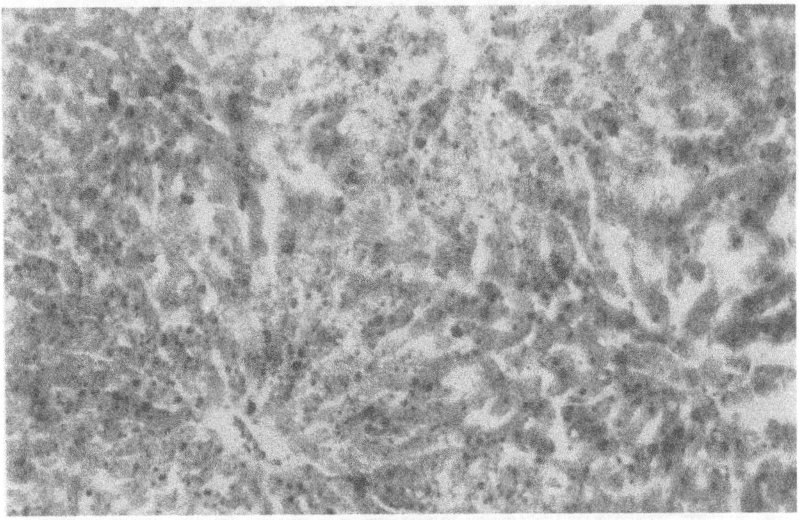

Fig. 9. Liver of rat fed hypolipotropic diet containing 10% xylitol for 195 days (Oil red O stain, ×140)

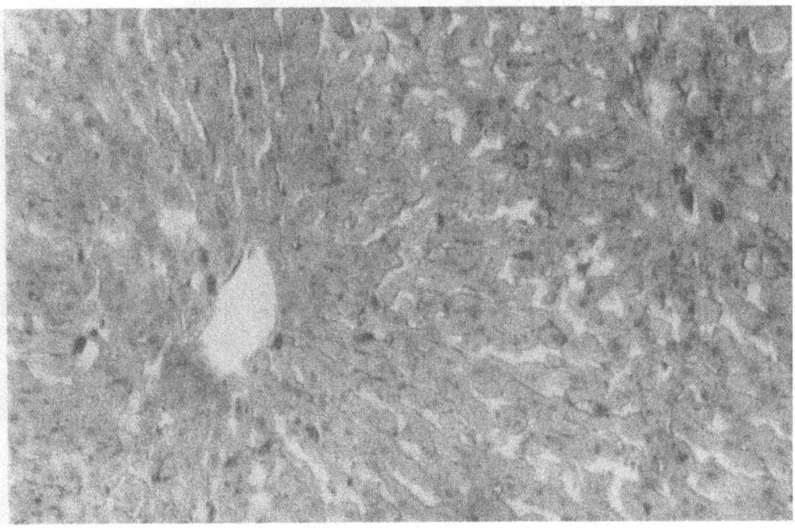

Fig. 10. Liver of rat fed hypolipotropic diet containing 20% xylitol for 124 days (Oil red O stain, ×140)

nutrition. The liver showed normal histology or minimum fat deposition in the centrolobular areas in 18 of 20 rats (Fig. 10). The hepatic cells looked slightly atrophic and were deposited with considerable amount of hemosiderin. The remaining 2 rats developed fatty cirrhosis. The

Table V. *Histology of the liver*

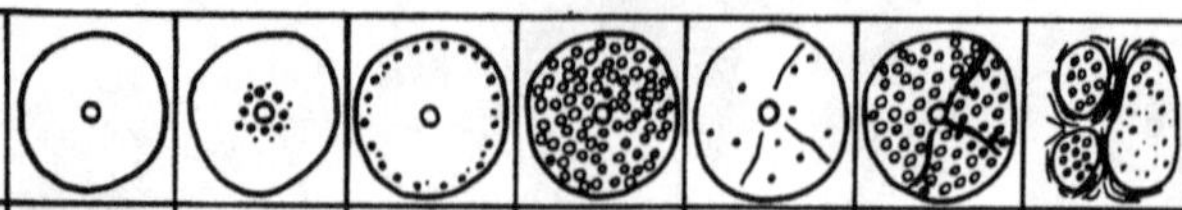

								Total
FoxChow	●●●●● ●●●●● ●●●●● XXXXXX							15 (6)
C. S. Suc.48%	●●●●● ●●●●● ● X		●●●●● ●●●●	●				21 (1)
C. D. Suc.48%		XXX		X		●●● X	●●●●● ●●●●● ●●●●● ●●● XX	21 (7)
C. D. Suc.48% Xyl. 10%	●●●●● ●●● X	●●●	●	●●●	●●	●●●	●●●● X	24 (2)
C. D. Suc.28% Xyl.20%	●●● XXX XXXXX XXXXX	● XXXXX	●●	●	●	● X	X	9 (20)

X Died ● Killed ( ) Died

Table VI. *Total lipid concentrations of livers*

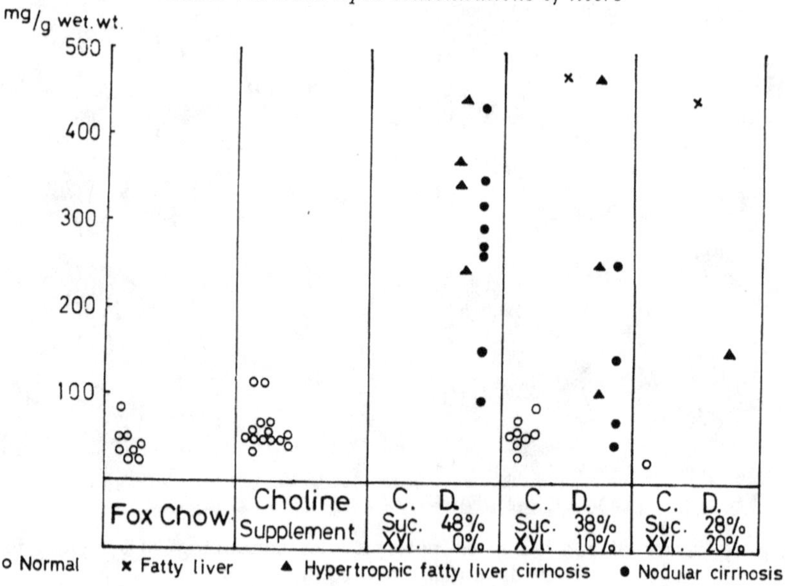

9 rats that survived for periods as long as 6 months were almost identical to the rats of group IV (Table V).

Hepatic Lipids. Total lipid concentration of the livers is shown in Table VI. Hepatic lipids changed in close relation to the pathological

Table VII. *Fatty acid composition of livers*

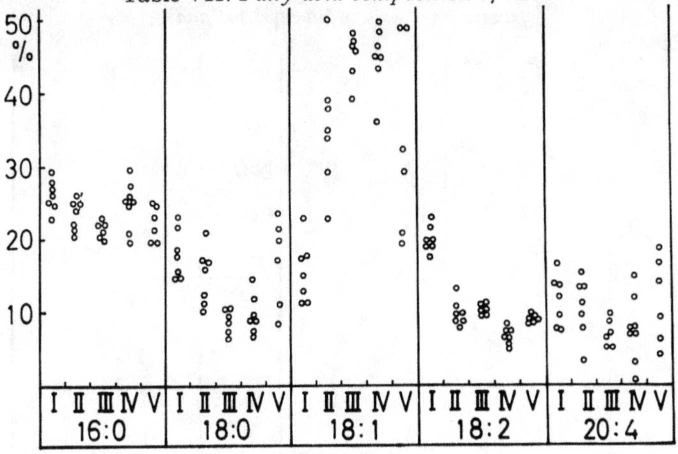

Table VIII. *Fatty acid composition of adipose tissue*

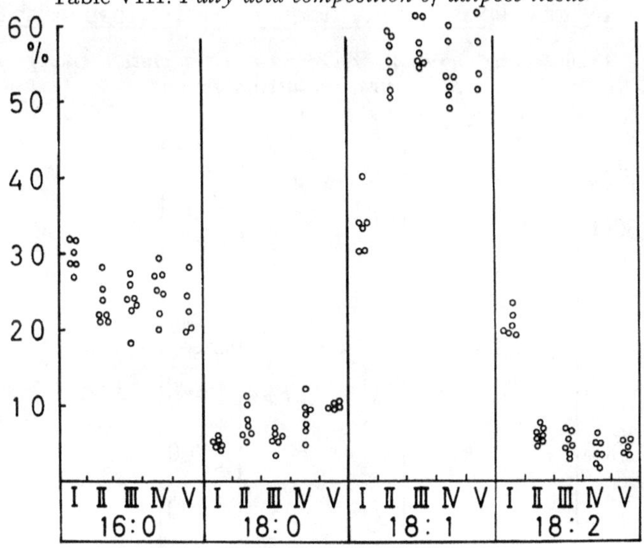

state of the livers, and the difference between Group III and IV was statistically significant ($P<0.01$). The fatty acid composition of liver tissues showed a decrease in stearic acid ($P<0.01$) and an increase in oleic acid ($P<0.05$) in the rats on the standard choline-deficient diet (Group III). The change was prevented by the addition of either choline or xylitol in the diet (Table VII).

Epididymal Fat Pad. The total fatty acid concentration and fatty acid composition of the epididymal fat pad was different in each experimental group (Table VIII). The synthesis of fatty acids in the

Table IX. *Absolute incorporation of* C^{14} *from 1-*C^{14} *acetate into gross lipid fractions of rat epididymal fat pad*

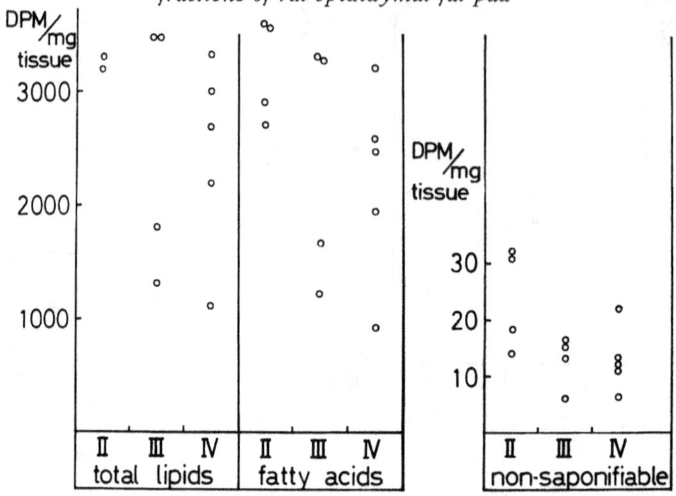

Table X. *Absolute incorporation of* C^{14} *from 1-*C^{14} *glucose into gross lipid fractions of rat epididymal fat pad*

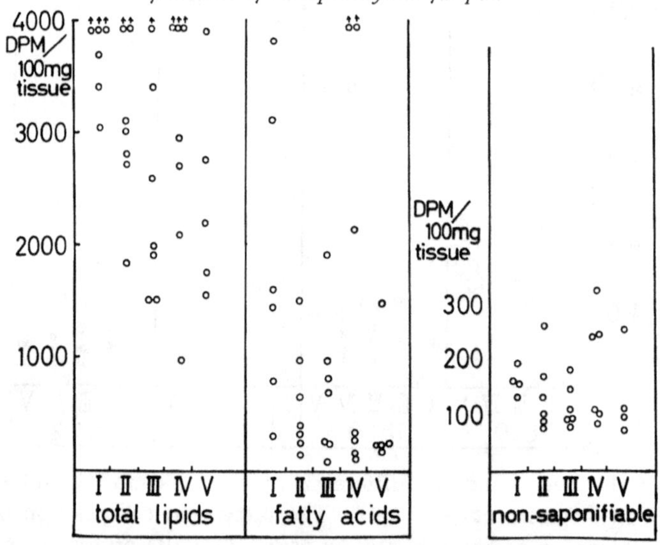

epididymal adipose tissue was calculated from the absolute incorporation of C^{14} from either 1-C^{14}-glucose or 1-C^{14}-acetate into gross lipid fractions. A reduction was detected in the rats on the standard choline-deficient diet, in comparison to the rats maintained on the diet supplemented with choline. Addition of xylitol to the diet did not prevent this reduction (Tables IX and X). The uptake of C^{14} into the individual

Table XI. *Relative incorporation of C¹⁴ from 1-C¹⁴ glucose and 1-C¹⁴ acetate into individual fatty acids of rat adipose tissue in vitro*

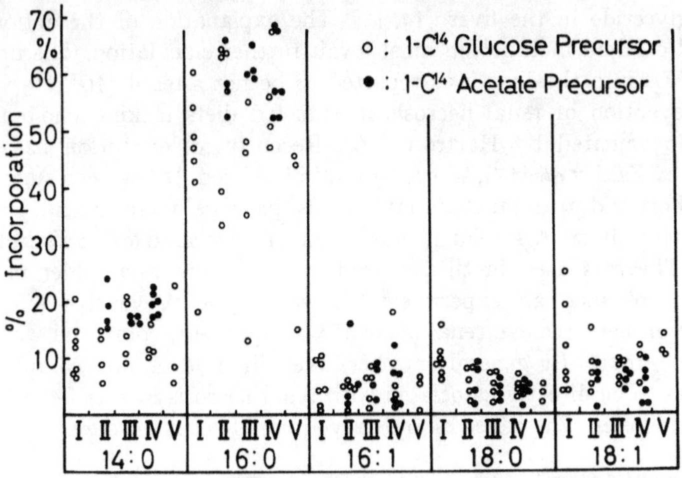

fatty acids showed no significant difference, with or without choline and xylitol in the experimental diet (Table XI).

Discussion

The absorption of xylitol from gut was carefully studied by Schmidt *et al.* (1964) [10] and Bässler *et al.* (1966) [11]). Its rate is dependent on the osmotic pressure induced by xylitol in the intestinal canal, and there is no evidence for active transport of xylitol. The maximum tolerable rate is 10 to 20% xylitol in the experimental diet. Bässler *et al.* [11] also observed an adaptation phenomenon during the dietary experiment. In our study, rats on the experimental diets exhibited diarrhea only at the beginning. The animals tolerated the amounts of xylitol in the experimental diets although the average weekly food intake was reduced to approximately two thirds of the control rats for the first 3 weeks. The nature of feces at this stage was neither loose nor greasy. The rats showed slowness in growth rate and a decrease in survival rate, notably in the rats on the 20% xytilol diet. The animals exhibited malnutrition, in contrast to rats on the 10% xylitol diet.

The rats of both Group IV and V, receiving xylitol, revealed a marked reduction in hepatic lipids, and the fatty acid composition of their livers was similar to that of the rats maintained on the diet supplemented with choline. The effect of xylitol on the synthesis of fatty acids in the epididymal adipose tissue was shown to be negligible.

The pathogenesis of choline-deficiency fatty liver is characterized by increased synthesis of triglyceride [12, 13] and impaired utilization of triglyceride in the liver [14, 15]. The explanation of the actions of xylitol on lipid metabolism must await further elucidation. The uptake of NEFA into the liver was reported to be not altered [16].

Prevention of renal necrosis in rats fed diets lacking choline was well documented by Hartroft [16]. Recently an exception has been noted by Zaki *et al.* [17], where coconut oil is used. It has been suggested that short and medium chain fatty acids, principally lauric acid, exhibit protective effect. A similar phenomenon was observed when xylitol was used. The rats used in this experiment were somewhat elder for the purpose of such an experiment. However, the preventive effect of xylitol on hemorrhagic renal necrosis was apparent, compared with the control group. In general, neither liver lipid level nor composition appears to be directly associated with renal necrosis in rats fed a hypolipotropic diet. The cause of the preventive effect is obscure.

Acknowledgement. Xylitol used in the present investigation was kindly supplied by the Eisai Company, Tokyo, Japan.

Summary

Xylitol (10 and 20%) in a choline-deficient diet was found to modify the production of fatty liver and fatty liver cirrhosis of the rat. The other notable finding was prevention of hemorrhagic renal necrosis with xylitol.

References

1. Patek, A. J., Jr., F. E. Kendall, N. M. de Fritsch, and R. L. Hirsch: Arch. Path. **79**, 494 (1965).
2. — — — —: Arch. Path. **82**, 596 (1966).
3. Zaki, F. G., F. W. Hoffbauer, and F. Grande: Arch. Path. **80**, 323 (1965).
4. — — —: Arch. Path. **81**, 78 (1966).
5. MacDonald, I.: J. Physiol. (Lond.) **162**, 334 (1962).
6. Kuo, P. T., and D. R. Bassett: Ann. intern. Med. **62**, 1199 (1965).
7. Folch, J., M. Lees, and G. H. Sloane Stanley: J. biol. Chem. **226**, 497 (1957).
8. Bjorntorp, P.: Scand. J. clin. Lab. Invest. **12** (Suppl. 52), 1 (1960).
9. Metcalfe, L. D., and A. A. Schmitz: Analyt. Chem. **33**, 363 (1961).
10. Schmidt, B., M. Fingerhut u. K. Lang: Klin. Wschr. **42**, 1073 (1964).
11. Bässler, K. H., G. Stein u. W. Belzer: Biochem. Z. **346**, 171 (1966).
12. Rosenfeld, B., and J. M. Lang: Proc. Canad. Fed. Biol. Sci. **1**, 43 (1958).
13. Yoshida, A., and A. E. Harper: J. biol. Chem. **235**, 2586 (1960).
14. Recknagel, R. O., and A. K. Ghoshal: Fed. Proc. **23**, 243 (1964).
15. Haines, D. S. M., and S. Mookerjea: Canad. J. Biochem. **43**, 507 (1965).
16. Hartroft, W. S.: Brit. J. exp. Path. **29**, 483 (1948).
17. Zaki, F. G., F. W. Hoffbauer, and F. Grande: Arch. Path. **81**, 85 (1966).

Discussion

Dr. Takagi: Have you also tested the effect of xylitol-feeding on the fatty liver caused by the administration of other compounds such as orotic acid?

Dr. Ohta: No, I have not.

Dr. Bässler: I think you should have adapted your rats by increasing the xylitol content of the diet successively up to 20%. We have never lost rats on a diet with 20% xylitol.

Dr. Touster: Do other polyols have a similar effect as xylitol in preventing choline deficient fatty liver?

Dr. Ohta: I have not yet tested that point.

Dr. MacDonald: It is known that bacterial flora can influence the development of choline deficient fatty liver. Do you think this may have played some part in your results?

Dr. Ohta: I have no idea about it.

Dr. Lang: In your groups of animal receiving 20% xylitol mortality was extremely high and death occured in the beginning of the feeding period, but in the 10% group it was very low. What can be the cause? Have you made histopathological examinations? Can it be possible, that diarrhea or exsiccosis or caloric undernutrition was the cause?

Dr. Ohta: I made the histological examinations of dead rats maintained on 20% xylitol. The liver cells were not fatty, but they appeared to possess a vacuolated cytoplasm with increased hemosiderin and a pycnotic nucleus. The electron microscopic examination of such hepatic cells revealed reduced glycogen particles and giant mitochondria. These findings could be interpreted as rats on starvation for several days. In this sense, undernutrition could be a cause of death. The low growth rate also supports this assumption.

Session III

Clinical Use of Pentoses and Pentitols

The Effect of Intravenous Administration of Xylitol Solutions in Normal Persons and in Patients with Liver Diseases and Diabetes mellitus

H. MEHNERT, H. FÖRSTER, and K.-H. DEHMEL

IIIrd Medical Department of the Städtische Krankenhaus München-Schwabing, West Germany

Medizinische Poliklinik, University of München, West Germany

I. Introduction

The biochemical studies of Touster, Hollmann, and co-workers and of Lang, Bässler, and associates showed that xylitol is a metabolite in the glucuronic acid-xylulose-cycle in mammals. Clinical and experimental studies were carried out to state whether these findings could also be applied to human metabolism and whether under various pathological conditions a utilization of xylitol may be expected. For this reason, we have examined normal subjects of various age groups, patients with liver diseases showing a liver damage of various degrees and diabetic persons. In addition, we gave an extremely high dosage of xylitol to normal persons and examined the reactions after fasting and after a high carbohydrate meal.

II. Methods

The experiments were carried out in 6 groups, altogether 60 subjects. 5 groups were given an infusion of 10% xylitol (0.5 g/kg body weight) in 90 minutes; another group received the double amount of xylitol in 30 minutes (20% solution of xylitol, 1.0 g xylitol/kg body weight/30 minutes).

The groups were classified as follows:

Group 1: 10 normal young persons;

Group 2: 8 older persons without any liver damage and without diabetes mellitus;

Group 3: 6 older patients with slight liver damage;

Group 4: 6 older patients with severe liver damage (5 liver cirrhoses, 1 severe epidemic hepatitis);

Group 5: 15 patients with diabetes mellitus of various degrees;

Group 6: 15 young normal persons receiving the above mentioned extremely high dosis of xylitol.

The purpose of this study was to find out if there is any difference in the utilization of xylitol between persons with normal metabolism and patients with liver diseases or diabetes mellitus and also between younger and older persons.

Xylitol in blood was determined according to Rapoport and West and urinary xylitol according to Malaprade. Blood glucose was determined according to Hagedorn-Jensen, blood lactate by the enzymatic method according to Boehringer. Glucose in urine was assessed enzymatically or by polarimetry. The reducing substances — with the exception of glucose — as e.g. xylulose, were determined by Fehling's test.

Results

The results of our experiments are shown in the following 7 figures:

Fig. 1 presents the results of the intravenous infusion of xylitol (0.5 g per kg body weight in 90 minutes) in 10 normal fasting subjects

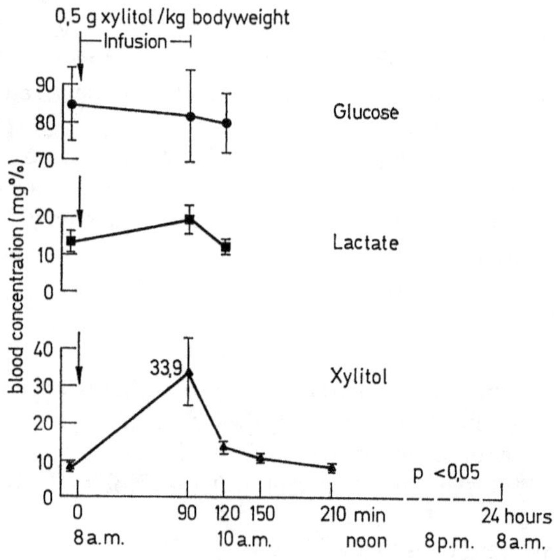

Fig. 1. Results of intravenous infusion of xylitol (0.5 g/kg bodyweight/90 min) in 10 normal fasting persons. Average age: 25 years

Average of infused xylitol: 31.2 g

Average of excreted xylitol in urine during 24 hours after begin of infusion: 2.7 g (= 8.8%)

	8 a.m. — noon	noon — 8 p.m.	8 p.m. — 8 a.m.
Excretion of xylitol in urine	1.8 g = 5.9%	0.5 g = 1.5%	0.4 g = 1.4%
Reducing substances in urine	∅	∅	∅

with an average age of 25 years. The amount of xylitol infused averaged 31.2 g. Within 24 hours after the beginning of the infusion, 2.7 g, i.e. 8.8% were excreted in the urine; 1.8 g were excreted in the first 4 hours, 0.5 g in the following 8 hours and 0.4 g in the following 12 hours. The blood xylitol level was highest at the end of the infusion averaging

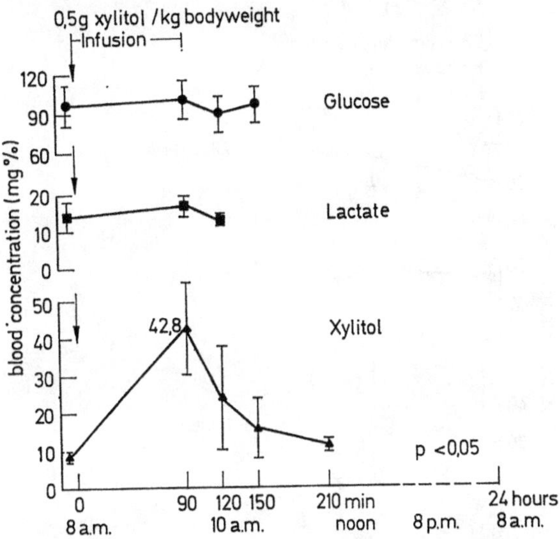

Fig. 2. Results of intravenous infusion of xylitol (0.5 g/kg bodyweight/90 min) in 8 fasting persons without diabetes or liver diseases. Average age: 63 years

Average of infused xylitol: 33 g
Average of excreted xylitol in urine during 24 hours after begin of infusion: 4 g (= 12.1%)

	8 a.m. — noon	noon — 8 p.m.	8 p.m. — 8 a.m.
Excretion of xylitol in urine	2.3 g = 7.0%	1.4 g = 4.2%	0.3 g = 0.9%
Reducing substances in urine	∅	∅	∅

33.9 mg-%. 120 minutes after the end of the infusion, the xylitol level had returned to normal, but already 30 to 60 minutes after the end of the infusion most of the xylitol had disappeared from the blood. The initial slight rise and subsequent fall of lactate may be an indication for the rapid utilization of xylitol in peripheral tissue.

Fig. 2 shows the results of the intravenous administration of xylitol in 8 older fasting persons without diabetes mellitus or any liver damage. The average age of this group was 63 years. The procedure of the test was the same as in group 1. Following the values of the xylitol levels, one can see that at the end of the infusion the xylitol level in blood with 42.8 mg-% has gone higher up than in the younger persons' group. Furthermore, the elimination from blood is also slower and a comparable decline is evident only after 60 to 90 minutes. The slower utilization of xylitol is also indicated by the smaller rise and delayed decline of

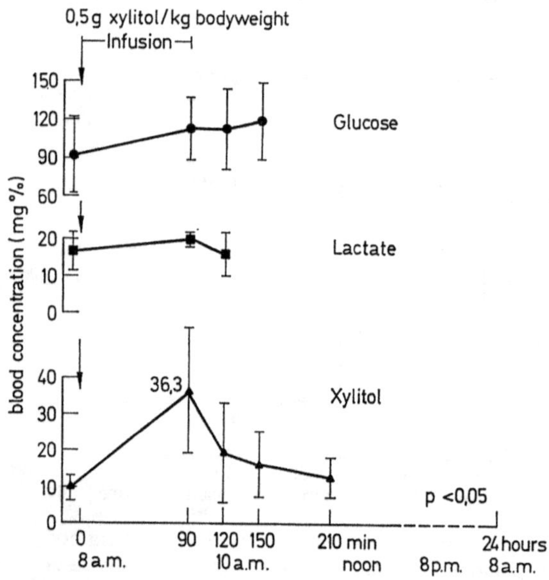

Fig. 3. Results of intravenous infusion of xylitol (0.5 g/kg bodyweight/90 min) in 6 fasting persons with slight liver disease. Average age: 63 years

Average of infused xylitol: 32.5 g

Average of excreted xylitol in urine during 24 hours after begin of infusion: 3.3 g (= 10.1%)

	8 a.m. — noon	noon — 8 p.m.	8 p.m. — 8 a.m.
Excretion of xylitol in urine	1.6 g = 5.0%	0.9 g = 2.7%	0.8 g = 2.4%
Reducing substances in urine	∅	∅	∅

lactate levels. The slightly higher blood glucose level in the older age group is not significantly influenced by the xylitol infusion.

The following figure (Fig. 3) gives the results of the same test in 6 fasting subjects with slight liver damage. In this group, the average age is also 63 years. The highest xylitol level in blood was 36.3 mg-%. As far as rise and elimination of lactate are concerned, there is no difference compared with the same age group of normal persons.

In contrast, patients with severe liver damage show different results (Fig. 4). The six persons tested (5 cirrhoses, 1 severe epidemic hepatitis) had an average age of 52 years. It can be seen that the peak xylitol levels are comparable with those of the older age group (Fig. 2)

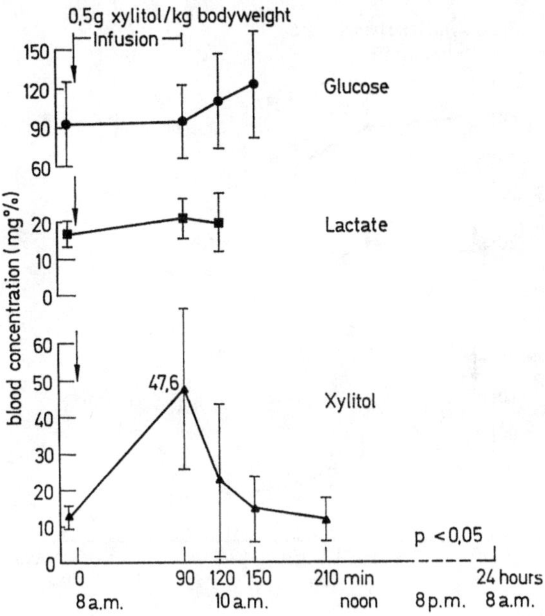

Fig. 4. Results of intravenous infusion of xylitol (0.5 g/kg bodyweight/90 min) in 6 fasting persons with severe liver disease (cirrhosis or acute hepatitis). Average age: 52 years

Average of infused xylitol: 32 g
Average of excreted xylitol in urine during 24 hours after begin of infusion: 2.5 g (= 7.8%)

	8 a.m. — noon	noon — 8 p.m.	8 p.m. — 8 a.m.
Excretion of xylitol in urine	1.7 g = 5.3%	0.5 g = 1.5%	0.3 g = 1.0%
Reducing substances in urine	Ø	Ø	Ø

the mean value being higher than in the young group as indicated in Fig. 1. Blood glucose shows a slight but not significant rise.

Because of the therapeutical consequences, the results of the intravenous administration of xylitol in diabetics are of interest and importance. We have tested 15 fasting persons with diabetes mellitus of an average age of 45 years (Fig. 5). Comparing the results of this group with group 1, the young normal persons, one can see that the xylitol level rises much higher with an average of 43.7 mg-% but declines as rapidly. The blood glucose levels do not rise and even fall during xylitol infusion. This becomes even more striking when compared with the glucose levels of the previous day.

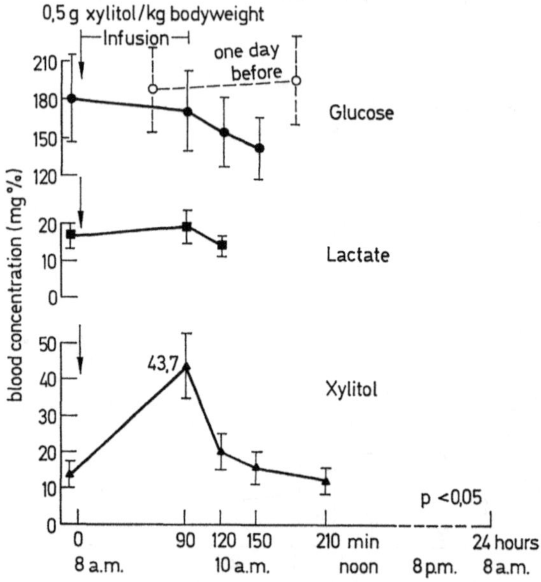

Fig. 5. Results of intravenous infusion of xylitol (0.5 g/kg bodyweight/90 min) in 15 fasting persons with diabetes mellitus of different severity. Average age: 45 years

Average of infused xylitol: 37 g
Average of excreted xylitol in urine during 24 hours after begin of infusion: 3.4 g (= 9.2%)

	8 a.m. — noon	noon — 8 p.m.	8 p.m. — 8 a.m.
Excretion of xylitol in urine	2.5 g = 6.8%	0.5 g = 1.3%	0.4 g = 1.1%
Reducing substances in glucose free urine	Ø	Ø	Ø

The good results concerning the utilization of xylitol encouraged us to perform the following test. We loaded a group of 15 normal fasting subjects of an average age of 25 years with an extremely high quantity of xylitol: instead of 0.5 g/kg in 90 minutes we gave 1.0 g/kg in 30 minutes (Fig. 6). We came to rather high blood levels for xylitol (average: 238 mg-%) and the elimination from blood was also obviously delayed if compared with the elimination of similar doses of glucose.

After 4 to 5 hours, the value returned to normal. There is again no significant increase of blood glucose. It is of interest that inspite of the high load with xylitol the average excretion of xylitol in the urine is not higher than under the previous conditions of the test.

After a carbohydrate rich breakfast when the glycogen stores of the liver should be replenished, there is still a good utilization of xylitol

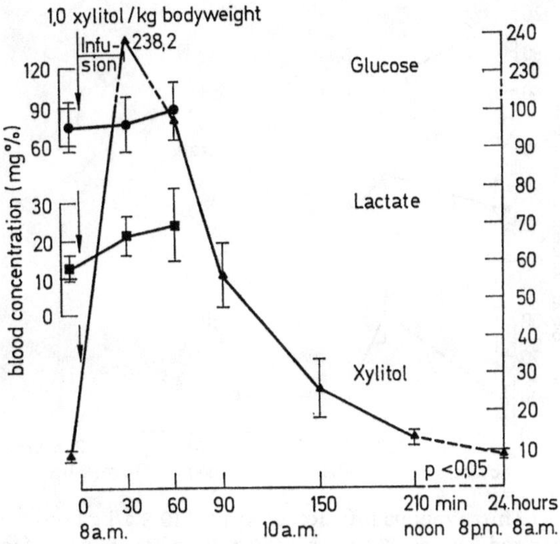

Fig. 6. Results of intravenous infusion of xylitol (1.0 g/kg bodyweight/30 min) in 15 normal fasting persons. Average age: 25 years

Average of infused xylitol: 64.1 g
Average of excreted xylitol in urine during 24 hours after begin of infusion: 5.2 g (= 8.1%)

	8 a.m. — noon	noon — 8 p.m.	8 p.m. — 8 a.m.
Excretion of xylitol in urine	3.7 g = 5.8%	1.2 g = 1.9%	0.3 g = 0.4%
Reducing substances in urine	∅	∅	∅

without any rise of blood glucose (Fig. 7). These experiments were carried out in five normal subjects of an age of 20 to 30 years.

In all test groups, we could not find any striking difference in the excretion of xylitol in the urine which was 8 to 12% of the xylitol given. These results agree with those of Lang who found also a constant xylitol excretion independent of quantity and mode of administration. Even after application of 1 g of xylitol per kg body weight in 30 minutes as xylitol excretion did not rise significantly. We could also in group 6 confirm the results of Bässler *et al.* who even in low blood levels still found some xylitol in the urine. As the tests took place in the morning, the blood levels of xylitol had returned to almost normal in the afternoon and during the night, but there was still some xylitol in the urine.

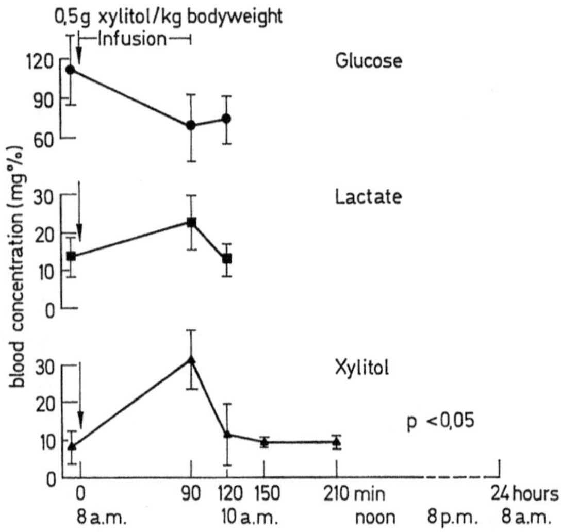

Fig. 7. Results of intravenous infusion of xylitol (0.5 g/kg bodyweight/90 min) in 5 persons aged 20 to 30 years (60 min after a breakfast with 50 g carbohydrates and 10 g fat)

Average of infused xylitol: 34 g
Average of excreted xylitol in urine during 24 hours after begin of infusion: 3.2 g (= 9.3%)

	8 a.m. — noon	noon — 8 p.m.	8 p.m. — 8 a.m.
Excretion of xylitol in urine	7.0%	0.0%	2.3%
Reducing substances in urine	∅	∅	∅

As in all groups there was no significant influence of xylitol infusions upon the blood glucose levels, we do not feel that there is a stimulation of insulin secretion. This is in contrast to the results of other authors and was confirmed by serum insulin determinations after xylitol administration which will be reported by Geser of our group.

I should be mentioned that an excretion of metabolites of the uronic acid pathway (e.g. D-xylulose or L-xylulose) could be ruled out by controlling the urine on its content of reducing substances.

Summary

The utilization of intravenously administered xylitol was tested in 60 subjects. These subjects included younger and older metabolically normal persons as well as patients with diabetes mellitus of various degrees and patients with slight or severe liver damage. It could be shown that there are only slight differences in the utilization of xylitol of the various groups. Older persons as well as persons with liver diseases seem to have a lower utilization of xylitol. This phenomenon can be seen with all carbohydrates (monosaccharides, polyols). It is remarkable that in diabetic persons blood glucose levels did not rise as happens after administration of fructose or sorbitol. However, in insulin-deficient diabetics treated temporarily without insulin, it should be considered that xylitol might contribute to glucose formation and to hyperglycemia. Even when the glycogen stores of the liver are replenished there is no rise in blood glucose levels after xylitol infusion suggesting various possibilities of the organism to utilize this polyol.

References

1. Bässler, K. H., W. Prellwitz, V. Unbehaun u. K. Lang: Xylitstoffwechsel beim Menschen. Zur Frage der Eignung von Xylit als Zucker-Ersatz beim Diabetiker. Klin. Wschr. **40**, 791 (1962).
2. Geser, C.-A., H. Förster, H. Pröls u. H. Mehnert: Zur Frage einer Wirkung von Xylit auf die Insulinsekretion des Menschen. Klin. Wschr. **45**, 851 (1967).
3. Hirata, Y., M. Fujisawa, H. Sato, T. Asano, and S. Katsuki: Blood Glucose and plasma insulin responses to xylitol administered intravenously in dogs. Biochem. biophys. Res. Commun. **24**, 471 (1966).
4. Hollmann, S.: Nicht-glykolytische Stoffwechselwege der Glucose, S. 95. Stuttgart: Georg Thieme 1961.
5. Kuzuya, T., Y. Kanazawa, and K. Kosaka: Plasma insulin response to intravenously administered xylitol in dogs. Metabolism **15**, 1149 (1966).
6. Lang, K.: Xylit als Nahrungskohlenhydrat. Med. u. Ernähr. **4**, 45 (1963).
7. Malaprade, L.: Bull. chim. **43**, 683 (1928). Zit. nach M. Sickert, Med. Inaug.-Diss. München 1961.
8. Mehnert, H., u. H. Förster: Zur Prüfung der Resorption von Zuckern und Zuckeralkoholen bei Mensch und Tier. Vortrag, Kongr. der Dtsch. Ges. f. Verdauungs- und Stoffwechselkrankheiten, Wiesbaden, 1964.
9. — K. Stuhlfauth, B. Mehnert, R. Lausch u. W. Seitz: Vergleichende Untersuchungen zur Resorption von Glucose, Fructose und Sorbit beim Menschen. Klin. Wschr. **37**, 1138 (1959).

10. Stuhlfauth, K., H. Mehnert u. Ch. Pette. Das Verhalten des Glucose-, Fructose- und Sorbitstoffwechsels bei leberkranken und lebergesunden Patienten vor, während und nach intravenöser Infusion von Sorbit. Med. Welt **25**, 1367 (1960).
11. Summa, J. D.: Med. Inaug.-Diss. München 1964.
12. Touster, O., and D. R. D. Shaw: Biochemistry of the acyclic polyols. Rev. Physiol. **42**, 181 (1962).
13. West, C. D., and S. Rapoport: Modification of colorimetric method for determination of mannitol and sorbitol in plasma and urine. Proc. Soc. exp. Biol. (N.Y.) **70**, 141 (1949).

The Effect of Xylitol on Activities of Hepatic Enzymes in Normal and Alloxanized Rats

KIYOSHI FUJISAWA, KIYOSHI OHKAWA, TERUJI TANAKA, and TADAO TAKAHASHI

Department of Medicine, The Jikei University, School of Medicine, Tokyo, Japan

Since the function of the Pentose Phosphate Pathway and the Glucuronate-Xylulose-Pathway in the intermediary metabolism of carbohydrates has been established, xylitol has become a subject of our attention in relation to the coupling of these two pathways and their alternative mechanisms. It also has been applied to clinical use; i.e., its biological role as a new type of energy source, or as a sugar substitute, especially in the diabetic state has been mainly focused.

In this report, we will discuss the effect of xylitol with the emphasis upon how orally administered xylitol modifies carbohydrate metabolism at the level of hepatic enzymes in normal and alloxan diabetic rats.

Materials and Methods

Male Wistar rats around 140 g body weight were fed on a basal diet, CE-2, provided by Japan Clea Co., containing 10% xylitol, while control rats were fed on a diet containing 10% glucose. A diet containing 10% xylitol brought about a normal gain in weight, and caused no diarrhea. Alloxan diabetic rats were induced by a subcutaneous injection of 15 mg of alloxan per 100 gm body weight, and 10 days after alloxanization treatment with 10% xylitol was given to rats whose blood sugar level was elevated steadily over 200 mg/dl. In this series of experiments, rats were sacrificed 6 hours after fasting.

Among the various hepatic enzymes, we have chosen glucose-6-phosphatase (G-6-Pase) [1], fructose-1,6-diphosphatase (FDPase) [2], glucose-6-phosphate dehydrogenase (G-6-PDH) [3], 6-phosphogluconate dehydrogenase (6-PGDH) [3], uridine diphosphate glucose dehydrogenase (UDPG-DH) [4], NAD dependent xylitol dehydrogenase (NAD-Xyl-DH) [5], NADP dependent xylitol dehydrogenase (NADP-Xyl-DH) [5], and L-glutamine hexose-6-phosphate transamidase (H-6-P-Tr) [6, 7]. At the same time the content of hepatic glycogen [8], lipid [9], protein [10] and the activities of transaminases (GOT and GPT) [11] also were determined.

For the determination of activities of UDPG-DH, NAD-Xyl-DH, and NADP-Xyl-DH of the Uronic Acid Pathway, the water extract of acetone dried liver was used as an enzyme source. The activities of such crude enzyme preparation were rather unstable, and reactions were of the first order. Activities were determined, therefore, at the very initial reaction phase in the presence of sufficient substrates. For example, the activity of NADP-Xyl-DH was determined by measuring the initial rate of decrease in E_{340} in the following system;0.3 ml of 1% L-xylulose, 0.3 ml of 0.5 M Tris buffer, pH 8.1, 0.3 ml of 0.08 M $MgCl_2$, 0.1 ml of NADPH (10 mg/ml), 0.3 ml of enzyme extract and water up to 3.0 ml.

Results

The Effect of Xylitol on Normal Rat Liver. First, the effect of orally administered xylitol on normal rat liver was studied at the enzymatic level. In the preliminary experiment, we failed to get any information except the increase of hepatic glycogen, since the rats were sacrificed for the experiment at 7-day intervals. Therefore closer observations were made at 2-day intervals.

As shown in the table, the ratio of liver weight to body weight as well as protein content in the liver remained constant during xylitol treatment, and liver glycogen increased gradually up to 50% above normal on the 21st day, while lipids in the liver increased only 10%. On the other hand, a significant rise or fall in the activity of hepatic enzymes was found temporarily in the Uronic Acid Pathway as well as in the Pentose Phosphate Pathway, while levels of transaminases and NADP-Xyl-DH, which catalyze the conversion of xylitol to L-xylulose, were almost constant in the course of treatment. These results are summarized in Fig. 1. The activities of NAD-Xyl-DH, UDPG-DH and NADP-Xyl-DH, which catalyze the conversion of L-xylulose to xylitol, rose rapidly soon after the start of xylitol administration and reached maximum level after 5-day treatment, followed, after a short lag phase, by an increase in the activities of G-6-PDH and 6-PGDH. By contrast, a rapid fall in the activities of G-6-Pase and FDPase was found soon after xylitol treatment.

These significant changes in enzyme activities returned to normal levels within one to two weeks, despite a continued administration of xylitol, whereas glycogen content and H-6-P-Tr activity increased steadily up to 50% above normal.

It is important to note that all of the liver enzymes do not show the same pattern of response to xylitol, and three groups of enzymes can be differentiated clearly. First, a group of enzymes participating in the Uronic Acid Pathway, such as UDPG-DH, NAD-Xyl-DH and

Table I. *Changes in constituents and enzyme levels in the liver during xylitol treatment*

$n=6$	Control	Days treated with xylitol				
		3rd	5th	7th	14th	21th
Liver weight/100 gm (body weight)	3.7 ± 0.43	3.8 ± 0.12	3.7 ± 0.07	3.9 ± 0.07	3.6 ± 0.28	3.8 ± 0.04
Protein (mg/gm)	98 ± 9.4	99 ± 12.0	105 ± 5.7	99 ± 11.3	94 ± 9.1	92 ± 9.1
Lipid (mg/gm)	52 ± 1.8	53 ± 3.4	54 ± 3.2	55 ± 3.2	57 ± 3.0	58 ± 2.9
Glycogen (mg/gm)	20 ± 3.0	21 ± 3.1	26 ± 2.2	27 ± 3.0	32 ± 3.1	32 ± 2.9
NAD-Xyl-DH (mM/gm/hr)	30 ± 3.5	45 ± 2.0	38 ± 9.6	33 ± 1.9	32 ± 0.7	32 ± 3.4
NADP-Xyl-DH (mM/gm/hr) (L-xylulose→xylitol)	56 ± 7.5	72 ± 2.2	81 ± 9.1	74 ± 1.0	69 ± 3.4	67 ± 5.7
NADP-Xyl-DH (mM/gm/hr) (xylitol→L-xylulose)	13 ± 1.9	13 ± 1.3	13 ± 0.5	12 ± 1.4	14 ± 1.3	13 ± 1.7
UDPG-DH (mM/gm/hr)	14 ± 1.6	19 ± 2.9	23 ± 2.7	17 ± 3.9	15 ± 1.0	13 ± 1.7
G-6-PDH (mM/gm/hr)	30 ± 2.8	32 ± 4.9	47 ± 3.5	43 ± 4.0	40 ± 5.4	36 ± 4.1
6-PGDH (mM/gm/hr)	72 ± 5.0	75 ± 3.5	100 ± 4.9	83 ± 9.9	78 ± 6.9	72 ± 4.0
G-6-Pase (PμM/gm/hr)	750 ± 43.6	574 ± 54.2	638 ± 97.6	680 ± 74.8	698 ± 85.3	702 ± 47.6
FDPase (PμM/gm/hr)	274 ± 17.9	193 ± 12.3	215 ± 13.4	237 ± 18.6	232 ± 23.6	235 ± 15.0
H-6-P-Tr (μM/gm/hr)	1.2 ± 0.85	1.3 ± 0.10	1.4 ± 0.53	1.6 ± 0.10	1.9 ± 0.48	1.9 ± 0.49
GOT (U)	99 ± 13.3	105 ± 5.2	102 ± 5.2	98 ± 9.5	101 ± 5.8	104 ± 14.4
GPT (U)	55 ± 5.7	57 ± 8.9	57 ± 11.6	54 ± 9.2	55 ± 5.5	58 ± 4.6

NADP-Xyl-DH, shows a rapid increase in activity, as shown in Fig. 1. These changes suggest that the administration of xylitol stimulates the Uronic Acid Pathway. Second, a group of enzymes participating in the Glycolytic Pathway, such as G-6-Pase and FDPase, shows a rapid fall, suggesting the maintenance of the concentration of glucose-6-phosphate at a certain level. Then after a short lag period, the third group of

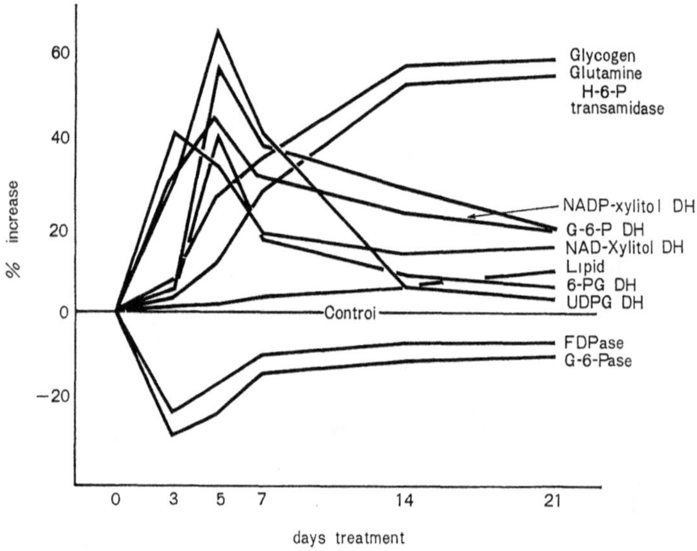

Fig. 1. Changes in enzyme activities in rat liver by treatment with xylitol

enzymes participating in Pentose Phosphate Pathway is induced and their activities elevate, suggesting that the intermediates of xylitol metabolism (for example, xylulose-5-phosphate) formed in the Uronic Acid Pathway may stimulate the Pentose Phosphate Pathway. These initial changes in enzyme activities return to normal levels within two weeks and the metabolic phase in liver enters a new steady state. Furthermore, a steady increase of glycogen and the activity of hexosamine phosphate-forming enzyme also suggests an extension of the sequential induction caused by xylitol.

It is probably excessive to attempt to learn everything about the metabolic effect of xylitol in liver only by the observation on hepatic enzyme levels. There is, however, one important reaction sequence which is markedly affected by xylitol: the accelerated Uronic Acid Pathway of direct xylitol metabolism is coupled with the activated Pentose Phosphate Pathway of the NADPH-generating system, which is also required for the complete cycle of the Uronic Acid Pathway. Conse-

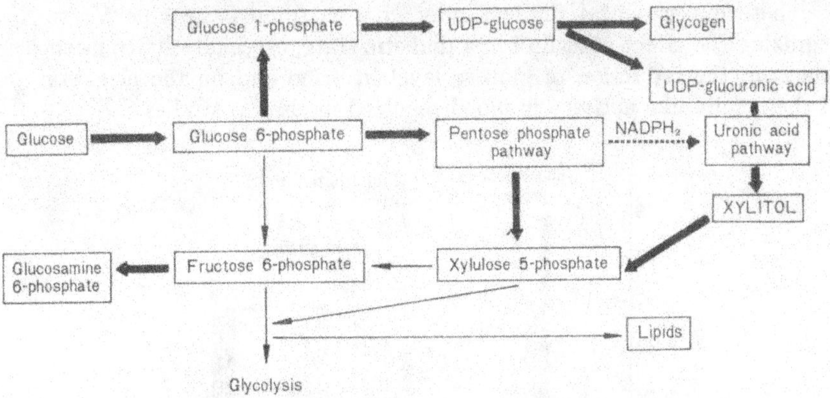

Fig. 2. A possible metabolic flow in G-6-P utilizing system stimulated by xylitol

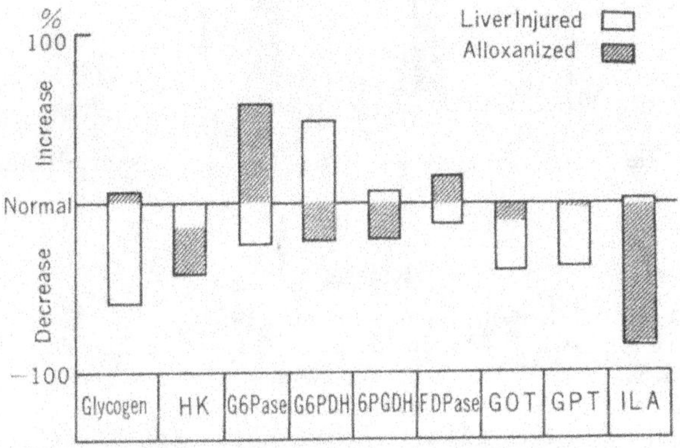

Fig. 3. Activities of hepatic enzymes in alloxanized rat and liver injured rat

quently, xylitol may accelerate the utilization of glucose in liver as shown in Fig. 2.

The Effect of Xylitol on Alloxan Diabetic Rat Liver. The changes in enzyme levels in alloxan diabetic rat liver are characterized by the decrease of hexokinase, G-6-PDH, and 6-PGDH, and the increase of G-6-Pase and FDPase, as shown in Fig. 3. Then the effect of xylitol on alloxan diabetic liver was studied at the enzyme level by feeding alloxan diabetic rats on a diet containing 10% xylitol. The increased activities of G-6-Pase and FDPase and the decreased activities of G-6-PDH and 6-PGDH, found in diabetic rat liver 2 weeks after alloxanization, return to normal levels by xylitol treatment; and it should be emphasized that the phases of the improvement coincided well with those found in normal rat liver treated with xylitol (Fig. 4).

These experimental findings may give us some basic clue for recognizing the effect of xylitol in a diabetic state, especially its transient effect on the reduction of glucose level in urine and on the elevation of the insulin-like activity in blood reported in clinical studies.

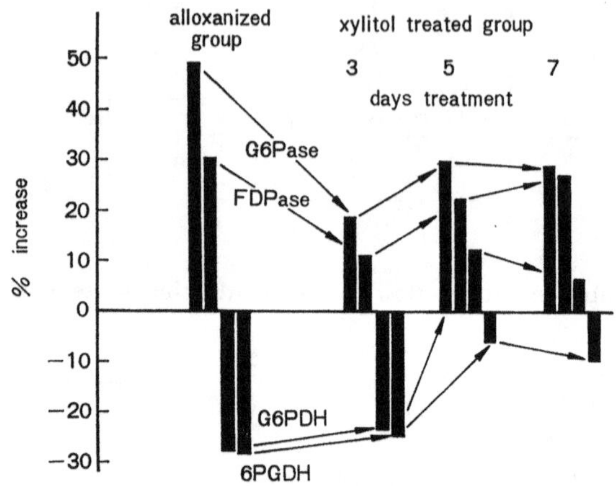

Fig. 4. The effect of xylitol on some hepatic enzymes in alloxanized rats

References

1. Swanson, N. A.: J. biol. Chem. **184**, 647 (1950).
2. Pogell, B. M., and R. W. McGilvery: J. biol. Chem. **197**, 293 (1952).
3. Glock, G. E., and P. McLean: Biochem. J. **56**, 171 (1954).
4. Strominger, J. L.: J. biol. Chem. **224**, 79 (1957).
5. Hollmann, S., and O. Touster: J. biol. Chem. **225**, 87 (1957).
6. Pogell, B. M., and M. G. Bosa: J. biol. Chem. **228**, 701 (1957).
7. Blix, G.: Acta chem. scand. **2**, 467 (1948).
8. Good, C. A., H. Kamer, and M. Somogyi: J. biol. Chem. **100**, 485 (1933).
9. Folch, J., M. Lees, and G. S. Stanley: J. biol. Chem. **226**, 479 (1957).
10. Gornall, A. G., C. J. Bardawill, and M. M. David: J. biol. Chem. **177**, 751 (1949).
11. Reitman, S., and S. Frankel: Amer. J. clin. Path. **28**, 56 (1957).

Comment

Some Effects of Xylitol on Carbohydrate Metabolism

H. Ishii and K. Sambe

Department of Internal Medicine, Keio University, School of Medicine, Tokyo, Japan

Hollmann and Touster [1] discovered that the polyalcohol xylitol was a regular intermediary product of carbohydrate metabolism (Glucuronate-Xylulose Cycle). There have been many reports since then on the effects of xylitol on carbohydrate metabolism in animals and human beings. Lang and his co-workers [2, 3] could demonstrate the almost complete utilization of xylitol given orally or intravenously to normal or alloxan diabetic rats and normal persons or diabetic patients, but the physiological significance of the Glucuronate-Xylulose Cycle still remains obscure.

We performed some experiments in normal adults and in patients with diabetes mellitus or cirrhosis of the liver in order to study metabolic changes induced by administration of xylitol. Animal experiments were also performed in order to clarify some aspects of metabolism of xylitol in carbon tetrachloride (CCl_4)-poisoned or alloxanized rats.

Methods

Clinical experiments were performed in 5 control, 10 diabetic and 10 cirrhotic patients. All subjects were studied at postabsorptive state. 500 ml of a 5 per cent solution of xylitol or glucose was administered intravenously over 60 minutes. Blood samples were obtained by vein-puncture for the determination of blood glucose, lactate, and pyruvate levels before, during and after the administration of xylitol or glucose. 500 ml of a 5 per cent solution of xylitol was given parenterally over 90 minutes in another series of experiments. Blood samples were obtained for the analysis of blood glucose, xylitol, and serum electrolytes before, during and after the administration of xylitol.

Urine specimens were also collected in order to determine the xylitol level in urine for 5 hours and for the next 19 hours after xylitol administration.

Blood glucose was determined by the Somogyi-Nelson's method. Blood pyruvate and lactate levels were determined enzymatically. The concentration of blood and urine xylitol was determined by the chromotropic acid method of Otsu and Kinoshita.

Serum potassium was measured flamephotometrically and serum phosphate was determined by Fiske-Subbarow's method.

Animal experiments were carried out by using male rats of Wistar strain, weighing about 150 g. All animals were fed a standard laboratory diet and were deprived of food for 24 hours before they were killed by decapitation. Alloxanized rats were produced by administering alloxan intraperitoneally (50 mg per kg body weight). CCl_4-poisoned rats were produced by injecting subcutaneously 0.2 ml of 50 per cent CCl_4 (in olive oil) per 100 g body weight twice a week up to 14 days.

The *in vitro* study using labelled sugars was carried out according to the procedure of Isselbacher [4].

The liver tissue to be studied was weighed and homogenized in 4 volumes of 0.15 M potassium chloride in 0.05 M nicotinamide with a Potter-Elvehjehm glass homogenizer. The incubation medium constisted of the following: 0.05 ml (0.5 μc) of either glucose-u-C^{14} or xylitol-u-C^{14}; 0.65 ml of 0.1 M potassium chloride; 0.2 M potassium phosphate buffer (pH 7.4); and 0.5 ml of homogenate.

The incubations were made in modified Warburg flasks with a center well, and the reactions were carried out for 60 min at 37° C in a shaking incubator. At the end of incubation 1 ml of Hyamine-10X was injected into the center well of each flask to trap the produced CO_2. The reactions were stopped by injecting 0.2 ml of 10 N H_2SO_4 into the main part of the flasks. The flasks were then shaken in the incubator for an additional 30 minutes to collect the CO_2. Hyamine-10X was transferred to counting bottles and the radioactivity measured in a liquid scintillation spectrometer.

The specific activity of the C^{14} used was 3.0 mc per mMole for glucose-u-C^{14} and 3.2 mc per mMole for xylitol-u-C^{14}. For the study on liver glycogen, another group of CCl_4-poisoned rats fasted for 24 hours received oral administration of 2.0 ml of 50 per cent glucose solution and, after approximately 15 minutes, intraperitoneal administration of 10 μc of xylitol-u-C^{14} or glucose-u-C^{14} per 100 g body weight. The animals were killed by decapitation 3 hours after injection, and the livers were extirpated and plunged into 30 per cent potassium hydroxide.

Glycogen was isolated by the method of Good, Kramer, and Somogyi [5]. Radioactivity in isolated glycogen was measured in a liquid scintillation spectrometer.

Results

Clinical Experiments. As seen in Fig. 1, blood glucose levels after xylitol administration almost did not change for 2 hours in controls,

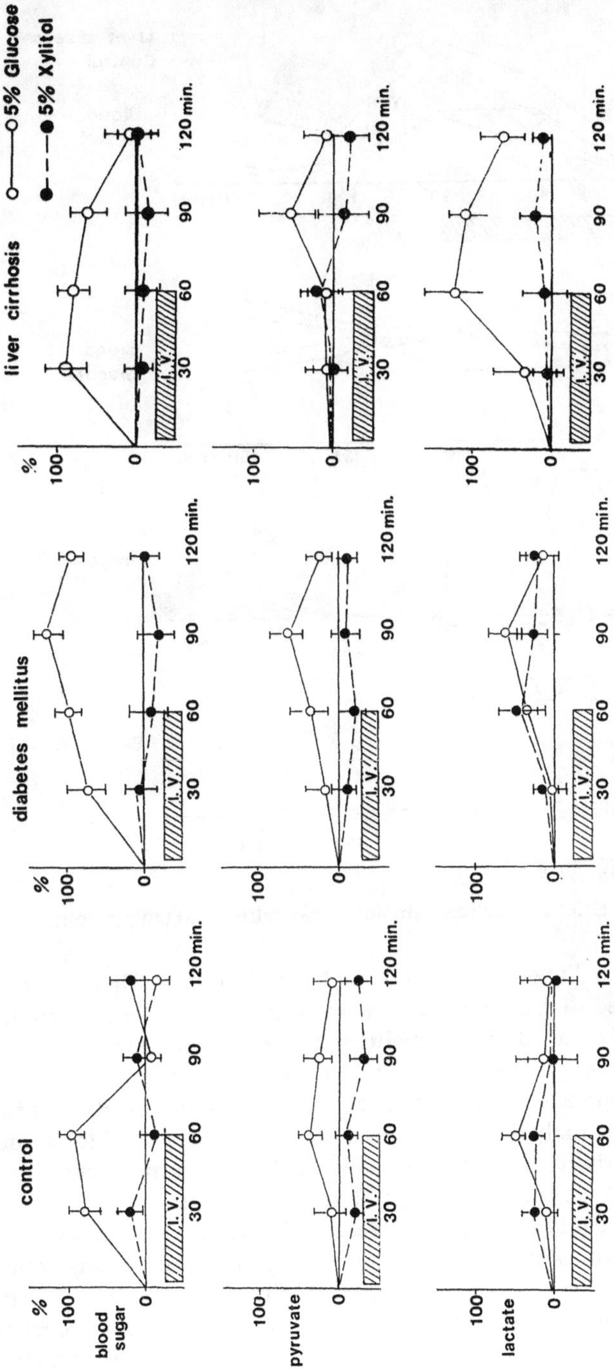

Fig. 1. Change of the level of blood sugar, pyruvate and lactate during and after intravenous injection of glucose or xylitol

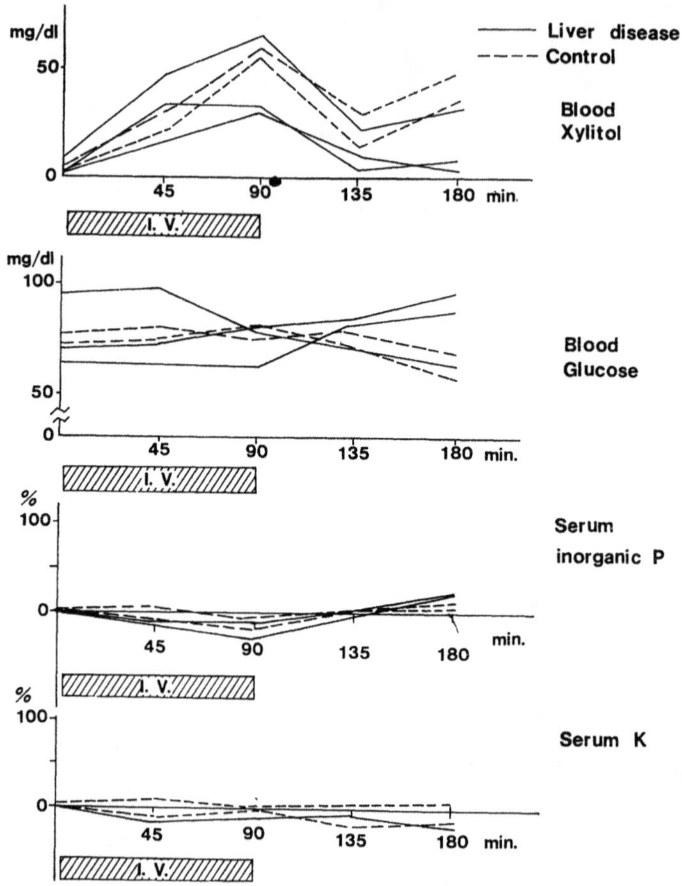

Fig. 2. Effect of xylitol infusion (5% xylitol 500 cm^3/90 min)

in patients with diabetes mellitus or cirrhosis of the liver. Blood pyruvate levels decreased slightly following xylitol administration both in controls and in patients with diabetes mellitus. There was no significant change of blood pyruvate in patients with cirrhosis of the liver.

Blood lactate levels rose at the end of xylitol administration both in controls and in diabetic patients and 30 minutes after xylitol administration in cirrhotic patients. As shown in Fig. 2, plasma xylitol levels reached a peak at the end of xylitol administration.

Serum inorganic phosphate decreased slightly at the end of the infusion but serum potassium levels showed no significant change during or after infusion. In Table I, it was seen that the mean urinary excretion of xylitol for 24 hours is 2,940 g in controls, and 3,749 g in the cirrhotic patients, which is about 12 per cent and 15 per cent of administered

Table I. *Xylitol excretion in urine following intravenous administration of xylitol solution to man (5% xylitol 500 cm³/90 min)*

	Urinary excretion of xylitol		
	5 hr-urine	19 hr-urine	24 hr-urine
Control [3]	2.835 g (11.34%)	0.105 g (0.42%)	2.940 g (11.76%)
Liver diseases [5]	3.152 g (12.60%)	0.597 g (2.38%)	3.749 g (14.98%)

Table II. *Recovery of C^{14} in CO_2 produced from xylitol-u-C^{14} and glucose-u-C^{14} by liver homogenate of CCl_4-treated rats*

	Animals No.	CO_2 production by liver incubated with	
		glucose-u-C^{14}	xylitol-u-C^{14}
Control	10	3.78 ± 1.04%	7.45 ± 2.20%
Time after first CCl_4 injection:			
22 hr	6	0.66 ± 0.07	0.71 ± 0.27
10 days	8	0.55 ± 0.46	0.59 ± 0.17
21 days	5	1.65 ± 0.86	6.64 ± 2.56
35 days	4	2.44 ± 0.48	7.70 ± 2.20

The data are expressed as % of administered C^{14} recovered in CO_2/100 mg of liver (mean ± S.D.).

doses, respectively. The difference between the two groups was not statistically significant.

Animal Experiments. Table II shows values for the per cent of the administered doses of xylitol-u-C^{14} or glucose-u-C^{14} recovered in CO_2 per 100 mg of the liver. CO_2 production from xylitol-u-C^{14} was more marked than that from glucose-u-C^{14} in controls. There was markedly decreased CO_2 production both from xylitol-u-C^{14} and glucose-u-C^{14} in CCl_4-poisoned rats.

However, CO_2 production was almost recovered to the control level 21 days after the first CCl_4 injection in the xylitol-u-C^{14} group. Table III indicates that CO_2 production from xylitol-u-C^{14} was not decreased in alloxanized rat liver. On the contrary, CO_2 production from glucose-u-C^{14} was markedly reduced.

Finally, Fig. 3 shows the data on the recovery of C^{14} in the glycogen of control and CCl_4-poisoned rat liver after administration of glucose-u-C^{14} and xylitol-u-C^{14}. Xylitol was found to be an efficient precursor of glycogen in the rat.

Table III. *Recovery of C^{14} in CO_2 produced from xylitol-u-C^{14} and glucose-u-C^{14} by liver homogenate of alloxan diabetic rats*

	Animals No.	CO_2 production by liver incubated with	
		glucose-u-C^{14}	xylitol-u-C^{14}
Control rat liver	10	3.78 ± 1.04%	7.45 ± 2.20%
Alloxan diabetes rat liver	10	1.70 ± 0.95% [a]	6.77 ± 3.28%

[a] $P<0.01$.

The data are expressed as % of administered C^{14} recovered in CO_2/100 mg of liver (mean ± S.D.).

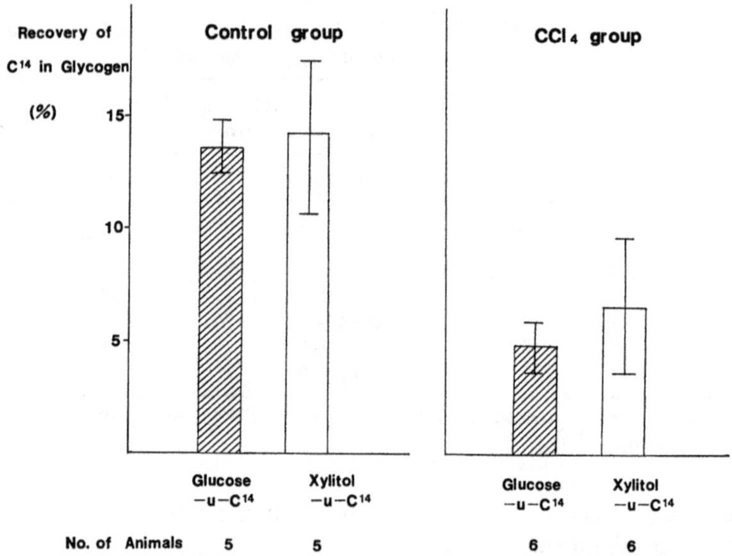

Fig. 3. Incorporation of xylitol-u-C^{14} and glucose-u-C^{14} into rat liver glycogen

Summary

We performed experiments in normal adults and in patients with diabetes mellitus or cirrhosis of the liver in order to investigate whether xylitol given intravenously could be utilized under these conditions. Good utilization of xylitol was indicated by a rapid decrease of the blood xylitol level following its administration, and by the small urinary loss.

There was no marked elevation of blood glucose or pyruvate levels during or following xylitol administration. Animal experiments using labelled sugars revealed that xylitol is incorporated easily into liver glycogen and oxidized to CO_2 in normal and CCl_4-poisoned or alloxanized rats.

Acknowledgement. The authors wish to express their thanks to Eisai Co., Ltd., Tokyo, Japan, for a gift of xylitol-u-C^{14}.

References

1. Hollmann, S., and O. Touster: J. Amer. chem. Soc. **78**, 3544 (1956).
2. Lang, K.: Med. u. Ernähr. **2**, 45 (1963).
3. Prellwitz, W., u. K. H. Bässler: Klin. Wschr. **41**, 196 (1963).
4. Isselbacher, K. J., and W. A. Jones: Gastroenterology **46**, 424 (1964).
5. Good, C. A., H. Kramer, and M. Somogyi: J. biol. Chem. **100**, 485 (1933).

Discussion

Dr. Fukui: We also investigated the effect of xylitol on hepatic enzymes in normal and alloxanized rats, and we obtained results similar to yours. However there is one difference. That is the level of G-6-Pase activity. Following our studies, the level of G-6-Pase activity showed higher elevation after xylitol administration in alloxanized rats, although that of yours decreased.

Dr. Fujisawa: We repeated these experiments and we have never found elevation of G-6-Pase activity after xylitol treatment.

Dr. Touster: How were various liver fractions prepared for analysis? Since many of the enzymes are greatly influenced by metabolites, it would be advisable to check the ones which changed significantly by employing specific subcellular fractions. For example, if the UDPG dehydrogenase increase is valid, then it means that xylitol is exerting a positive feedback action of considerable interest.

Dr. Fujisawa: We have not employed any specific subcellular fractions for analysis. Actually we prepared two kinds of liver fractions for these experiments: the supernatant of whole homogenate and water extract of acetone dried liver.

In the case of the determination of UDPG-dehydrogenase in whole homogenate, we found a continuous gradual increase in activity up to 100% above normal on the 21 days of xylitol treatment in normal rats. When the activity was determined using water extract of acetone dried liver as the enzyme source, the maximal elevation of the activity of 60% above normal was found on the 5th day and it returned normal, as shown in the paper. I think it should be analysed in the future on dialysed liver homogenates to avoid probable contamination of xylitol metabolism intermediates.

Dr. Horecker: I would like to know if you studied the levels of any glycolytic enzymes, such as aldolase or triosephosphate dehydrogenase, in these experiments.

Dr. Fujisawa: We didn't analyse the levels of glycolytic enzymes other than G-6-Pase and FDPase.

Clinical Application of Xylitol in Diabetes

SHOICHI YAMAGATA, YOSHIO GOTO, AKIRA OHNEDA, MORITOSHI ANZAI, SATOSHI KAWASHIMA, JIN KIKUCHI, MAMORU CHIBA, YOSHISUKE MARUHAMA, YUICHI YAMAUCHI, and TAKAYOSHI TOYOTA

Department of Internal Medicine, Tohoku University School of Medicine, Sendai, Japan

Introduction

As a result of the pioneering experiments of Dr. Lang and his coworkers [1], and the subsequent clinical investigations of Prellwitz and Bässler [2], and Mehnert, Summa, and Förster [3], it became apparent that the disappearance of intravenously administered xylitol was essentially the same in normal and diabetic subjects. The production of xylitol in large quantities since 1964, made it possible to explore the use of this pentitol in clinical conditions, particularly diabetes mellitus, in which animal studies suggested a possible beneficial effect.

Our studies were concerned with the acute and chronic effects of xylitol administration on carbohydrate and fat metabolism in normal adults and diabetic patients.

Experiments

Acute Intravenous Administration. First we examined the effects of acute intravenous administration of xylitol on blood sugar, pyruvate, and lactate, and on plasma free fatty acids and insulin levels in normal adults and diabetic patients. The studies were carried out on six healthy subjects and eleven mild adult-onset diabetics.

A blood specimen was drawn after an overnight fast and 30 g of xylitol in 10% solution was injected intravenously over 90 minutes. Blood specimens were then collected at 90, 120, and 180 minutes after the start of the injection. Urine specimens were collected at 90 and 180 minutes. In normal subjects, the mean plasma xylitol level was 4.1 ± 1.2 mg-% under fasting, 40.1 ± 5.5 mg-% at 90 minutes, and 6.8 ± 1.9 mg-% at 180 minutes. The urinary excretion of xylitol over the three hour period was 2.5 ± 0.5 g. These results suggest that 90% of the injected xylitol was utilized in the body. In diabetic patients, the plasma xylitol level was 7.7 ± 0.9 mg-% under fasting, 65.2 ± 6.3 mg-% at 90 minutes and 8.7 ± 1.5 mg-% at 180 minutes. The urinary excretion over the three hour period was 2.2 ± 0.6 g. These results indicate that intravenously administered xylitol was utilized as well in diabetic patients as in normal subjects.

Changes in blood metabolites after the xylitol injection were described in our previous report [4]. Xylitol infusion produced no change of blood sugar in normal subjects and a slight increase in diabetic patients. No change in blood pyruvate was seen in either group. Blood lactate showed a significant increase in the diabetic group but only a slight increase in the normal group. Blood ketone bodies decreased in both groups following xylitol administration. Plasma free fatty acid levels decreased markedly in normal subjects as well as in diabetic patients. Plasma insulin, as estimated by the method of Morgan and Lazarow, increased slightly in 5 of the 6 normal subjects and in 4 of the 9 diabetic patients.

Kusumoto *et al.* [5] observed an increase in blood sugar after xylitol injection, but Horiuchi *et al.* [6], Yoshitoshi *et al.* [7] and Ishii *et al.* [8] reported no change or a decrease. The findings on the change of blood pyruvate have varied from laboratory to laboratory. Most of the workers have reported a slight fall of pyruvate after xylitol injection [5, 9]. Fujisawa *et al.* [10] have reported a prompt and transient fall in the dog following rapid injection of xylitol. There is agreement that a rise in blood lactate is seen after xylitol injection. Masuda *et al.* [9] reported an increase in plasma α-ketoglutarate and Kusumoto *et al.* [5] and Ishii *et al.* [8] observed a fall of serum inorganic phosphate.

An antiketogenic action of xylitol was reported by Bässler [11] and this has been confirmed by Horiuchi *et al.* [6] and in our laboratory [4]. Since this effect is very important with regard to the value of xylitol administration in clinical medicine, it was studied using liver slices as the experimental system. Fig. 1 shows the effect of xylitol added to the medium on the production of ketone bodies by liver slices from normal and alloxan diabetic rats.

The decrease in plasma free fatty acids following xylitol administration has been observed by Horiuchi *et al.* [6], Masuda *et al.* [9] and Yoshitoshi *et al.* [7]. Fig. 2 shows an inhibitory effect of xylitol and glucose on free fatty acid mobilization from rat adipose tissue *in vitro*. In normal rats glucose and xylitol showed similar effects, but in diabetic rats xylitol exerted more inhibitory effect than glucose.

In the next experiments, a 10 per cent solution of xylitol was given as drinking water to normal and alloxan diabetic rats for one month. The oxidation of xylitol-U-C^{14} to CO_2 and the incorporation of xylitol-U-C^{14} into triglyceride in epididymal adipose tissue of normal and diabetic rats *in vitro* was significantly greater in the rats taking a xylitol solution than in the control rats (Fig. 3).

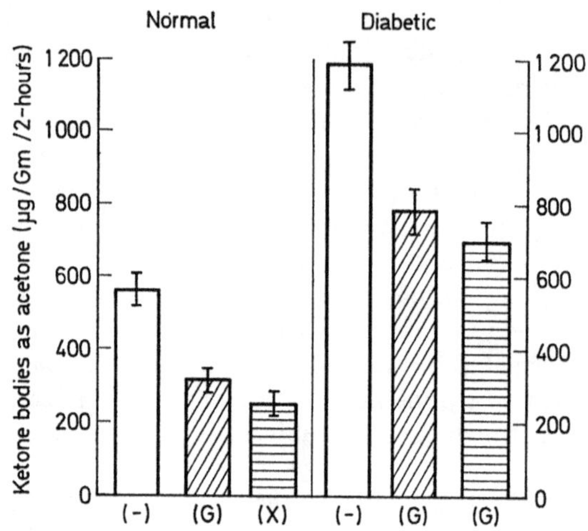

Fig. 1. Effects of glucose and xylitol *in vitro* on the production of ketone bodies by liver slices from normal and diabetic rats. Twenty four-hour fasting rats were used. The liver slice was incubated in Krebs-Ringer phosphate buffer at 37° C for 2-hours. Gas phase was 95% O_2 and 5% CO_2. The medium contained no carbohydrate-(—), 5 mM glucose-(G) and 5 mM xylitol-(X).

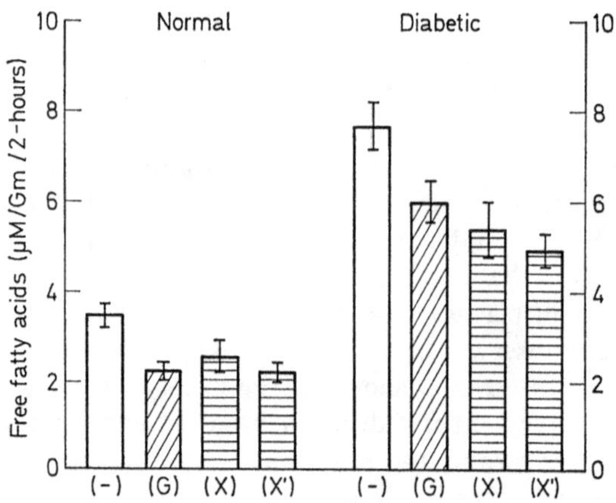

Fig. 2. Effects of glucose and xylitol *in vitro* on fatty acid mobilization from adipose tissue of 24-hour fasting normal and diabetic rats. Epididymal adipose tissue was incubated in Krebs-Ringer bicarbonate buffer containing 5% bovine albumin at 37° C for 2-hours with gas phase of 95% O_2 and 5% CO_2. The medium contained no carbohydrate-(—), 5 mM glucose-(G) and 5 mM xylitol-(X). The rat group previously given xylitol solution for a month was expressed by (X'). Net release of free fatty acids was calculated

To summarize the results of the acute experiments, xylitol infusion produces a decrease in plasma free fatty acids, serum inorganic phosphate and blood ketone bodies in normal subjects as well as in diabetic patients.

The changes obtained are very similar to those seen with insulin administration or after glucose administration. Hence, the results indicate that xylitol is metabolized via the Pentose Phosphate Cycle to

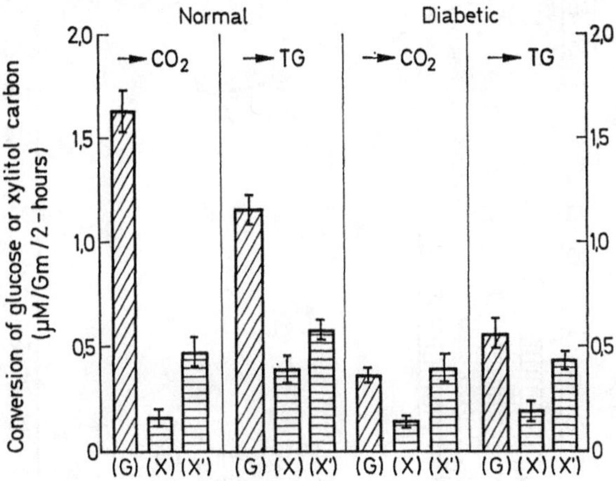

Fig. 3. Incorporation of glucose-U-C^{14} and xylitol-U-C^{14} *in vitro* into CO_2 and triglyceride (TG) in adipose tissue from fed normal and diabetic rats. Epididymal adipose tissue was incubated in Krebs-Ringer bicarbonate buffer containing 5% bovine albumin at 37° C for 2-hours with gas phase of 95% O_2 and 5% CO_2. The medium contained glucose-U-C^{14}-(G) and xylitol-U-C^{14}-(X). Non-labelled glucose or xylitol was added to the medium to make a final concentration of 5 mM. The rat group given xylitol solution for a month was expressed by (X′)

fructose-6-phosphate or glyceroaldehyde-3-phosphate, resulting in a decrease in plasma free fatty acids and a slight increase in blood sugar. In addition it may stimulate the secretion of insulin from the pancreas, which might account for a slight decrease in blood sugar, in plasma free fatty acids and in serum inorganic phosphate.

Chronic Intravenous Administration. Twelve mildly diabetic inpatients were used for this study. Their diet was kept constant both in calories and composition. Thirty grams of xylitol were infused intravenously every morning for 7 days. The diet and dosage of insulin or oral drugs which they were recieving were kept unchanged during and after this infusion period.

The fasting blood sugar did not change significantly. However, 4 cases indicated the marked decrease of urine sugar excretion during the study period. No significant change was seen in serum triglycerides, total cholesterol, serum electrolytes or ketone bodies. The noteworthy finding in this series was that urine excretion decreased during the infusion period in spite of the extra calories provided by the 30 g of xylitol. Those findings are illustrated in Figs. 4 and 5. Similar results

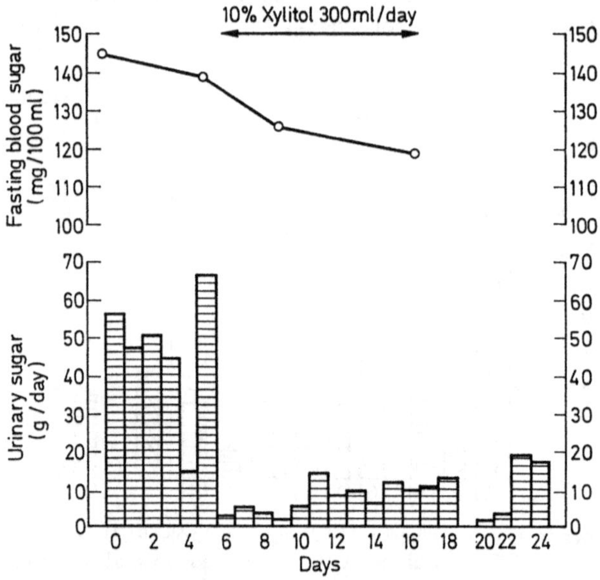

Fig. 4. Effect of xylitol infusion on blood sugar and urine sugar in a male diabetic. 30 g of xylitol in 10% solution was injected intravenously every morning for one week

were also observed by Yoshitoshi and his co-workers [7]. The explanation for this effect is not readily apparent. Fukui found that the metabolism of alloxan diabetic rat liver is characterized by an increase in FDPase and a decrease in glucokinase, and is changed to a normal pattern when rats are fed diets containing xylitol [12]. This observation may provide an explanation for our finding.

Xylitol Infusion in Diabetic Ketosis. The results stated above suggest that one of the best indications of xylitol administration to diabetic patients is diabetic ketosis. Several cases of diabetic coma and precoma were treated with xylitol with or without additional insulin therapy. The following two mild ketogenic cases were most noteworthy.

Case 1 (Fig. 6) was a 29 year-old female. Her diabetes was found at 19 years old and she has been treated by insulin, but with interruption of therapy for one to three months on several occasion. She became pregnant and developed hyperemesis which lasted two weeks. For this reason she stopped her insulin injections for one week. When

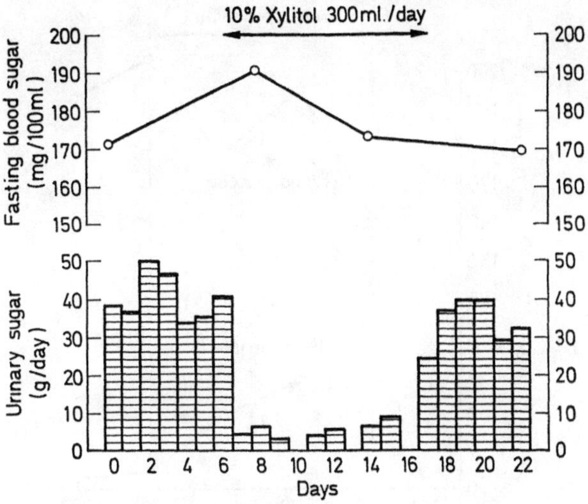

Fig. 5. Effect of xylitol infusion on blood sugar and urine sugar in a female diabetic. 30 g of xylitol in 10% solution was infused intravenously every morning for one week

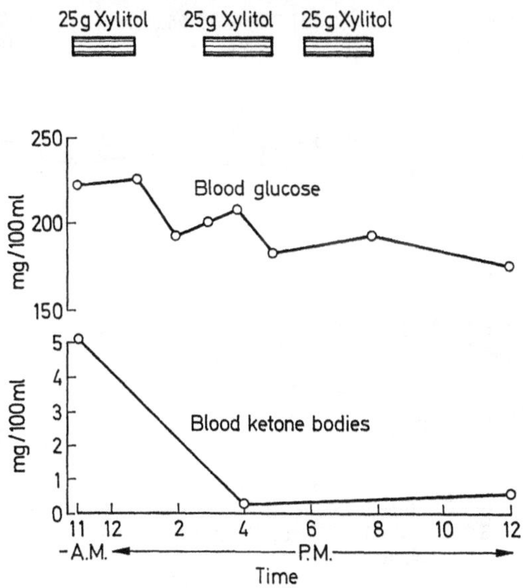

Fig. 6. Case 1, 29 year-old female. Xylitol in 5% solution was administered intravenously

she was admitted to our clinic, she was semicomatous with a blood sugar of 220 mg-%. Her urine gave a ++++ test for ketone bodies. Xylitol solution was infused immediately. Her blood ketone bodies decreased

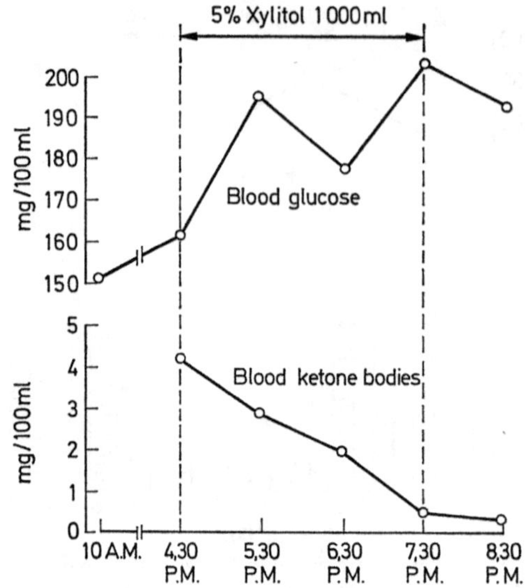

Fig. 7. Case 2, 51 year-old female

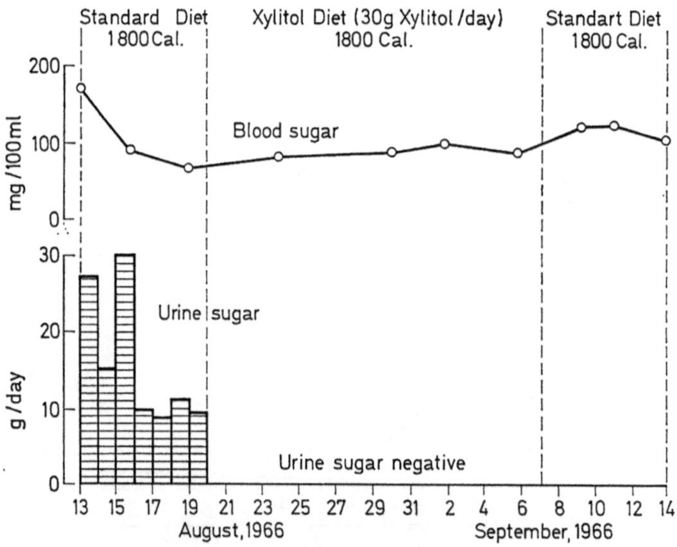

Fig. 8. Effect of oral administration of xylitol on blood sugar and urine sugar in a male diabetic

from 5.1 to 0.2 mg-% within four hours after the start of the infusion. No insulin had been given. In this case, the blood sugar did not increase after xylitol infusion, but decreased gradually.

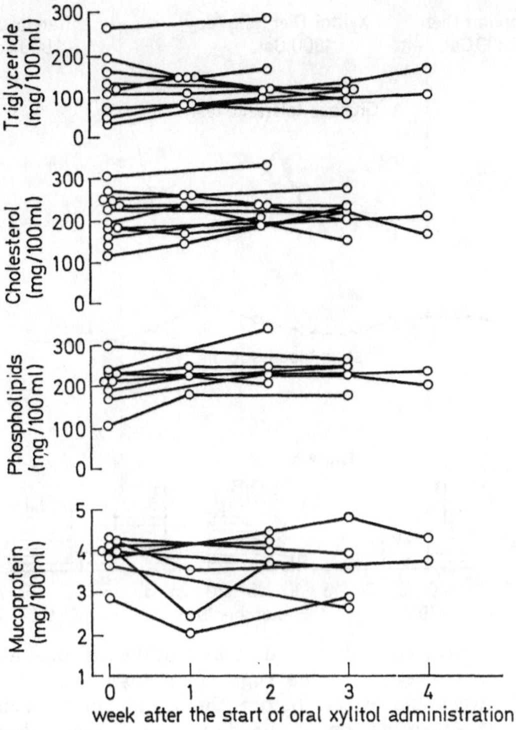

Fig. 9. Effect of oral administration of xylitol (30 g/day) on serum triglyceride, cholesterol, phospholipids and mucoprotein in insulin independent diabetics

Case 2 (Fig. 7) was a 51 year-old female. Her diabetes was found at the age of 38 and since that time she had received 20 units of NPH or Lente insulin daily. However, she stopped her injections for one week during a trip. She was semicomatous on admission. Her fasting blood sugar was 151 mg-% and urine gave a ++++ test for ketone bodies. Four hours after the start of xylitol infusion her blood ketone body level decreased from 4.2 to 0.6 mg-% and the urine gave a negative test for ketone bodies. In this case blood sugar increased after the xylitol infusion. The report of Horiuchi [6], of Fukiyama [13] and ofothers [14] confirmed the prompt disappearance of ketosis after xylitol injection.

Oral Administration of Xylitol. The oral administration of xylitol as a sweetener for diabetic patients was attemped by Mellingkoff in 1961 [15]. Our study was made to see whether the oral administration of xylitol would have an effect similar to that of intravenous infusion. Thirty grams of xylitol was given orally to 12 stable diabetics. The total caloric intake and the amount of carbohydrate were kept constant,

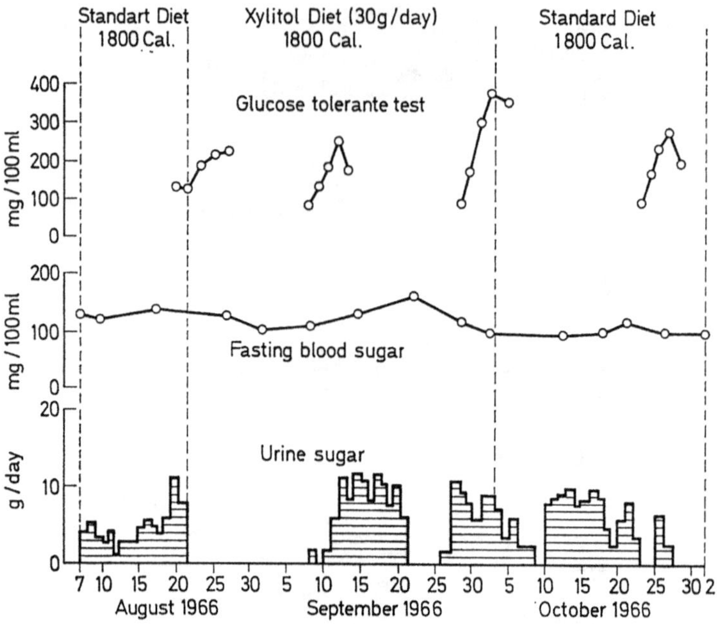

Fig. 10. Effect of oral administration of xylitol on fasting blood sugar, oral glucose tolerance curve and urine sugar in a male diabetic. 50 g glucose tolerance tests were performed before the initiation of xylitol diet, and during and at the termination of the diet, and 4 weeks thereafter. Blood sugar was estimated before and at 30, 60, 90, and 120 minutes after the glucose loading

but 30 g of carbohydrate was replaced with xylitol. This diet was given for 2—6 weeks and the changes in fasting blood sugar, urine sugar, serum lipids, serum mucoprotein, GOT and GPT were studied before, during and after the xylitol diet period.

In three of the twelve cases, urinary sugar excretion disappeared from the first day of xylitol administration as seen in Fig. 8. No change was observed in serum cholesterol, phospholipids, triglyceride or mucoprotein, as shown in Fig. 9. The glucose tolerance test was performed during the xylitol diet period. In some cases, a slight impairment of glucose tolerance was observed, which improved after termination of the xylitol diet (Fig. 10). A decrease in serum GOT and GPT was seen in most of the cases although it was not statistically significant. Serum mucoprotein as well as serum total cholesterol, triglyceride and phospholipids showed no apparent change following oral xylitol administration for four weeks. In a few cases, diarrhea appeared for a few days after the start of the xylitol diet, but it stopped without discontinuation of the diet.

Summary and Conclusion

Our results and those of other investigators show that xylitol is an important tool for the treatment of diabetic ketosis, and can be a good energy source when the patients need a parenteral supply of carbohydrate in certain circumstances such as during surgery. The present clinical studies also suggest the possibility that xylitol can be used as a substitute for sucrose, which does not affect plasma lipids or mucoprotein.

References

1. Kieckebuch, W., W. Griem u. K. Lang: Klin. Wschr. **39**, 447 (1961).
2. Prellwitz, W., u. K. H. Bässler: Klin. Wschr. **41**, 196 (1963).
3. Mehnert, H., J. D. Summa u. H. Förster: Klin. Wschr. **42**, 382 (1964).
4. Yamagata, S., Y. Goto, A. Ohneda, M. Anzai, S. Kawashima, M. Chiba, Y. Maruhama, and Y. Yamauchi: Lancet **1965 II**, 918.
5. Kusumoto, T., Y. Ohnishi, T. Takano, H. Kono, and T. Yamabuki: Tonyobyo (J. Jap. Diab. Soc.) **8**, 107 (1965).
6. Horiuchi, A., K. Tanaka, S. Kitamura, Y. Niki, and H. Kawara: Tonyobyo **9**, 266 (1966).
7. Yoshitoshi, Y., and T. Kaneko: Shinryo **20**, 597 (1967).
8. Ishii, H., N. Takahashi, M. Oda, T. Shishido, M. Tsuchiya, and K. Sambe: Tonyobyo **11**, 36 (1968).
9. Masuda, M., K. Kuzuya, K. Mori, J. Irino, S. Tomikawa, C. Nakano, and T. Kanetsuna: Tonyobyo **8**, 107 (1965).
10. Fujisawa, M., and Y. Hirata: Tonyobyo **9**, 131 (1966).
11. Bässler, K. H., u. G. Dreiss: Klin. Wschr. **41**, 593 (1963).
12. Fukui, S., T. Tsujii, Y. Fukuoka, M. Tamura, S. Sato, Y. Matsuoka, O. Makiura, Y. Funada, J. Sakamoto, and S. Fukuda: J. Jap. Soc. Int. Med. **56**, 805 (1967).
13. Fukiyama, K., S. Fujisawa, and Y. Hirata: Naika **16**, 587 (1965).
14. Akazawa, Y., T. Iehara, H. Mori, B. Hashimoto, and T. Tanaka: Iryo **21**, 891 (1967).
15. Mellinghoff, C. H.: Klin. Wschr. **39**, 447 (1961).

Comment

Difference in Metabolic Effects of D-Ribose and Xylitol in the Diabetic State

SHOICHI YAMAGATA, YOSHIO GOTO, AKIRA OHNEDA, MORITOSHI ANZAI, SATOSHI KAWASHIMA, JIN KIKUCHI, MAMORU CHIBA, YOSHISUKE MARUHAMA, YUICHI YAMAUCHI, TAKAYOSHI TOYOTA, NORIO FUKUHARA, and SHINICHIRO SATO

Department of Internal Medicine, Tohoku University School of Medicine, Sendai, Japan

It has been shown that the uptake and utilization of administered D-ribose or xylitol is considerably rapid in man [1, 2]. D-Ribose and xylitol are known to be converted to ribose-5-phosphate and xylulose-5-phosphate, respectively, in the Pentose Phosphate Pathway. Taking these facts, coulped with our previous studies on xylitol [3], the present study was performed to see the difference in metabolic effects of D-ribose and xylitol in normal subjects and in diabetic patients. A few animal experiments were carried out to observe the effects of pentose and pentitol on the metabolism of isolated tissues.

Clinical study

Thirty grams of D-ribose or xylitol in 10% solution were infused intravenously over 90 minutes in six healthy volunteers and nine mild diabetics after an overnight fast. The blood specimens were collected before and at 5, 15, 90, 120 and 180 minutes after the start of the infusion. Blood levels of glucose, ketone bodies, pyruvate and lactate and plasma levels of free fatty acids and immunoreactive insulin were measured in these specimens.

In normal subjects, D-ribose infusion produced a decrease in blood glucose and an increase in plasma free fatty acids. Xylitol produced no change in blood glucose but a significant decrease in plasma free fatty acids ($p<0.05$). No apparent change in plasma insulin was observed with ribose infusion but a significant increase was produced by xylitol infusion ($p<0.01$). These changes are shown in Fig. 1. Changes in blood ketone bodies, pyruvate and lactate are shown in Fig. 2. During D-ribose infusion, no definite changes in blood ketone bodies, pyruvate

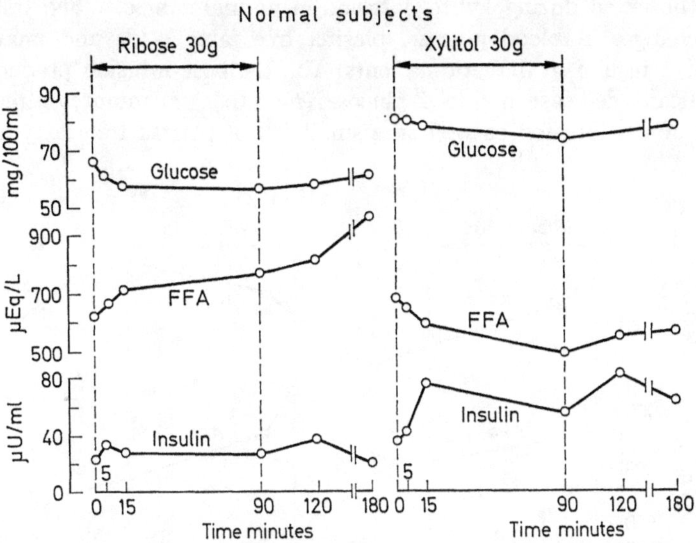

Fig. 1. Changes in blood glucose, plasma free fatty acids and immunoreactive insulin during and after intravenous infusion of D-ribose or xylitol in normal subjects

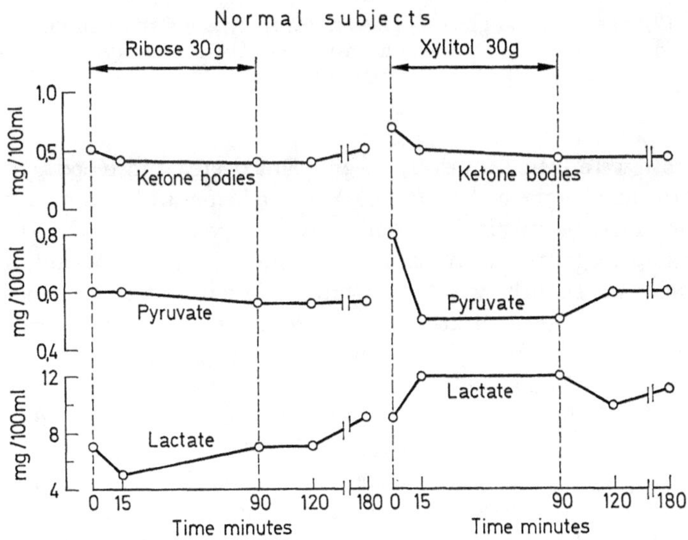

Fig. 2. Changes in blood ketone bodies, pyruvate and lactate during and after intravenous infusion of D-ribose or xylitol in normal subjects

were observed during xylitol infusion in normal subjects. Fig. 3 shows the changes in blood glucose, plasma free fatty acids and immunoreactive insulin in diabetic patients. The D-ribose infusion produced a significant decrease of blood glucose ($p<0.01$) 120 minutes after the start of the infusion as well as a small fall of plasma free fatty acids.

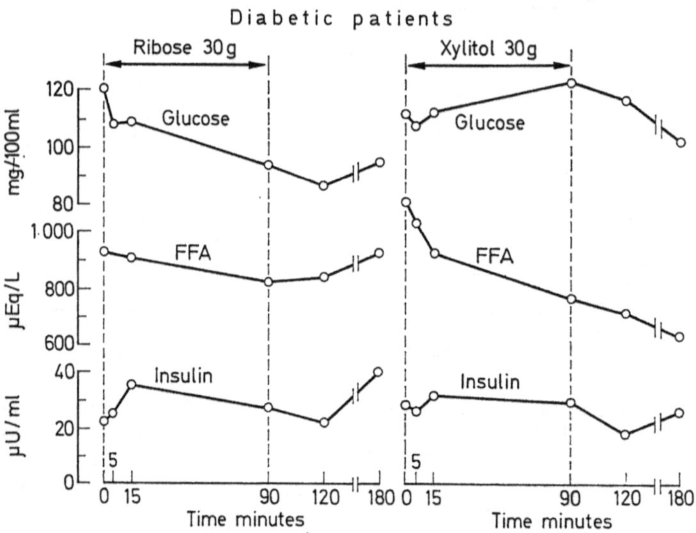

Fig. 3. Changes in blood glucose, plasma free fatty acids and immunoreactive insulin during and after intravenous infusion of D-ribose or xylitol in diabetic patients

or lactate were seen. However, a significant decrease in blood pyruvate ($p<0.01$) and ketone bodies ($p<0.01$) and an increase of lactate ($p<0.05$) A slight increase in plasma insulin was observed during the D-ribose infusion, though the increase was not statistically significant. Blood glucose roses slightly with the xylitol infusion and this change was different from that in normal subjects. Plasma free fatty acids decreased markedly during the xylitol administration ($p<0.01$), as described previously [3]. The change in blood ketone bodies, pyruvate or lactate during the D-ribose or xylitol infusion showed the same tendency as observed in normal subjects, as seen in Fig. 4.

A comparison was made between the metabolic effects of D-ribose and xylitol in patients with mild diabetic ketosis. A sharp drop in blood glucose was induced by the infusion of 100 gm of D-ribose and the same dose of xylitol infusion produced a rapid disappearance of hyperketonemia in these cases, as shown in Fig. 5.

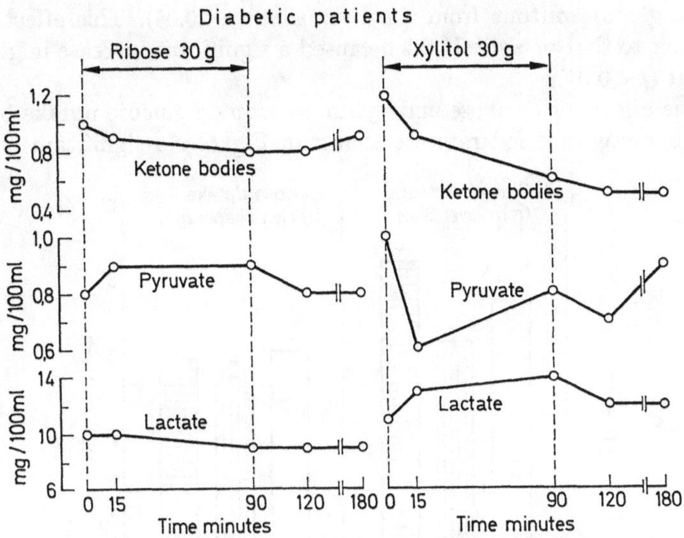

Fig. 4. Changes in blood ketone bodies, pyruvate and lactate during and after intravenous infusion of D-ribose or xylitol in diabetic patients

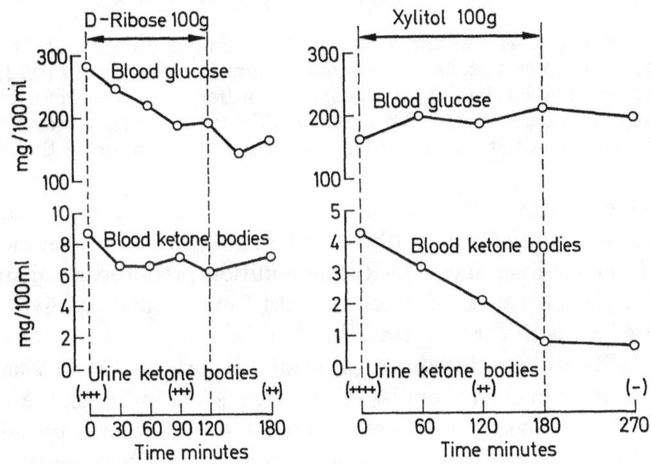

Fig. 5. Changes in blood glucose and ketone bodies as produced by intravenous infusion of D-ribose or xylitol in patients with mild diabetic ketosis

Animal Experiments

In vitro experiments were carried out using male albino rats of the Wistar strain. The effects of the addition of D-ribose and xylitol in the medium on glucose output from liver slices of fed rats are shown in Fig. 6. The addition of D-ribose *in vitro* produced a significant decrease in the glucose output from the liver slice ($p<0.05$). This effect is in contrast to that of xylitol which caused a significant increase in glucose output ($p<0.01$).

The effects of D-ribose and xylitol *in vitro* on glucose uptake by the hemidiaphragm of fed rats are shown in Fig. 6. No significant change

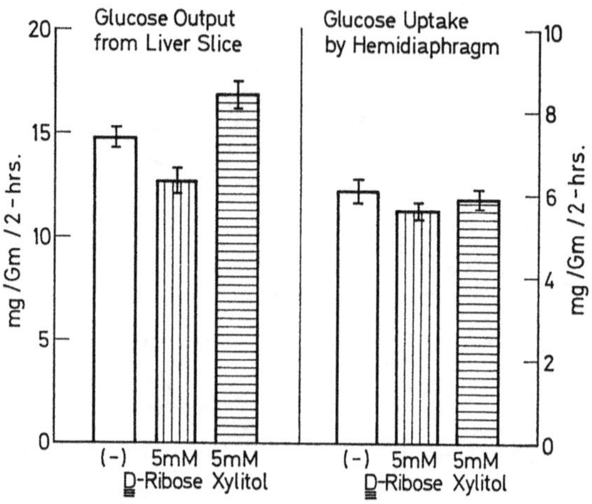

Fig. 6. Effects of D-ribose and xylitol *in vitro* on glucose output from liver slices and on glucose uptake by hemidiaphragm in fed normal rats. Incubation medium was Krebs-Ringer phosphate buffer, and was carried out at 37° C with gas phase of 95% O_2 and 5% CO_2 for 2 hours. Glucose uptake was calculated as a net uptake of glucose by hemidiaphragm in the medium

was produced either by addition of D-ribose or xylitol in the medium. Fig. 7 shows the effects of D-ribose and xylitol *in vitro* on ketone body release from rat liver slices. D-Ribose addition produced no significant change in the liberation of ketone bodies from normal rat liver slices, but caused a significant decrease in that from diabetic rat liver slices ($p<0.01$). Xylitol produced a significant decrease of the ketone body liberation either in normal or diabetic rat liver slices ($p<0.001$). A greater decrease in the liberation of ketone bodies was produced by addition of xylitol than by D-ribose *in vitro* and the difference between the effect of xylitol and that of D-ribose was statistically significant in normal ($p<0.001$) as well as in diabetic rats ($p<0.05$).

The effects of D-ribose or xylitol *in vitro* on the net release of free fatty acids from adipose tissue are shown in Fig. 8. D-Ribose addition showed no effect on fatty acid mobilization *in vitro* from adipose tissue in both normal and diabetic rats. Xylitol addition, on the other hand, produced a significant decrease of fatty acid mobilization from adipose tissue in normal ($p<0.05$) as well as in diabetic rats ($p<0.01$).

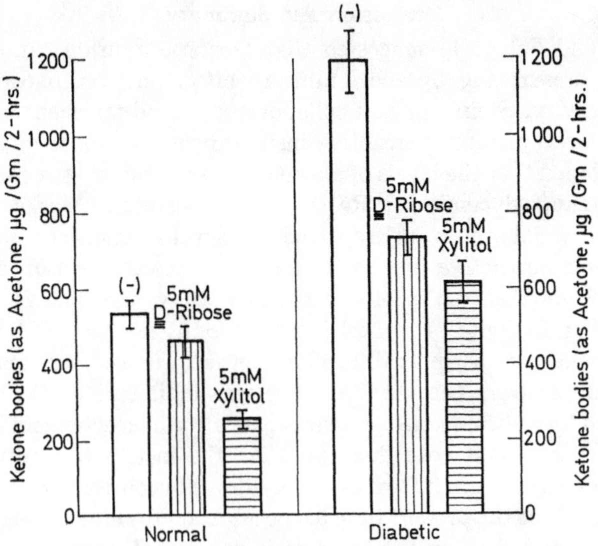

Fig. 7. Effects of D-ribose and xylitol *in vitro* on liberation of ketone bodies from liver slices of 24-hour fasting normal and diabetic rats. Incubation medium was Krebs-Ringer phosphate buffer. The first incubation was carried out for 5 minutes and the second incubation was performed in another incubation flask at 37° C for 2 hours with gas phase of 95% O_2 and 5% CO_2. The amount of ketone bodies liberated during the second incubation period was calculated per gram of wet liver

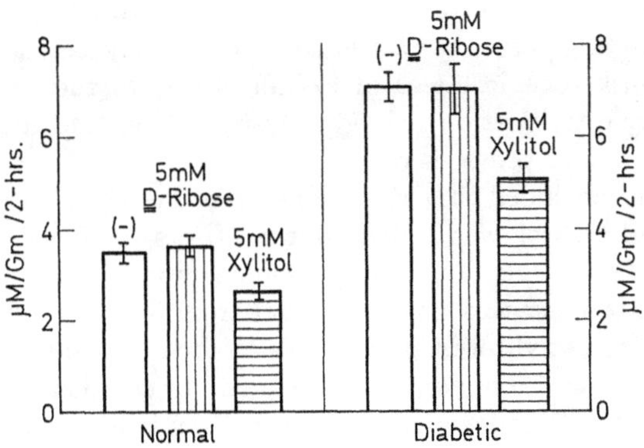

Fig. 8. Effects of D-ribose and xylitol *in vitro* on net release of free fatty acids from epididymal fat pad of 24-hour fasting normal and diabetic rats. Incubation medium was Krebs-Ringer bicarbonate buffer containing 5% bovine albumin, and was carried out at 37° C for 2 hours with gas phase of 95% O_2 and 5% CO_2

Discussion and Summary

Segal and Foley [1] suggested that D-ribose infusion caused hypoglycemia, presumably by the inhibitory effect on liver phosphoglucomutase activity. Steinberg and collaborators found transient elevations in serum levels of immunoreactive insulin during D-ribose infusion but they concluded that the levels of plasma insulin attained were insufficient to produce hypoglycemia [4]. Recently, Kuzuya and co-workers [5] and Hirata and collaborators [6] reported a sharp but transient increase of plasma immunoreactive insulin by the intravenous injection of xylitol in dogs. From the results of our present study, the change in blood glucose after D-ribose or xylitol infusion seems not to be attributable to the change in peripheral insulin concentrations. Our result that glucose output from the rat liver slice was inhibited by D-ribose and accelerated by xylitol *in vitro* explains partially the mechanism by which the level of blood glucose fluctuates. As the insulin concentration in the pancreatic vein after D-ribose or xylitol infusion was not measured, it is impossible at the present time to speculate on the effect of endogenous insulin on metabolism in the liver. It is presumable, however, that the major parts of the molecules of D-ribose or xylitol administered may be converted to members of the Pentose Phosphate Pathway, mainly in the liver cell, accompanied by consumption of equal amounts of ATP. The only difference between D-ribose and xylitol is that xylitol and not D-ribose produces $NADH_2$ during the reaction by NAD-linked dehydrogenase. The increase in blood lactate after xylitol infusion accounts for the increase of $NADH_2$. The inhibitory effect of xylitol on fatty acid mobilization from adipose tissue *in vitro* may be attributed to the increase in the ratio of $NADH_2$ to NAD which accelerates the formation of α-glycerophosphate from dihydroxyacetone phosphate. A strong antiketogenic effect of xylitol is due to the acceleration of fatty acid re-esterification and glycolysis in liver and adipose tissue, as suggested in our previous report [3].

It can be concluded that the difference in metabolic effects between D-ribose and xylitol was found in the effects on blood glucose and plasma free fatty acids. The hypoglycemic action of D-ribose is, at least in part, due to decreased glucose output from liver, and the fatty acid lowering effect of xylitol is attributable to the direct inhibition of fatty acid mobilization from adipose tissue. An interaction between these carbohydrates and endogenous insulin in liver is to be clarified.

References

1. Segal, S., and J. Foley: J. clin. Invest. **37**, 719 (1958).
2. Mehnert, H., J. Summa, u. H. Förster: Klin. Wschr. **42**, 382 (1964).
3. Yamagata, S., Y. Goto, A. Ohneda, M. Anzai, S. Kawashima, M. Chiba, Y. Maruhama, and T. Yamauchi: Lancet **1965 II**, 918.

4. Steinberg, T., P. Ortman, R. Poucher, B. Cochran, Jr., and G. Gwinup: Metabolism **16**, 40 (1967).
5. Kuzuya, T., Y. Kanazawa, and K. Kosaka: Metabolism **15**, 1149 (1966).
6. Hirata, Y., M. Fujisawa, H. Sato, T. Asano, and S. Katsuki: Biochem. biophys. Res. Commun. **24**, 471 (1966).

Discussion

Dr. Bässler: Can you explain the disappearance of urinary glucose during xylitol application?

Dr. Goto: A possible explanation for this phenomenon is a normalization of the activity of liver enzymes after xylitol administration, as reported by Fukui *et al.* and by Takahashi *et al.* A competition in reabsorption from the renal tubuli and an increase of insulin secretion from the pancreas in some cases may be another possibility. However, it is difficult to find an exact explanation at the present time.

Dr. Horecker: I believe that mannoheptulose is excreted with a very low renal threshold. It might be of interest to study the effect of xylitol on mannoheptulose excretion, in the hope of gaining some information pertaining to the mechanism of the effect of xylitol on glucose excretion.

Dr. Asakura: The increase of lactate and decrease of pyruvate in blood after xylitol administration may be due to the increase of NADH/NAD ratio by the oxidation of xylitol to xylulose. The accumulation of lactate at the same time indicates that NADH utilization is slower than its production or that lactate utilization in liver is smaller than its production. So that the measurement of lactate in the blood may be good criteria to evaluate whether the dosis of xylitol administered is suitable or not.

Dr. Horecker: I believe that it is necessary to insert a word of caution with respect to the significance of increases in insulin levels in response to xylitol infusion or oral administration. In the so-called maturity onset type of diabetes the level of insulin may be normal, but the response to a glucose load is distinctly different. The group in Stockholm under Prof. Rolf Luft has studied the time course of insulin levels in the blood in response to a glucose load. In "normal" individuals the response is biphasic — a rapid initial rise is followed by a slower secondary increase. In the diabetic or "prediabetic", the first peak may be absent, but the final insulin level may be in the normal range. Therefore, in studying the effect of xylitol on insulin levels in the blood, it may be important to evaluate the response to a single load, rather than the levels during continuous administration.

Preventing Effect of Xylitol on Suppression of Adrenocortical Function by Steroid Therapy

MINORU OHNUKI

Second Department of Internal Medicine,
Tokyo Medical and Dental University, School of Medicine,
Tokyo, Japan

In many biological areas, increasing attention has been payed to the Pentose Phosphate Pathway. This pathway is important as one of the supplying systems of $NADPH_2$ which is necessary for steroidgenesis, and also as a nucleic acid-pentose producing pathway through ribose-5-phosphate. On the other hand, it is generally recognized that the Pentose Phosphate Pathway is dominant in adrenal cortex [1, 2].

When the adrenocortical function is under the condition of suppression, the action of the Pentose Phosphate Pathway will be decreased not only in $NADPH_2$ production, but also in the nucleic acid synthetic process. Therefore, in this suppressed condition of the adrenal, it will be beneficial to administer substances which accelerate the current of the Pentose Phosphate Pathway.

It is presumed that xylitol, which is a physiological intermediate of the Uronic Acid Pathway, might play the role of accelerator of the Pentose Phosphate Pathway by conversion to D-xylulose-5-phosphate, and that adrenocortical function could be affected, especially under suppressed condition.

In this paper, I would like to present my investigation about the effect of xylitol administration on steroidgenesis in adrenal cortex of rabbits.

Materials and Methods

About a hundred male adult rabbits were divided into four groups as shown in Fig. 1. The first group was intact normal animal. The second was treated with prednisolone for four weeks, the third was the group treated with xylitol and prednisolone, and the fourth was the group with xylitol alone.

In these treatments, prednisolone was injected intramuscularly in doses of 0.5 mg per kg body weight every day, and 5 ml of 10 per cent xylitol solution was infused intravenously. After 4 weeks of the treatment, I studied the 5 following points: 11-OHCS output in the incubation medium from adrenal slices; 11-OHCS level in plasma; determination of corticosterone in adrenal homogenate by thin layer chromatographic technique; incorporation of ^{32}P-orthophosphate into RNA of adrenals; histological examinations of adrenal cortex.

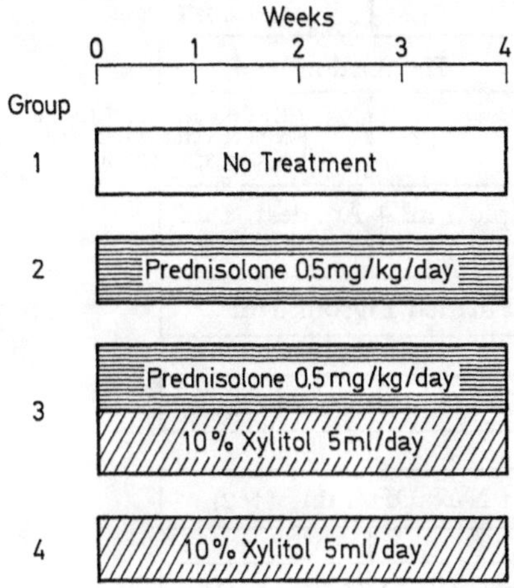

Fig. 1. Treatment of experimental animals

The procedure of determination of 11-OHCS secreted from adrenal slices is shown in Fig. 2. This method is a modification of Saffran's procedure [3]. Plasma 11-OHCS levels were determined by DeMoor's procedure [4].

Corticosterone, contained in adrenals of rabbits of each group, was isolated by thin layer chromatographic technique. Each adrenal was homogenized with water, and after extraction into dichlor-methane, the supernatant was evaporated to dryness. This sample was pointed on thin layer plates of Wakogel, chromatographed and developed with 3.5 per cent ethanol-chloroform solution. The area corresponding to corticosterone was dissolved into ethanol and corticosterone in this supernatant was determined by sulfuric acid fluorometry.

2 hours after intramuscular injection of 4 mC of ^{32}P-orthophosphate, RNA of adrenals was extracted by Schmidt-Thannhauser's method [5], and determined by Schneider's procedure [6]. The radioactivity of RNA was determined by a 2π gasflow counter.

The adrenals in each group were stained with hematoxylin and eosin for the purpose of histological examination. On the other hand, for the purpose of investigating the effect of xylitol on adrenal function in clinical cases, I observed cortical reserved capacity by plasma 11-OHCS

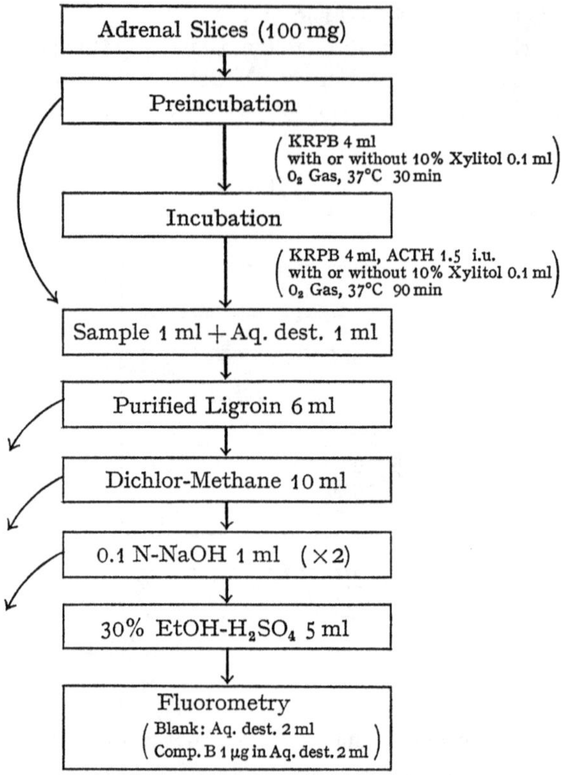

Fig. 2. Determination of 11-OHCS in adrenal slices

increasing rate under ACTH load test in 11 patients, receiving long steroid therapy with xylitol. They were all given orally 30 g of xylitol a day during steroid therapy.

Results

1. Fig. 3 shows the output of 11-OHCS from adrenal slices of rabbits under various conditions. In the normal group, addition of xylitol at incubation time significantly accelerated the output of 11-OHCS from adrenal slices into the incubation medium.

Adrenal corticoidgenesis of the prednisolone-treated group was very much suppressed compared with that of the normal group. On the other hand, combined treatment of xylitol with prednisolone prevented the adrenocortical suppression noted in the prednisolone treated group. In the group traeted with xylitol only, the output of 11-OHCS was identical to that in the normal group.

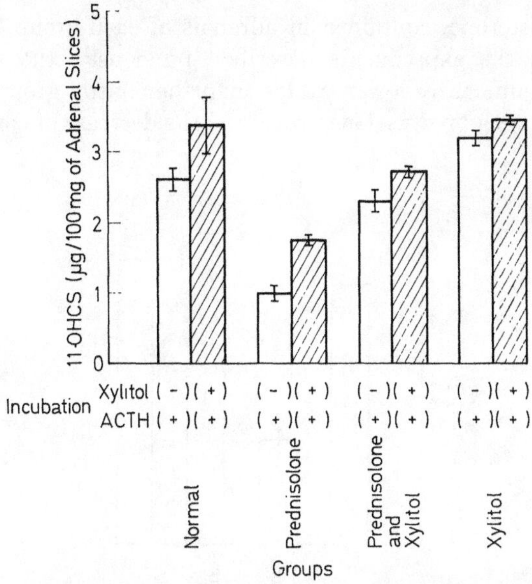

Fig. 3. Effect of xylitol administration on suppression of adrenal corticoidgenesis in prednisolone treated rabbits

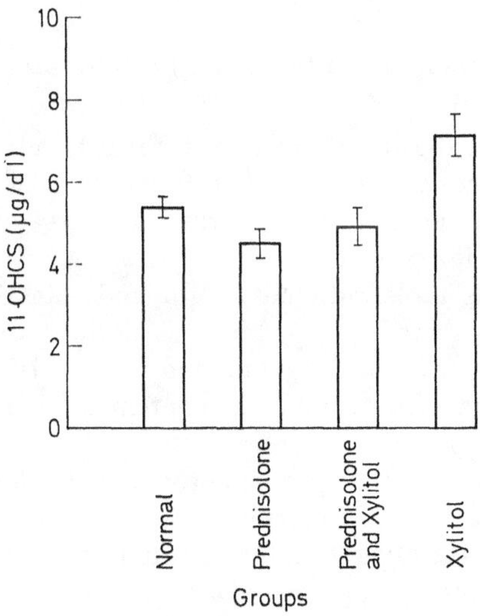

Fig. 4. Plasma 11-OHCS

2. Fig. 4 shows the plasma 11-OHCS levels in each group. The prednisolone group showed significantly lower values than the normal group. The xylitol group showed the highest level among those four groups.

3. Corticosterone, contained in adrenals of each group, is shown in Fig. 5. As in the experiments described previously, the prednisolone group had significantly lower values than the control group. Combined use of xylitol with prednisolone prevented the decrease of corticosterone.

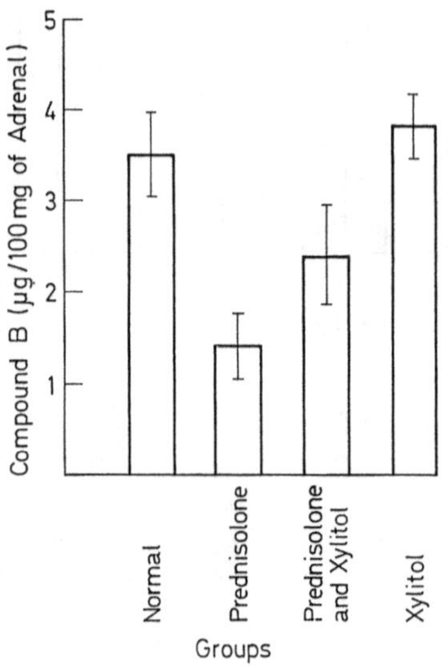

Fig. 5. Compound B (corticosterone) in adrenals by TLC

4. Table I shows the incorporation of ^{32}P into adrenal RNA, and shows also adrenal weight and RNA content of each group. In this experiment, I added one more group, which was treated with 10 units of ACTH every day for two weeks.

According to this table, adrenal weight and total RNA decreased in the prednisolone-treated group, but this decline was not observed in the group treated with prednisolone and xylitol. In the xylitol and ACTH groups, adrenal weight and RNA content were much higher than those in the normal group.

From measurement of the specific activity of ^{32}P incorporated into RNA, the ACTH-treated group showed rather low values, the prednisolone-treated group the highest and the group treated with prednisolone and xylitol a tendency to approach the normal group.

From the above results, it was noticed that xylitol not only augmentes the direct process of corticoidgenesis but also influences cell components in the adrenal cortex and when adrenal atrophy is inevitable due to steroid treatment, it has a marked preventing effect.

5. In histological examination, the adrenal cortex of the prednisolone group became thinner than that of the normal group, especially in the

Table I. *Weight of adrenal RNA and incorporation of* ^{32}P *into adrenal RNA*

Groups	Normal	Predni-solone	Predni-solone Xylitol	Xylitol	ACTH
Adrenal weight (mg)	221 ±25.1[a] (5)	158 ±8.6 (5)	222 ±30.1 (5)	298 ±21.8 (4)	315 ±28.9 (5)
RNA (μg)	1,631 ±212 (5)	1,066 ±72 (5)	1,675 ±356 (5)	2,349 ±306 (4)	2,710 ±255 (5)
RNA/adrenal weight (μg/mg)	7.3 ±0.26 (5)	6.7 ±0.24 (5)	7.3 ±0.64 (5)	7.9 ±1.27 (4)	8.4 ±0.51 (5)
Total count of RNA (cpm)	24,474 ±3,999 (5)	31,120 ±2,964 (5)	36,528 ±5,405 (5)	37,149 ±9,314 (2)	28,592 ±6,648 (5)
Specific activity (cpm/μg of RNA)	15.4 ±2.10 (5)	30.4 ±3.11 (5)	24.1 ±3.28 (5)	15.8 ±0.14 (2)	9.9 ±1.17 (5)

[a] Mean ± standard error. ( ): tested number.

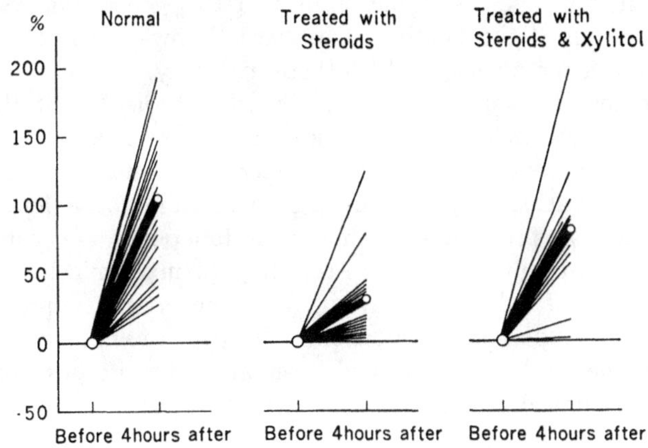

Fig. 6. Plasma 11-OHCS response by ACTH test

zona fasciculata. In the group with prednisolone and xylitol, the adrenal cortex retained normal width, or even seemed to be somewhat wider than that of the normal group.

6. Fig. 6 shows the cortical reserved capacity in 11 cases receiving long steroid therapy with xylitol, compared with that in 17 normal cases and 14 cases of steroid treatment without xylitol. In this figure, it was

revealed that in spite of the highly marked depression of cortical function in cases of xylitol non-administration, the depression of cortical reserved capacity in cases of combined xylitol treatment amounted to no more than a very scarce degree.

Discussion and Conclusion

Xylitol has an intimate relationship to the Pentose Phosphate Pathway which is dominant in adrenal cortex. Here, pursuit of the preventing effect of combined use of xylitol on suppression of rabbit adrenocortical function, caused by 4 weeks administration of prednisolone, was performed directly by measuring corticosterone content in adrenal homogenate, indirectly by measuring the plasma 11-OHCS level and further by measuring cortex reserved capacity with Saffran's procedure of adrenal slice incubation. Data also were obtained in which xylitol prevented abnormality in RNA content in the adrenal under steroid treatment. Furthermore, the same results were obtained also in histological examinations.

All these results indicated that xylitol has an augmenting effect of steroidgenesis in adrenal cortex, and that more clearly it has an effect of preventing suppression of cortical function caused by steroid treatment.

There are several interpretations of action of xylitol on the adrenocortical function; one is stimulation of $NADPH_2$-generating systems by the acceleration of the Pentose Phosphate Pathway. The other is the acceleration of the nucleic acid synthetic pathway, also by acceleration of the Pentose Phosphate Pathway. Further studies about these mechanisms are now under investigation in my laboratory.

Recently, in many clinical cases, steroid hormones are used for treatment of various diseases in large doses and for a long period. In these cases, suppression of adrenocortical function is a very important problem in iatrogenic disorders. According to my clinical experience, as much effect of xylitol on preventing adrenocortical suppression was obtained as in experiments on rabbits. However the number of clinical cases examined is not enough, and cases are not uniformly composed; thus further clinical examinations are expected.

References

1. Glock, G. E., and P. McLean: Biochem. J. **56**, 171 (1954).
2. Kelly, T. L., E. D. Nielson, R. B. Johnson, and C. S. Vestling: J. biol. Chem. **212**, 545 (1955).
3. Saffran, M., and M. J. Bayliss: Endocrinology **52**, 140 (1953).
4. De Moor, P., O. Steeno, M. Raskin, and A. Hendrikx: Acta endocr. (Kbh.) **33**, 297 (1960).
5. Schmidt, G., and S. J. Thannhauser: J. biol. Chem. **161**, 83 (1945).
6. Schneider, W. C.: Methods in enzymology, vol. III, p. 680. Academic Press Inc., N.Y. 1957.

Discussion

Dr. BÄSSLER: I already mentioned in my paper that McKerns [Biochim. biophys. Acta (Amst.) **121**, 207 (1966)] has shown the primary action of ACTH to be an activation of adrenal glucose-6-phosphate dehydrogenase for increased production of ribose 5-phosphate and RNA. In this respect xylitol can be expected to be able to replace ACTH to some extent, and your data are in complete agreement with this concept. In this very important function xylitol is unique and cannot be replaced by any other polyols.

Dr. OHNUKI: Recently, we had seen reports about adrenal nucleic acid, by Dr. Ferguson, Dr. Farese, Dr. Bransome, and so on. But in most of them the experiments were performed under the condition of ACTH loading, compared with normal animals, and the investigations studied under adrenal-suppressed condition are very few.

Therefore, the relationship between ACTH loading or xylitol loading and nucleic acid synthesis under the condition of suppressed adrenal function remains for future work.

Dr. FUKUI: Have you ever used xylitol for Addison's disease?

Dr. OHNUKI: No, I have not yet. But I have some cases that have adrenocortical insufficiency by long term steroid-therapy. In these cases, a slight improvement was found after two or three months administration of xylitol.

Dr. HORECKER: I find your results on RNA synthesis of great interest. We have been intrigued by the high levels of the Pentose Phosphate Pathway in the adrenal. We assumed that the requirement for TPNH would produce an excess of pentose phosphate over that required for nucleic acid synthesis, and that this organ could therefore have high levels of transaldolase and transketolase, in order to convert the excess pentose phosphate back to glucose phosphate. However, we found that the level of transaldolase, the limiting enzyme, was not unusually high. It would seem, therefore, that the pentose phosphate may be utilized for RNA synthesis. Your results, which suggest a high turnover of RNA in this organ, are consistent with this notion.

Dr. HALMAGYI: How early did you register an effect of xylitol on corticogenesis after the infusion of xylitol in prednisolone treated rabbits.

Dr. OHNUKI: The action of xylitol on the adrenocortical functions is not so acute, and continuous administration is required for the purpose of preventing adrenal suppression.

The Use of Polyols in Pediatrics

G. ERDMANN

From the Department of Pediatrics, University of Mainz,
Mainz, West Germany

Children have a high demand of sugars or similar substances. These ought to be tolerable and easily utilized, furthermore palatable if given by mouth, and apt to supply a sufficient caloric intake. Often, carbohydrates are used as the only source of calories for ill children, e.g. in case of acute disturbances of the intestinal tract (caused by infection or incompatibility of food ingested, by malabsorption syndromes etc.) and in stress situations.

After it had been demonstrated that the polyols sorbitol and xylitol enter the carbohydrate metabolism or even are natural links of it (Hollmann; Lang; McCormick and Touster), the pediatricians, too, were very interested in the problems arising from the use of these polyols in children.

Supply of fluid and electrolytes cannot be the only answer to meet the requirements of sick children in case of vomiting and/or diarrhea occurring during diseases of various origin. Fast growing infants especially need calories, given by parenteral route, if there is a lack of oral food supply. The additional use of dextrose proved to be of great benefit. But there are two reasons why glucose is not sufficient as the sole source of calories. One reason is that children have a high demand of protein which means amino acids in parenteral nutrition, and that amino acids cannot be sterilized together with glucose. Apart from this technical problem, the second reason is that children easily tend to disturbances of metabolism, as seen in ketonemic reactions, in infections, recurrent acetonemic vomiting or diseases of the intestine. Fructose seems to be advantageous in solving this second problem but its keto-group gives the same reaction with amino acids as does glucose.

Polyols appear to solve both these problems. So, in 1959 we started experiments with sorbitol. We could show that sorbitol did not produce any unwanted side effects apart from causing osmotic diuresis when given in high concentrations and very rapidly. This polyol did not cause any histopathological changes in liver and kidney of rabbits (Erdmann, 1960), which confirmed the results of Griem and Lang who proceeded in investigating the utilization of sorbitol in men. The promising results of Lang and co-workers using sorbitol together with balanced amino acid

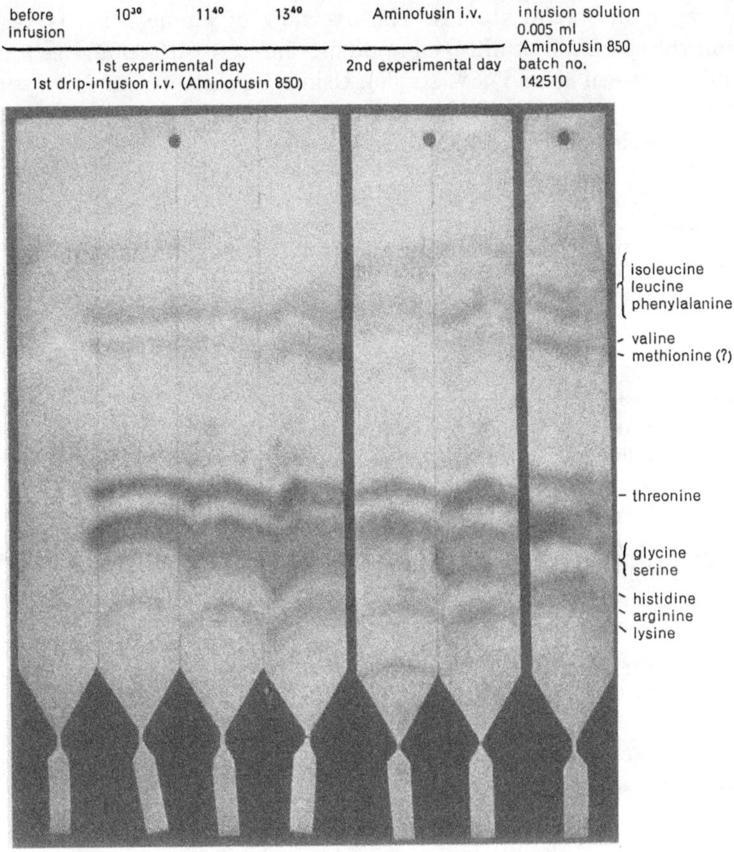

Fig. 1. Overflow-hyperaminoaciduria. Adult rabbit under drip-infusion (Aminofusin 850)

mixtures made us study the first industrially produced preparations of this kind, Aminofusin, for parenteral nutrition of severely ill children.

As utilization of intravenously administered amino acids partly depends on the concomitant carbohydrate or carbohydrate substitutes, the even more promising results gained with xylitol by Lang and coworkers — concerning its metabolisation rate, insulin-independent metabolism and low renal overflow — made us interested in comparing Aminofusin containing sorbitol with Aminofusin containing xylitol (Erdmann,1963, 1966; Erdmann and Heine, 1960).

We observed that following parenteral administration of Aminofusin blood amino nitrogen as well as urine amino nitrogen were rising. There was a special spectrum of urine amino acids under the infusion therapy (Fig. 1), but we have found a high retention of nitrogen and amino nitrogen in special studies. The overflow of amino acids is very low and the retention of nitrogen in infants infused with Aminofusin proved to be between 80 and 96% (table). Using the Matthias model in studying this overflow of amino acids under these conditions, a low overflow rate could be demonstrated under the infusion.

Table. *Amino-N-balance of infants under drip-infusion of aminofusin*

Child	Diagnosis	Weight (gm)	Age	Amino-fusin (ml/24 h)	N/min/kg (mg)	Amino-N-retention (%)
Pe.	state following otitis media	4,120	8 weeks	475	0.43	89.0
Schr.	state following parenteral dyspepsia	5,050	4 weeks	475	0.35	96.7
Wi.	premature birth	2,820	7 weeks	490 (!)	0.70 (!)	84.0
Be.	pretoxicosis	3,770	4 weeks	490	0.54	80.0
Str.	pretoxicosis	4,770	3 months	490	0.38	81.2
Wit.	toxicosis	4,020	2 months	490	0.45	85.6
Mö.	toxicosis	4,150	10 weeks	480	0.43	87.4

Later, we used amino acid solutions containing xylitol instead of sorbitol. The same low overflow of amino acids could be found, estimated in α-amino nitrogen balances: only 4% of the amino acids went into the urine under this condition. There is a difference between D- and L-forms of amino acids, as Bansi and co-workers have demonstrated. We could e.g. observe a retention of nearly 100% for L-lysine using a semiquantitative estimation in modification of the method of Waltz and Brenner, transferring the amino acids of the urine into nitro-phenylamino acids by means of nitrofluor-benzene (Erdmann, 1966). Therefore, we can state that there is no striking difference in using xylitol or sorbitol for addition to amino acid solutions concerning nitrogen and amino acid retention under the conditions of parenteral nutrition.

As for the excretion of xylitol in the urine, I am referring to the discussion of our last Session. Toussaint and Bässler have studied this problem in premature infants of our department. To these infants of a body weight between 2,000 and 2,600 g, xylitol solutions (0.5 to 1.0%) were administered intravenously. As we have heard this morning from the Munich group, an overflow of xylitol may take place. I like to lay stress on the important fact that even premature infants are able to retain xylitol by means of enzymes available in this state of development.

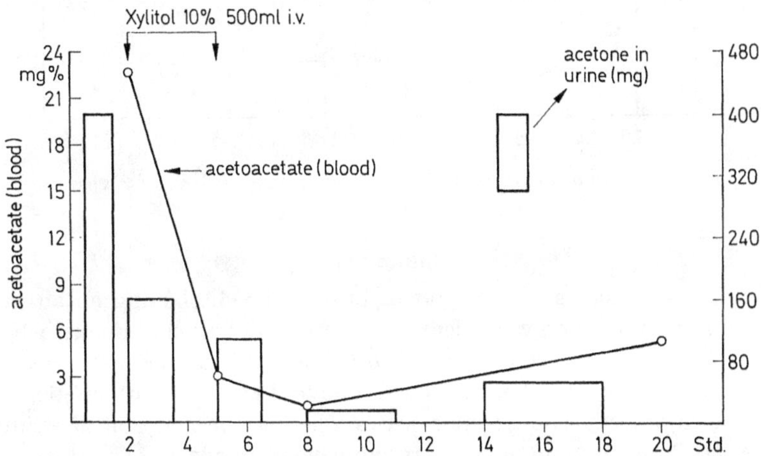

Fig. 2. Ketone bodies in diabetes. (Child of 10 years)

As poindet out already, pediatricians have often to deal with ketoses of various etiology. In our Department of Pediatrics, 33 children (3 to 14 years of age) were studied by Bässler in cooperation with Toussaint and Roggenkamp in this respect. Blood levels of acetoacetate, β-hydroxy-butyrate, glucose, pH and xylitol were measured. In all cases, the strong antiketogenic action of xylitol could be confirmed. Just let me demonstrate two cases chosen by random (by the courtesy of Bässler).

The first diagram shows the effect of 500 cm³ of a 10% xylitol saline solution in a 10 years old diabetic child. Acetoacetate is going down promptly and acetone excretion in the urine is decreasing, too (Fig. 2).

The other diagram shows a similar effect in a child of 8 years suffering from recurrent acetonemic vomiting. Blood levels of acetoacetate decrease, so does urinary excretion of acetone until about 7 hours following administration of 500 cm³ of a 10% xylitol saline solution (Fig. 3).

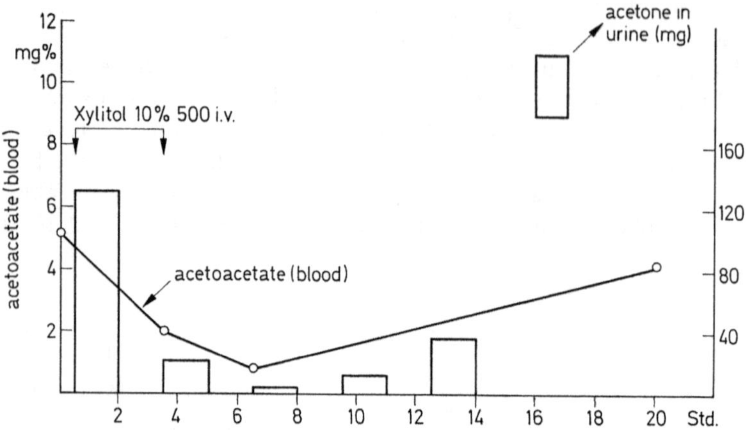

Fig. 3. Ketone bodies in acetonemic vomiting. (Child of 8 years)

Summary

For pediatricians, the polyols sorbitol and xylitol bring an answer for an important energy problem in parenteral nutrition of infants when balanced amino acid mixtures are used in combination with polyols. Furthermore, the treatment of ketosis due to diabetes and other etiology will be improved. The question arises whether the addition of xylitol to electrolyte solutions may be recommended in the treatment of dehydrated infants suffering from gastroenteritis or in newborn infants with similar metabolic derangements. Finally, the oral application of a certain amount of xylitol added to the diet of diabetic children may be discussed (cf. Lang; Mehnert *et al.*; Mellinghoff).

References

Bässler, K. H.: Stoffwechsel von Xylit. Scripta therapeutica **16**, 1—16 (1966).

— W. Prellwitz, V. Unbehaun u. K. Lang: Xylitstoffwechsel beim Menschen. Zur Frage der Eignung von Xylit als Zucker-Ersatz beim Diabetiker. Klin. Wschr. **1962**, 791—793.

— W. Toussaint u. G. Stein: Xylitverwertung bei Frühgeborenen, Säuglingen, Kindern und Erwachsenen. Klin. Wschr. **1966**, 212—215.

Diehl, U.: Diss. Mainz (is just prepared).

Erdmann, G.: Zur Nierenverträglichkeit parenteral verabreichter Sorbitlösungen. Klin. Wschr. **1960**, 1002—1003.

— Anwendung intravenös gegebener Aminosäurengemische in der Pädiatrie. Wiss. Veröff. Dtsch. Ges. Ernährung, Bd. 11. Darmstadt: Steinkopff 1963.

— Verwendung von Aminosäurengemischen in der Pädiatrie. In: Anaesthesiology and resuscitation, vol. 6, p. 114—128. Berlin-Heidelberg-New York: Springer 1966.

Erdmann, G., u. W. Heine: Aminostickstoff-Bilanz bei der Aminofusinbehandlung schwerer akuter Ernährungsstörungen im Säuglingsalter. Klin. Wschr. **1960**, 1001—1002.

Griem, W., u. K. Lang: Ein Beitrag zur parenteralen Ernährung mit Aminosäuren-Sorbit-Lösungen. Klin. Wschr. **1960**, 951.

Hollmann, S.: Nicht glykolytische Stoffwechselwege der Glukose. Stuttgart: G. Thieme 1961.

Lang, K.: Xylit als Nahrungskohlenhydrat. Med. u. Ernähr. **4**, 45—50 (1963).

Mehnert, H., J. D. Summa u. H. Förster: Untersuchungen zum Xylitstoffwechsel bei gesunden, leberkranken und diabetischen Personen. Klin. Wschr. **1964**, 382—387.

Mellinghoff, C. H.: Über die Verwendbarkeit des Xylit als Ersatzzucker bei Diabetikern. Klin. Wschr. **1961**, 447.

Toussaint, W., H. Roggenkamp u. K. H. Bässler: Behandlung der Ketonämie im Kindesalter mit Xylit. Z. Kinderheilk. **98**, 146—154 (1967).

Waltz, D., A. R. Fahmy, G. Pataki, A. Niederwieser u. M. Brenner: Dünnschicht-chromatographischer Nachweis der Aminosäuren im Urin. Experienta (Basel) **19**, 213—217 (1963).

Comment

The Clinical Use of Xylitol in Pediatrics *

MITIO KUMAGAI

Department of Pediatrics, School of Medicine Keio University,
Tokyo, Japan

It is generally accepted that xylitol administered to adult in various conditions including normal and diseased state, such as diabetes mellitus and hepatic disease, is rapidly metabolized and effectively utilized as a source of energy and that it exhibits a strong antiketogenic action. However, only few studies have been reported in the pediatric age group. The present studies were undertaken to examine whether xylitol administered to children is utilized as effectively as observed in adults and if there is any advantage in using this sugar alcohol for treatment of certain diseases of children.

Clinical Studies. Xylitol was administered intravenously to 16 healthy children aged 4 to 15, and blood glucose, xylitol, pyruvic acid, NEFA and ketone bodies were determined before, during and after xylitol infusion. Xylitol was given by one of the following two ways: a) rapid method by which xylitol was infused quickly by push at a level of 0.2 g/kg body weight, b) continuous drip infusion in a dose of 0.5 g/kg body weight for 90 minutes.

The results are indicated in Figs. 1 and 2, in which xylitol level is expressed in actual concentration in mg-% and others are in % change to the initial level. In rapid infusion group the glucose level was not significantly influenced, whereas the xylitol concentration rised rapidly and returned to initial level within 60 minutes after infusion was completed. The responses of blood pyruvate and lactate to xylitol infusion were similar to those reported in adults. Pyruvate fell sharply and returned to initial level by 60 minutes, while lactate rised initially and fell thereafter. The levels of both blood NEFA and ketones droped, and the latter more markedly. These changes were observed by both the rapid and drip infusion, but they were less remarkable in drip

* Data presented here are from: Department of Pediatrics, Tokyo Medical and Dental University; Department of Pediatrics, School of Medicine Keio University; Department of Pediatrics, Juntendo University; Department of Pediatrics, Koseikai Hospital.

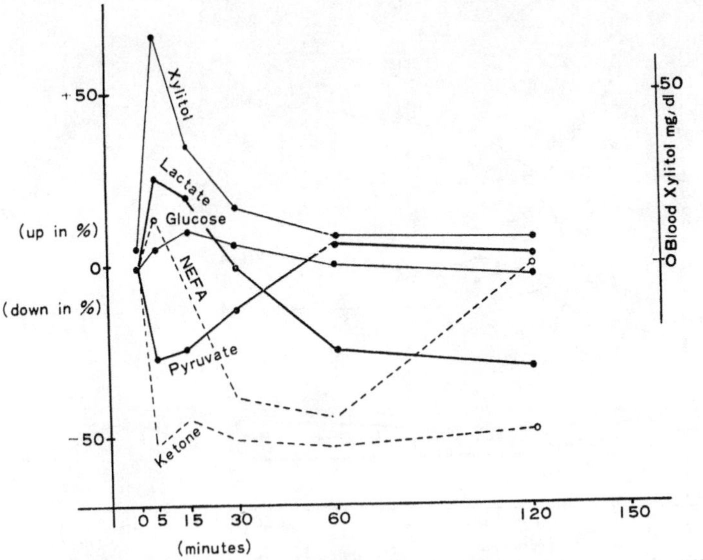

Fig. 1. The effect of xylitol infusion on blood chemistry in normal children (0.2 g/kg body weight by push)

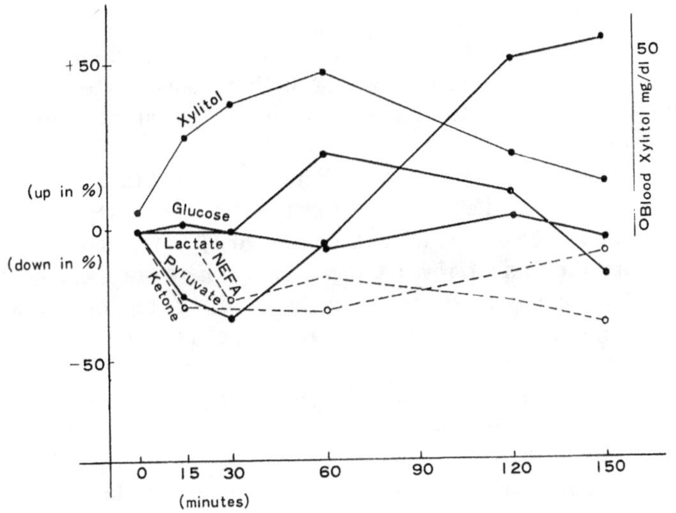

Fig. 2. The effect of xylitol infusion in blood chemistry in normal children (0.5 g/kg body weight by drip/90 min)

infusion group. The results are in accord with those reported in adults. From our results and Dr. Bässler's excretion studies in children, we assume that xylitol can be utilized in normal children as effectively as in adults.

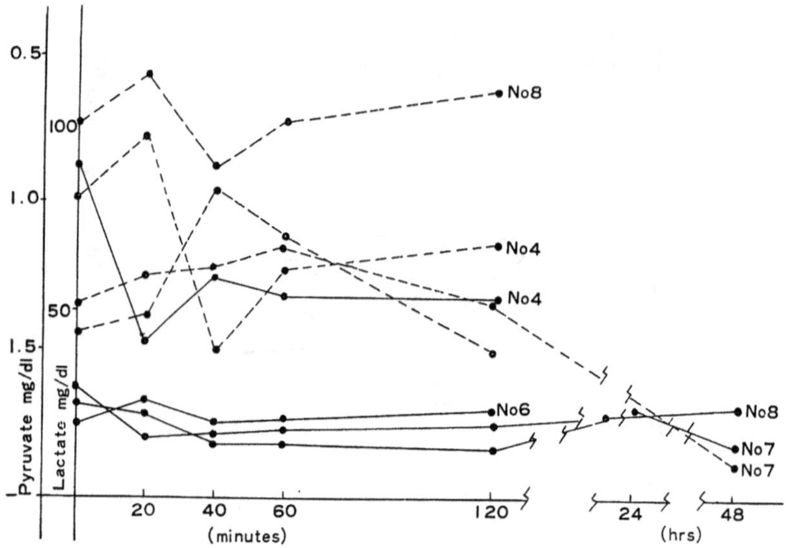

Fig. 3. Xylitol infusion effect on plasma pyruvate and lactate in children with cyclic vomiting (—— pyruvate, ----- lactate)

The same type of study was carried out in 4 children with cyclic vomiting, 5 with liver disease and 2 with juvenile diabetics. Figs. 3, 4 and 5 show some of the results, which are similar to those seen in the healthy children.

Furthermore we induced hyperketonemic condition in 6 normal children by placing them on a ketogenic diet for 2 days which was composed of 1,200 cal/1.73 m² and contained 67% fat and 16% carbohydrate. At the end of the 1st day blood NEFA and ketone bodies level rised as high as those seen in cyclic vomiting (Fig. 6) and actually some of children developed the clinical signs of this disease in 2nd day of the experiment.

Since this procedure could be used as an experimental model of cyclic vomiting, we produced this condition in 9 children. Then, glucose was administered to 3 of these children and xylitol to 6 of them. Both compounds were given in a dose of 2.5 g/kg body weight over a period of 12 hrs.

The fall of the blood NEFA and ketone bodies was observed in both cases, as indicated in Figs. 7 and 8, and there was very little difference. In all except one case blood pyruvate in the xylitol group fell and remained below the initial level during infusion, then it showed a gradual rise parallel to the level of glucose, whereas blood lactate was elevated during infusion, then reduced to the initial level.

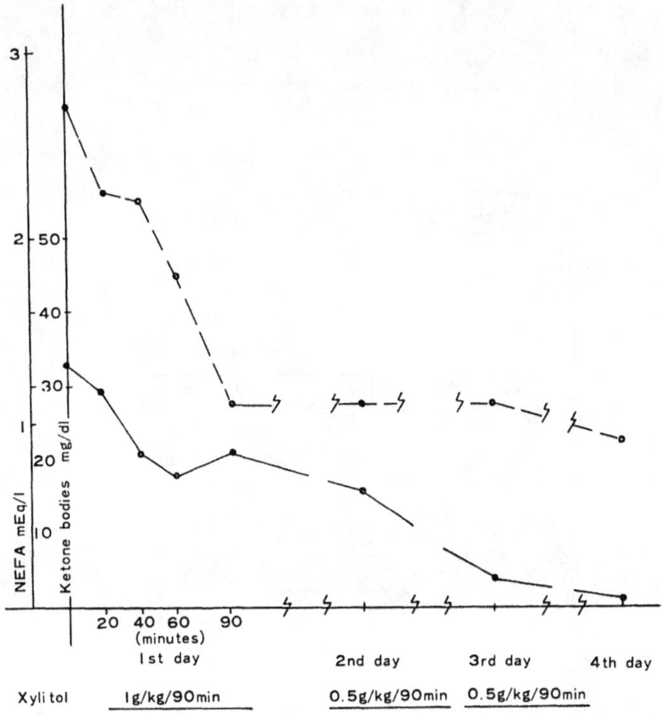

Fig. 4. Xylitol infusion effect on plasma NEFA and ketone bodies in children with cyclic vomiting [— — NEFA, —— ketone bodies (as acetone)]

In the glucose group pyruvate fell in the first 3 hrs, but thereafter rised to higher level than the initial, and appeared to run parallel to glucose and lactate levels. This initial fall in pyruvate was not seen by glucose infusion in normal children on normal diet.

At present we do not fully understand the significance of these results, but we can say that xylitol is as effective as glucose in correcting the ketonemia.

Clinical Application of Xylitol for Certain Pediatric Conditions

1. Cyclic Vomiting. Because of its strong antiketogenic effect, we used xylitol for the treatment of 32 cases with cyclic vomiting, and compared its effect with glucose treatment. As seen in Fig. 9, and table, in most of the cases clinical improvement was seen in 2 days after the treatment had started. However acetonuria remained few days thereafter in some cases. Thus we could not conclude that xylitol is more effective than glucose in treatment of this condition.

2. Prolonged Neonatal Jaundice. Fig. 10 illustrates the results of another clinical application of xylitol. Xylitol was given intravenously

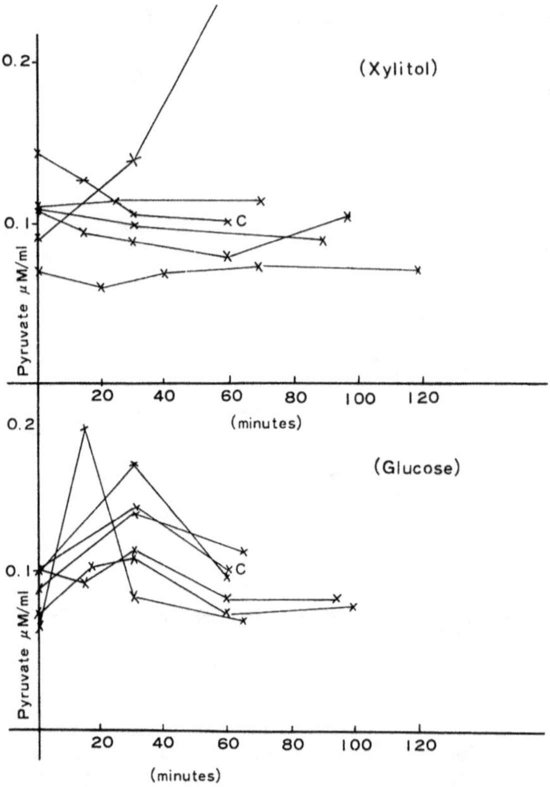

Fig. 5. The effect of xylitol infusion on plasma pyruvate in children with chronic hepatitis (0.5 g/kg body weight by push)

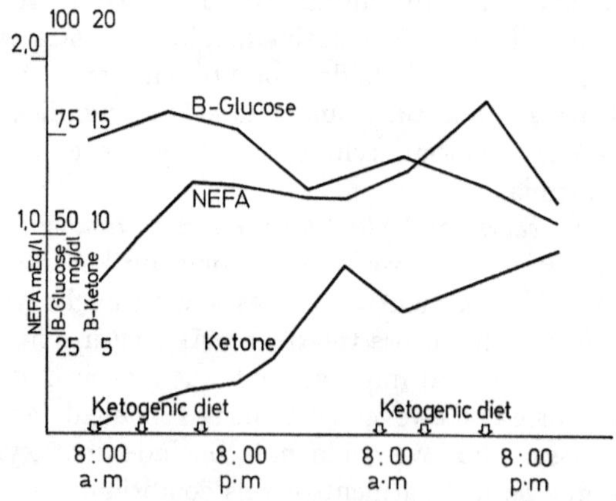

Fig. 6. Ketogenic provocative test. 6 normal children were placed on ketogenic diet (1,200 cal/1.73 m² body surface, 67% fat, 16% CHO, 17% protein)

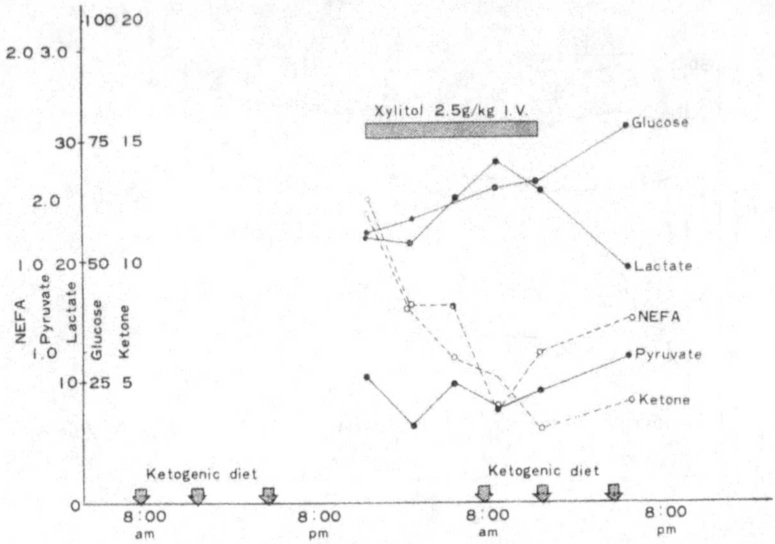

Fig. 7. The effect of continuous xylitol infusion on ketonemic condition induced by ketogenic diet

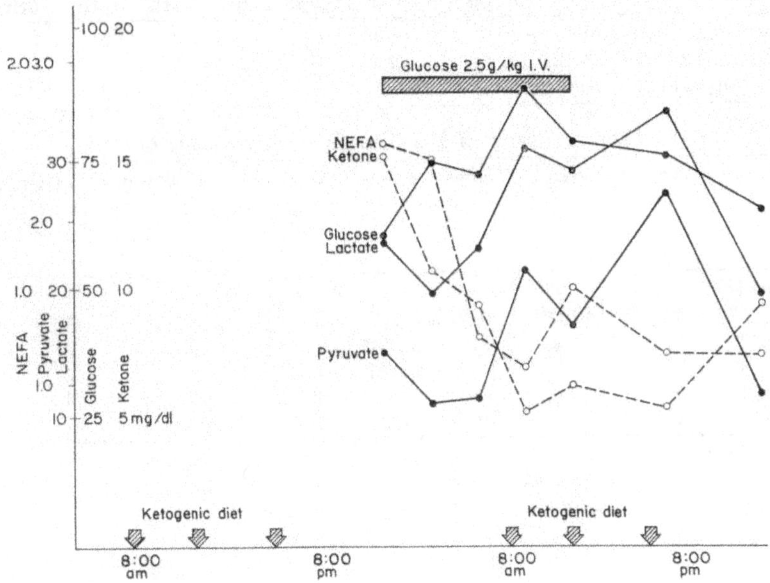

Fig. 8. The effect of continuous glucose infusion on ketonemic condition

in a dose of 0.5—1.0 g/kg body weight to 9 infants with prolonged jaundice. We were clinically impressed that xylitol has some beneficial influence on the clinical course of this type of infantile jaundice as judged by the fall of serum bilirubin.

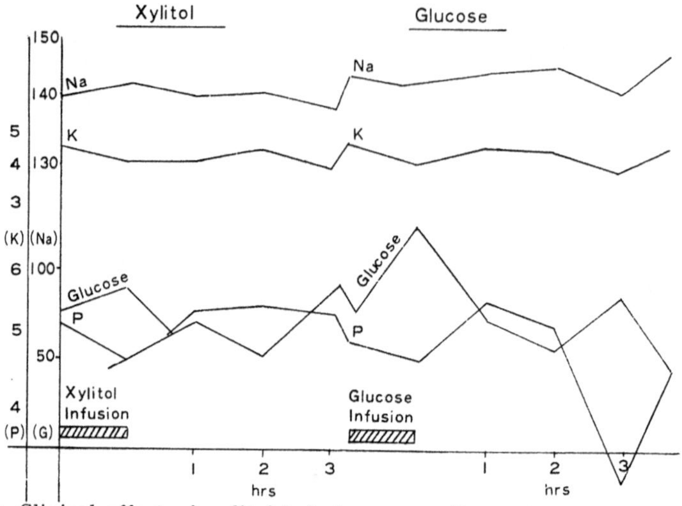

Fig. 9. Clinical effect of xylitol infusion on cyclic vomiting-(II) (21 cases). 10% xylitol sol. 80—100 ml by slow drip only

	1st	2nd	3rd	4th	5th	6th
General improvement	10	4	1	4	2	
Disappearance of vomiting	20		1			
Disappearance of acetonuria	2	2	7	5	3	2

Table I. *The effect of xylitol on cyclic vomiting (I)*

	Age	Severity	Doses of xylitol given	General improvement (day)	Disappearance of vomiting (day)	Disappearance of acetonuria (day)
Group I						
No. 1	5	mild-	10% sol	1	1	3
No. 2	7	moderate	0.2—0.3 g/kg	1	1	2
No. 3	7		body weight	1	1	2
No. 4	6		1 ~ 2 ×	1	1	3
No. 5	3		by i.v. push	1	1	2
Group II						
No. 4	4	moderate-	5% sol	2	2	3
No. 4	4	severe	0.5 ~ 1.0 g/kg	2	2	3
No. 7	7		body weight	2	2	2
No. 9	9		1 ~ 2 ×	2	2	2
			by i.v. drip			
Group III						
No. 10	7	moderate-	10% sol	2 ~ 3	2	3
No. 11	3	severe	0.2 ~ 0.3 g/kg	2 ~ 3	2	4
			body weight			
			2 ~ 3 ×			
			by i.v. push			

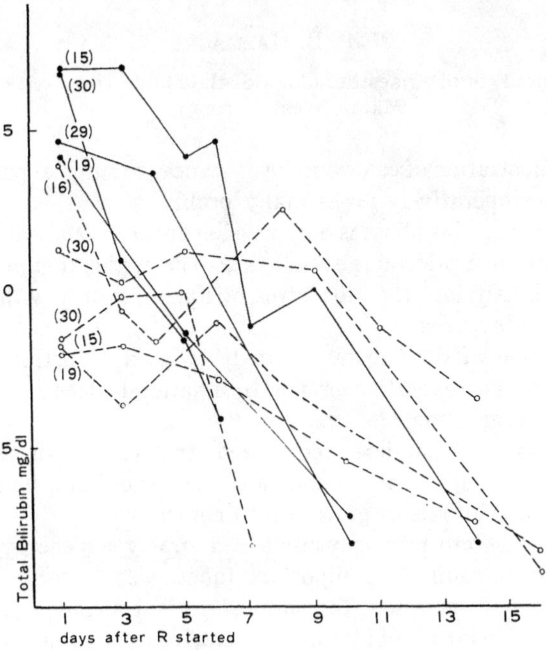

Fig. 10. Xylitol infusion effect on plasma bilirubin in infants with prolonged jaundice (0.5—1.0 g/kg body weight/i.v. push, —— bottle fed, — — breast fed. ( ) the days

Discussion

Dr. Fukui: Dr. Yoshikawa reported the preventing effect of xylitol on hemolysis. Have you investigated any cases of hemolytic jaundice?

Dr. Kumagai: No, I have not. The cases presented here are what we call prolonged physiological jaundice, which is mainly due to immature liver function.

Dr. Touster: Does sorbitol have an effect similar to xylitol in increasing bilirubin conjugation?

Dr. Kumagai: I have no experience with sorbitol, but glucose appears to have some effect also. However, these results are based only on clinical impression and blood bilirubin change.

Dr. Touster: Do you have an explanation for the difference between breast fed and bottle fed babies?

Dr. Kumagai: It is said that some breast milk contains high enough maternal steroid hormones to interfer bilirubin conjugation.

Carbohydrates and Polyols for Energy Supply of the Surgical Patient

M. P. B. HALMÁGYI

Institute of Anaesthesiology of the Mainz University, Mainz, West Germany

The administration of calorie-rich substances to surgical patients pre-, intra- and postoperatively poses many problems.

Feeding during this phase is only possible intravenously and therefore the conditions are a priori unphysiological. The application of nourishing substances directly into the bloodstream is possible only with low molecular-weight substances.

The selection of the substances to be infused is further influenced by the specific postoperative or posttraumatic changes in metabolism of the human organism.

As early as 1887 Landerer recognized, that carbohydrates are particularly well-suited for intravenous administration, and he introduced the infusion of solutions of glucose into clinical use.

Carbohydrates are primarily used as a supply for energy. However they also have to fulfil other important functions.

At the time of operation, the surgical patient already suffers a lack of glycogen, which is caused by the preoperative restriction of oral intake. The organism has approximately 400 mg of freely available glycogen-reserves, which will be first consumed for energy production. After this reserve has been exhausted the metabolism of fat leads to ketoacidosis.

Under the conditions of a complete turnover of fat to ketoacidosis the effective caloric value of fat amounts to about 30% of the original, as has been pointed out recently by Bässler. In addition to this, the regeneration of glycogen from protein constitutes an additional load on the negative protein-balance of the surgical patients.

These problems cannot be solved by the sole application of glucose. Not only narcotic substances, such as ether, cyclopropan and halothane, but also the surgical trauma per se, leads to disturbances of the carbohydrate metabolism, as has been pointed out by Seelig, Annamunthodo, and coworkers, Foldes and coworkers, Nakamura and many other authors. Documented findings repeatedly state that hyperglycaemia, glucosuria, insulin-resistance and diabetic-like glucose-tolerance always accompany trauma.

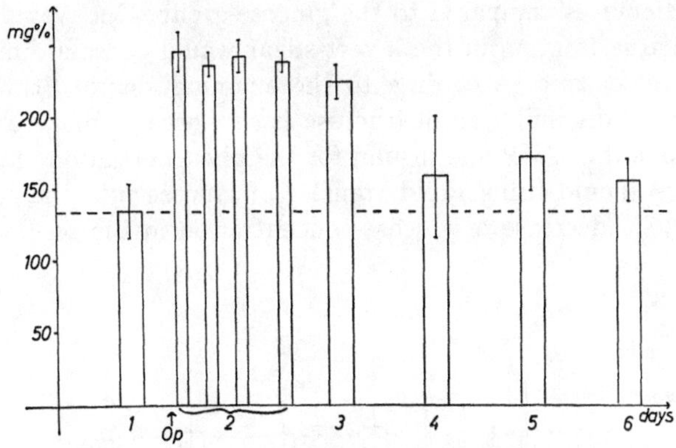

Fig. 1. Bloodsugar-level (glucose)

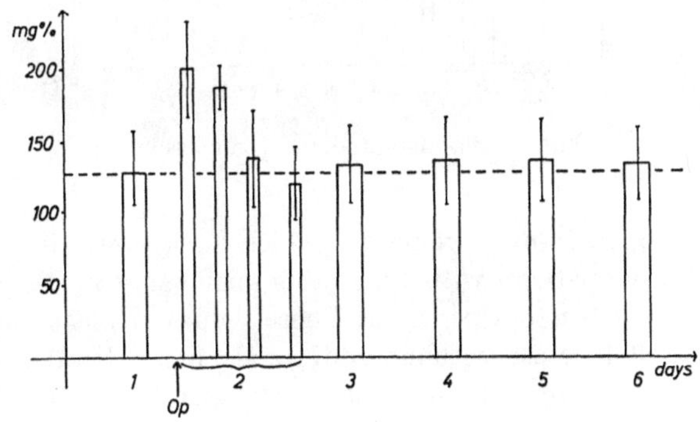

Fig. 2. Bloodsugar-level (honey)

Fig. 1 summarizes the results of measuring the glucose-concentration in the serum of 10 patients with upper-abdomen-operations. Infusion of 3 g glucose per hour resulted in hyperglycaemia, which remained for several days under this condition.

The strong initial hyperglycaemia is as result of the glycogenolysis, brought about by the secretion of Epinephrin. The stimulation effect of the 17-hydrocorticosteroids on the glycogenesis are further responsible for the hyperglycaemia. Glycocorticoids bring about a retardation of the peripheral consumption of glucose. Hypoxia also leads to an elevation of the blood-sugar level.

Fig. 2 shows the pre-, intra- and postoperative glucose-concentration in the serum of 10 patients with upper-abdomen surgery. In this group, an infusion of honey 10%, that is invert-sugar, was administered. The infused glucose-content also amounted to 3 g per hour. There were noticable differences compared to the glucose-group. One would expect that the fructose fraction in the invert-sugar would cause a diminution in the hyperglycaemia. According to the examinations of Strauss and Hiller the rapid dissimilation of fructose is supposed to make available increased amount of ATP and insulin for the phosphorisation of glucose, and therefore should bring rapid transfer of glucose into the cell with a resulting 30% decrease of glucose-concentration in the serum.

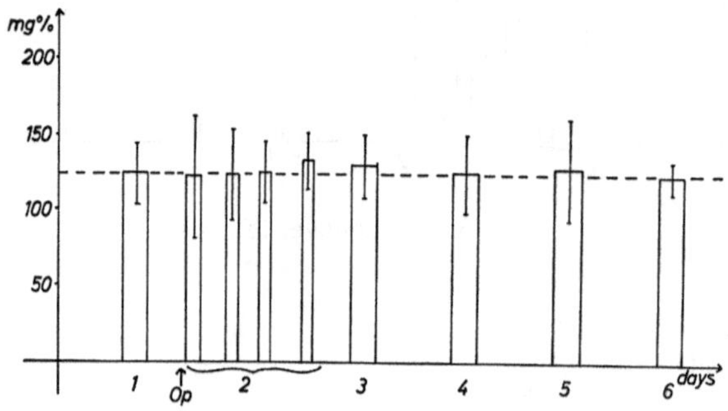

Fig. 3. Bloodsugar-level (fructose)

The serum-glucose-measurements, which are shown in Fig. 3 we obtained in 7 other patients with upper-abdomenal-surgery, they received an infusion of 3 g fructose per hour. These values of blood-sugar show that the infused fructose without additional glucose-load was able to prevent intraoperative and immediate postoperative hyperglycaemia via the above-mentioned "fellow-Traveller-effect".

Fructose shows further advantageous properties as well as sorbitol as far as the metabolic situation during the intra- and postoperative phase is concerned. Fructose is a rapid glycogen-producer, and the non-insulin-dependent utilization and the rapid glycogen-formation are advantageous during this phase. Fructose causes a larger increase in the ATP-concentration in blood than glucose, as has been shown by the investigations of Scheibe. This is important, since ATP-concentrations in blood are lowered in the intra- and postoperative phase. Also fructose shows during the postoperative phase a better potassium-saving-effect than glucose.

Fig. 4 shows the daily potassium-excretion in urin in milli-equivalents, which were determined in the glucose-group in 10 and in the laevulose-group in 7 patients following upper-abdominal-surgery. These results significant at 1% level show, without any doubt, that a higher potassium-excretion during the intra- and postoperative phase can be prevented by administration of laevulose.

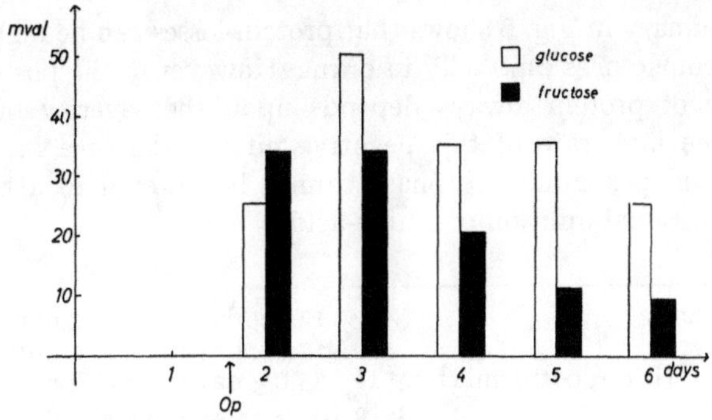

Fig. 4 Potassium loss in urin

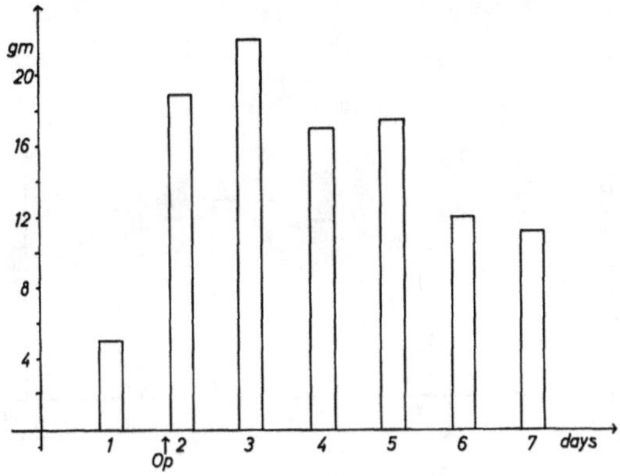

Fig. 5. Nitrogen loss in urin

Stuhlfauth and coworkers have observed that the postoperative nitrogen-loss could be diminished by administration of fructose. The reason for this can be seen in the fact that fructose forms a larger amount of pyruvic-acid, and therefore the hydrogen of the free sulfhydryl group, which is necessary for the activation of proteases, is used by the organism in increased amounts for the formation of lactic-acid from pyruvic-acid. This may explain a decrease in the activity of the endocellular proteases as Carstensen said. In spite of this protein-saving-effect of fructose, the postoperative and posttraumatic losses of protein are still too high.

Fig. 5 shows the daily nitrogen-elimination in urin in 10 patients following gastrectomy in gram. These patients received 3 g of fructose every hour for three days. A significant elevation of the postoperative nitrogen-losses up to 18—22 g per day, could be seen immediately following operation. An increased nitrogen-excretion remained during the postoperative phase of observation.

The summary in Fig. 6 shows that protein-losses can be higher under certain circumstances especially in burns. However in the postoperative phase, loss of protein always depends upon the severity of surgical trauma. The high rate of this negative nitrogen-balance in the postoperative and posttraumatic phase cannot be lowered to a minimum, without additional infusion of amino-acids.

Cholecystectomy:	12.0 g/day =	300 g muscle
Gastrectomy:	17.5 g/day =	440 g muscle
Headinjuries with hypothermia:	up to 35.0 g/day =	875 g muscle
Severe burns:	up to 120.0 g/day =	3,000 g muscle

Fig. 6. Loss of nitrogen after operation and trauma

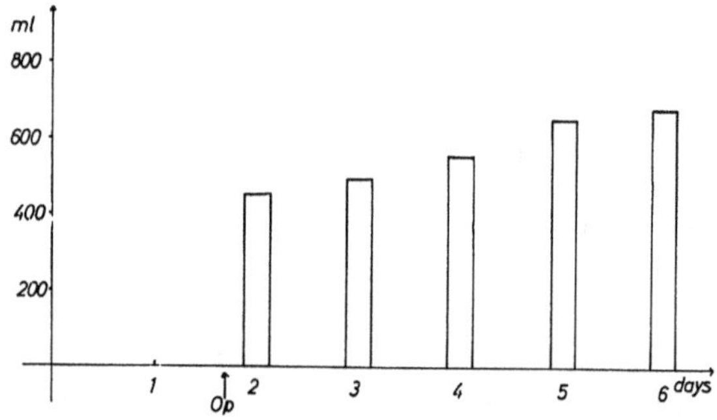

Fig. 7. Urinary output

In Fig. 7 we see the mean values of the daily urinary output in 27 patients after upper-abdominal-surgery. The retention of water after operation and trauma is a regular finding, which has been often observed. The retention of water is caused in this phase by the ADH-mechanism and by the increased incretion of the adrenal-cortical hormones. In this investigation the postoperative values of urinary output amounted to 400—700 ml per day, as compared to a normal excretion from 1,200—1,500 ml per day, were therefore considerable lower. The retention of water because of the danger of edema, calls for a restriction of water-intake, during the postoperative phase. This means that a combination of amino-acids and sugar in the same infusion-solution is necessary. Furthermore, the economic use of the amino-acids also calls for this combination. Without the concomitant supply of calorie-rich substances, the amino-acids would be used for the maintenance of metabolism and therefore burned up. However for the infusion therapy it would be a very serious mistake to sterilize reducing sugars such as glucose and fructose together with the amino-acid mixtures.

When sterilizing reducing sugars, together with free amino-acid in water, one obtains numerous reaction-products, according to the Maillard's reaction as was mentioned by Prof. Lang. The danger of the Maillard's reaction is not only the decrease in biological value, but this reaction also leads to the occurrence of toxic substances, such as those shown by investigations ın the institute of Prof. Lang. All these difficulties can be circumvented by use of polyalcohols, such as sorbitol and xylitol. These substances also have a better antiketogenic effect

Operation	Number of patients	Number of 500 ml-bottles	
		10% xylitol	50% xylitol
Head	3	2	1
Neck	85	59	26
Chest	75	48	27
Abdomen	574	401	455
Urology	199	130	72
Extremities	253	185	91
Total	1,189	825	672

Fig. 8. Groups of patients and numbers of infused 500 ml-bottles of xylitolsolutions

than fructose. The reason for this effect can be found, according to Bässler and coworkers, in the dehydrogenation, which is the first step in the metabolism of the two polyalcohols.

The investigations of Emerson, Fuss and several other authors show that anaesthesia and surgery, the disturbances in the utilization of glucose and the increased fat-metabolism during the postoperative phase, lead to a metabolic acidosis and to an increase in the ketogenic substances in the blood.

The use of an antiketogenic substance such as xylitol during this phase, would certainly be an advantage. We have therefore investigated between 1963 and 1965 xylitol-containing infusion-solutions in a group of surgical patients, first of all with the regard to general tolerance. We infused in 1189, cases 1,497 500 ml-bottles xylitol-containing-solution in form of Tutofusin EX 5 and Tutofusin EX 10. Fig. 8 shows each of the surgical groups and the amount of the infused xylitol-containing solution. The patients were between the age of one day and 80 years. Clinical tolerance could be established without any doubt. The rate of utilization of xylitol, during the postoperative phase, is presently under investigation and since these investigations have not yet been completed

I cannot report in detail about the results. The results obtained to date, however, suggest that there is no disturbances or utilization of xylitol during the postoperative phase.

If one looks at the rates of utilization of various sugars and polyalcohols published to date, one can see that none of the mentioned substances alone would be sufficient to cover the energy-demands in the intra- and postoperative phase. For this reason one should in my opinion combine the carbohydrates—as sugars and polyalcohols—for infusion-therapy. In so doing one would of course have to consider the investigations of Bässler and coworkers which were made with regard to the additive formations of lactate of these substances. The results of the investigations showed that the combination of xylitol—glucose or xylitol—fructose and sorbitol—glucose are applicable.

I wanted especially to discuss some of the important aspects of the infusion-therapy with sugars and polyols in the intra- and postoperative phase. All substances which I have mentioned have to fulfil specific metabolic requirements, beyond the requirements for energy-supply. They also have different organic affinities. Only skilled combinations of these substances, taking into account of technical problems, such as for instance the Maillard's reaction and the specific metabolic conditions of the human organism, during the intra- and postoperative phase, allows an optimal supply of the organism with these substances intravenously. I am convinced that the youngest member in the family of parenteral food-substances, as xylitol, constitutes an essential contribution with new possibilities in the infusion-therapy of patients.

References

1. Annamunthodo, H., J. V. Keating, and S. J. Patrick: Liver glycogen alterations in aneasthesia and surgery. Anaesthesia **13**, 429 (1958).
2. Bässler, K. H., W. Prellwitz, V. Unbehaun u. K. Lang: Xylitstoffwechsel beim Menschen. Zur Frage der Eignung von Xylit als Zuckerersatz beim Diabetiker. Klin. Wschr. **40**, 791 (1962).
3. — Die Rolle der Kohlenhydrate in der parenteralen Ernährung. In: Parenterale Ernährung. Berlin-Heidelberg-New York: Springer 1966.
4. Carstensen, E.: Infusionstherapie und parenterale Ernährung in der Chirurgie. Stuttgart: Schattauer 1964.
5. Emerson, G. A., S. J. Klyza, B. Abreu, and N. M. Phatak: Hyperglycemia and ketonemie with ether and divinyl oxide. Anesth. Analg. Curr. Res. **16**, 85 (1937).
6. Foldes, F. F., A. J. Murphy, and B. C. Wilson: The effect of various anesthetic agents on inorganic serum phosphorus levels. J. Pharmacol. exp. Ther. **100**, 14 (1950).
7. Fuss, H.: Kohlenhydrathaushalt bei der Narkose. Bedeutung der bei der Äthernarkose des Hundes auftretenden Säuren für das Säurebasengleichgewicht. Z. ges. exp. Med. **76**, 731 (1931).

8. Halmágyi, M.: Die Bedeutung der klinischen Vollbilanzierung bei der parenteralen Ernährung. In: Fettstoffwechsel, 2. Lochham bei München: Pallas 1967.
9. Landerer, D.: Über Transfusion und Infusion .Langenbecks Arch. klin. Chir. **34**, 807 (1887).
10. Lang, K., R. Frey u. M. Halmágyi: Parenterale Ernährung. Berlin-Heidelberg-New York: Springer 1966.
11. Nakamura, T.: Klinische Beobachtungen des Einflusses der Avertinrektalnarkose auf die Alkalireserve des Blutes und den Blutzucker. Tohoku J. exp. Med. **26**, 450 (1935).
12. Scheibe, O.: Langenbecks Arch. klin. Chir. **292**, 124 (1959).
13. Seelig, A.: Über Ätherglykosurie und ihre Beeinflussung durch intravenöse Sauerstoffinfusionen. Naunyn-Schmiedebergs Arch. exp. Path. Pharmak. **52**, 481 (1905).
14. Strauss, E., u. J. Hiller: Die Laevuloseverwertung des Diabetes mellitus und ihre biochemischen Grundlagen. Ärztl. Forsch. (Wörishofen) **4**, 326 (1952).
15. Stuhlfauth, K., A. Engelhardt-Gölkel, K. Plessner, B. Steinhuber, D. Wollheim, V. Struppler, F. Pirner, C. D. Netzer u. L. Tusban: Untersuchungen über die Störungen des Stoffwechsels nach Operationen und deren therapeutischen Beeinflussung durch Kohlenhydrate. Z. klin. Med. **153**, 287 (1955).

Discussion

Dr. Macdonald: In view of the ability of fructose to form more triglycerides than does glucose, did your patients who were given intravenously fructose have a higher incidence of fat embolism than patients receiving intravenously glucose?

Dr. Halmágyi: We did not control it, and so did not register in our studies presented now.

Xylitol Infusion Related to Pathophysiology in Anesthesia *

K. Yoshikawa

Department of Anesthesiology, Osaka University School of Medicine, Osaka, Japan

Infusion therapy is an important and interesting problem to anesthesiologists. Infusion therapy is based on correction of water depletion including electrolyte imbalance, metabolic disorders and other physiological disturbances during anesthesia and operation. I am not trying to debate fluid therapy in general extensively in this paper. I shall limit the problem to fluid therapy of xylitol as a source of energy and control of disorders of the metabolism of carbohydrate and fat, as found in diabetes.

The toxic effects of anesthetic agents and their effects on the metabolism of carbohydrates are well recognized; i.e., barbiturates change the rate of carbohydrate metabolism [1].

Recently, pentitol has become available for fluid therapy. The present experiments were carried out to determine the effect of xylitol on patients under general anesthesia, during and after operation. The effects of xylitol on enzyme activities after hemorrhagic shock in dogs were also investigated.

The Effect of Xylitol Infusion on Glucose, Lactate, Pyruvate, Inorganic Phosphate (Pi), Non-Esterified Fatty Acid (NEFA) in Blood in the Absence of Anesthesia. After the infusion of 5% xylitol in a dose of 0.5 g/kg body weight for 30 minutes to patients without any metabolic disorder, arterial blood was sampled. The levels of glucose [2], pyruvate [3] and lactate [4] in whole blood and the concentrations of Pi [5] and NEFA [6] in serum were measured. Glucose, fructose and sorbitol were infused in the same dose to compare the blood levels of the above substances with those obtained after the infusion of xylitol. The level of glucose in the blood after infusion of glucose, fructose, sorbitol or xylitol is shown in Fig. 1a. The lowest level of glucose was observed after the infusion of xylitol.

---

* Data presented here were from Department of Anesthesiology, Osaka University; Department of Anesthesiology, Kyushu University; Department of Anesthesiology, Juntendo University; Department of Anesthesiology, Kyoto University; Central Operation Room, Kagoshima University.

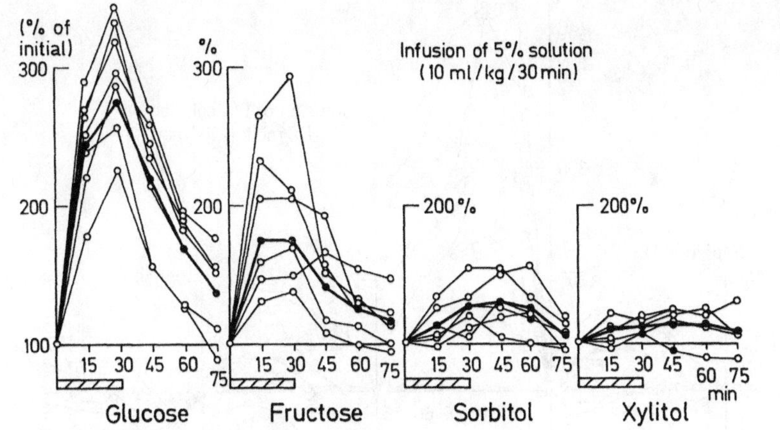

Fig. 1a

Fig. 1. Changes of glucose (a), inorganic phosphate (b), pyruvate (c), lactate (d) and non-esterified fatty acid (NEFA) (e) level in blood after infusion of various compounds without anesthesia. (Kyushu Univ.)

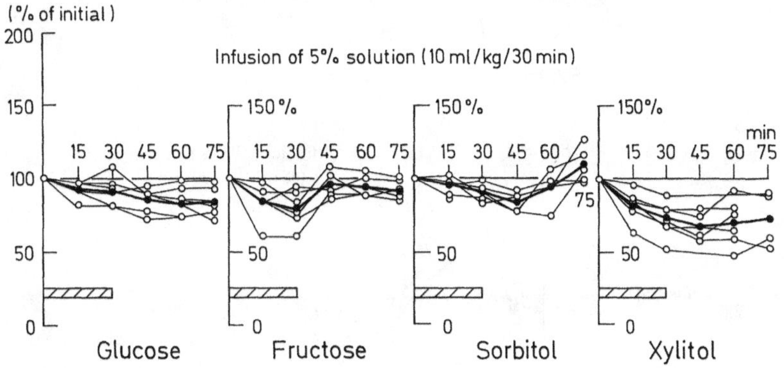

Fig. 1b

Fig. 1b shows the level of Pi in serum. Considerable decrease of phosphate was found after the infusion of xylitol. The levels of pyruvate and lactate are shown in Figs. 1c and d. The increase of lactate concentration was of almost the same extent as that after glucose, sorbitol or xylitol infusion, but the highest increase was found after fructose infusion. Pyruvate rather decreased after xylitol infusion. The infusion of those carbohydrates to patients decreased the concentration of NEFA almost to the same extent, as shown in Fig. 1e.

Change of Metabolism after Xylitol Infusion under General Anesthesia. The levels of glucose, lactate and pyruvate after infusion of glucose or

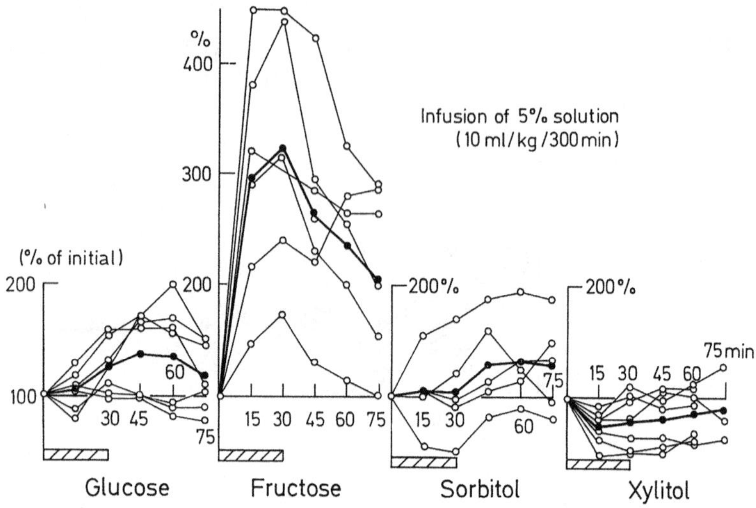

Fig. 1c

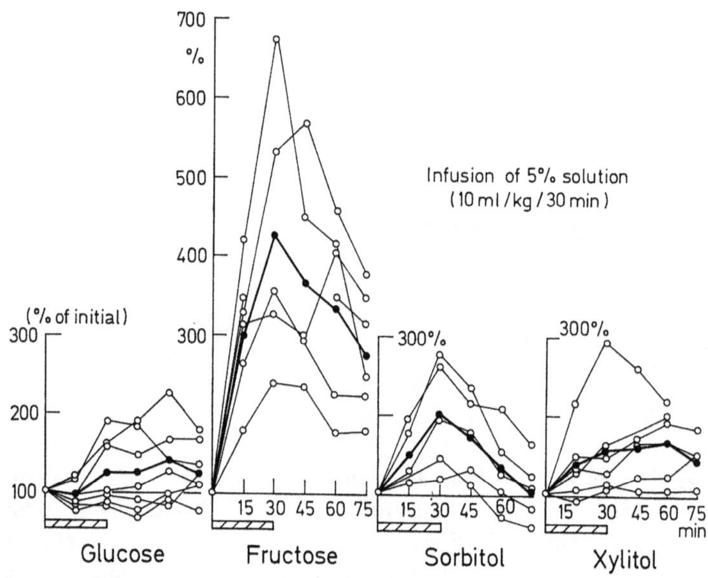

Fig. 1d

xylitol under anesthesia with ether and halothane are shown in Fig. 2a. It has been reported that the glucose concentration in blood is elevated in man during ether anesthesia [7, 8]. The cause of hyperglycemia by ether anesthesia is not clear, but it should be hepatic in origin, because

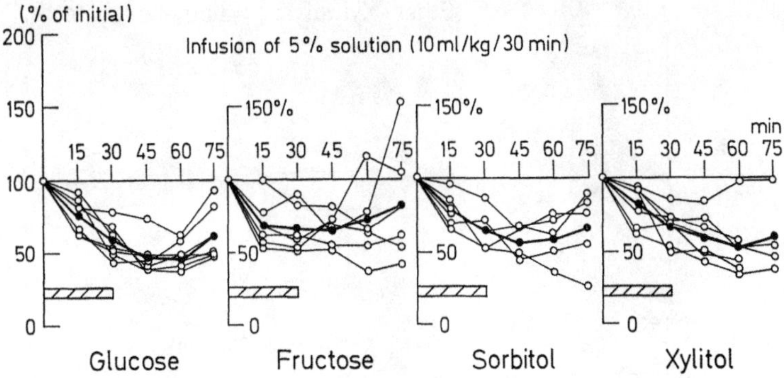

Fig. 1e

hepatectomized animals does not show hyperglycemia by ether anesthesia [9]. The change in muscle glycogen during ether anesthesia was not conclusive, but some reports have revealed no change or slight elevation [10, 11]. In our experiments, ether anesthesia increased the blood glucose level after glucose infusion, compared to the level without anesthesia. However, xylitol infusion lowered the blood glucose level. By the infusion of glucose under ether anesthesia, the lactate level in blood was increased, but the infusion of xylitol reduced this increase. Similar results were obtained under halothane anesthesia. The concentration of pyruvate in blood after xylitol infusion was lower than that after infusion of glucose.

The changes in concentration of Pi are shown in Fig. 2b. The infusion of xylitol tended to decrease the level under ether or halothane anesthesia, but glucose infusion did not. Acidosis was seen during the operation under anesthesia with glucose infusion, but no remarkable acidosis was found during halothane or ether anesthesia when xylitol was infused. It has been reported that xylitol has an antiketogenic action [12, 13]. Sato and his coworkers found that ketone bodies in the blood and urine decreased after xylitol infusion during and after surgery (Table I).

The Infusion of Xylitol to Diabetic Patients during Anesthesia and Surgery. A typical case of xylitol infusion during the operation on a diabetic patient is shown in Fig. 3. The concentration of glucose in the blood decreased gradually during xylitol infusion under cyclopropane anesthesia. After administration of 5% glucose with 4 units of regular insulin, the concentration of Pi increased gradually, but NEFA remained at a low level. Antiketogenic action of xylitol was also observed.

The changes of lactate, pyruvate and NEFA in blood during general anesthesia when xylitol was infused to diabetic patient are shown in

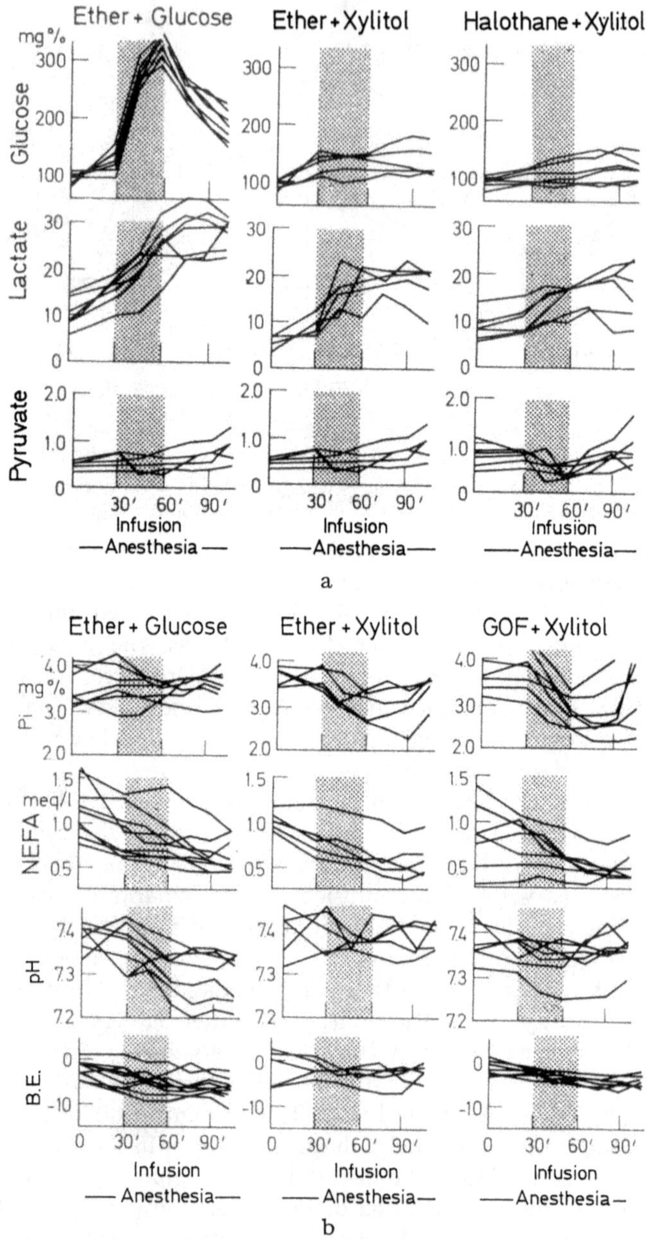

Fig. 2. The effects of glucose and xylitol infusion on the level of glucose, lactate, pyruvate, inorganic phosphate (Pi), NEFA, pH and base excess during general anesthesia including ether and halothane-oxygen-nitrous oxide. (Kyushu Univ.)

Table I. *The effect of xylitol infusion on the concentration of ketone bodies in blood and the amount excreted into the urine (Juntendo Univ.)*

Cases	Ketone bodies in blood (mg/dl)					Ketone bodies in urine (mg/day)		
	before operation	during operation	immediately after operation	5 hours after operation	24 hours after operation	op. day	1 day	2 day
				Group 1 (none)				
1	0.65	0.67	0.83	—	—			
2	0.07	0.03	0.40	8.80	0.00	42.46	120.07	5.44
3	0.05	0.07	0.07	0.54	2.63	5.53	28.84	5.00
4	—	—	5.55	9.15	9.80	48.27	0.03	1.85
5	1.98	1.72	2.30	13.37	3.37	93.48	136.26	0.00
6	0.42	0.56	0.49	0.89	4.12	0.00	65.25	0.08
7	0.03	0.05	0.02	1.02	4.55			
				Group 2 (5% glucose)				
8	0.82	0.50	0.42	0.21	—			
9	0.50	—	0.25	2.02	2.61	9.43	5.85	—
10	0.54	0.30	0.53	4.18	9.59			
11	1.30	0.95	0.80	2.65	0.91			
12	0.93	0.91	0.16	4.80	1.44	24.69	53.19	0.00
13	0.36	0.36	0.43	2.92	0.60			
14	—	—	—	—	—	0.00	37.35	23.52
15	—	—	—	—	—	14.03	4.84	0.02
				Group 3 (5% xylitol)				
16	0.00	0.00	0.00	0.69	0.25	4.68	16.66	5.30
17	0.38	0.25	0.49	1.60	1.40	4.00	38.20	—
18	0.78	—	0.91	0.91	1.12			
19	1.26	—	0.73	1.28	—			
20	0.23	—	0.64	1.51	0.81			
21	1.53	—	0.21	1.40	—			
22	0.00	0.67	0.81	1.16	0.29	1.17	4.90	0.00
23	0.00	0.00	3.47	2.00	0.00	0.00	9.12	13.33

Fig. 4. The concentration of lactate increased in all cases, but pyruvate and NEFA were constantly maintained at the control level.

Xylitol concentration after its administration in a dose of 0.4 g/kg body weight is shown in Fig. 5. The elimination curve of xylitol from blood of diabetic patients was almost identical to that of non-diabetic patients.

During anesthesia, the glucose level in blood could be controlled easily by the infusion of xylitol without addition of insulin. However, in brittle diabetics, it was difficult to control it with xylitol alone as

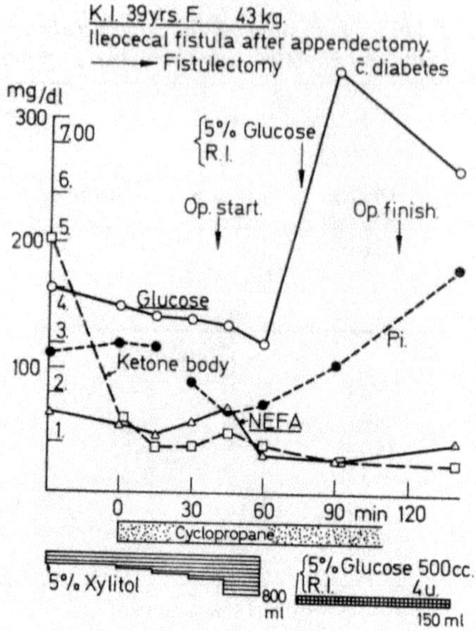

Fig. 3. A case of the surgical operation of a diabetic patient under cyclopropane anesthesia. (Kyushu Univ.)

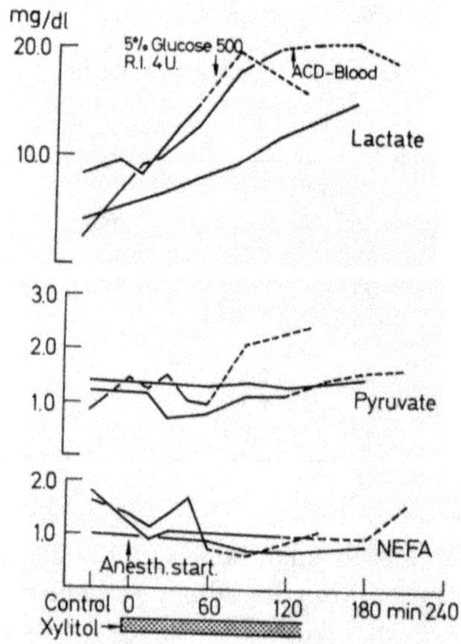

Fig. 4. Changes of blood lactate, pyruvate and NEFA concentration of diabetic patients during anesthesia after xylitol infusion. (Kyushu Univ.)

shown in Fig. 6. It might be considered that in the brittle type of diabetics, it is desirable to administer insulin even if xylitol is infused.

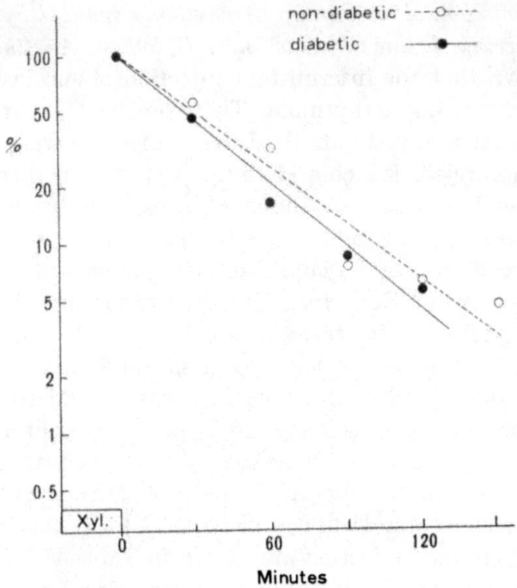

Fig. 5. The elimination curve of xylitol from blood. (Osaka Univ.)

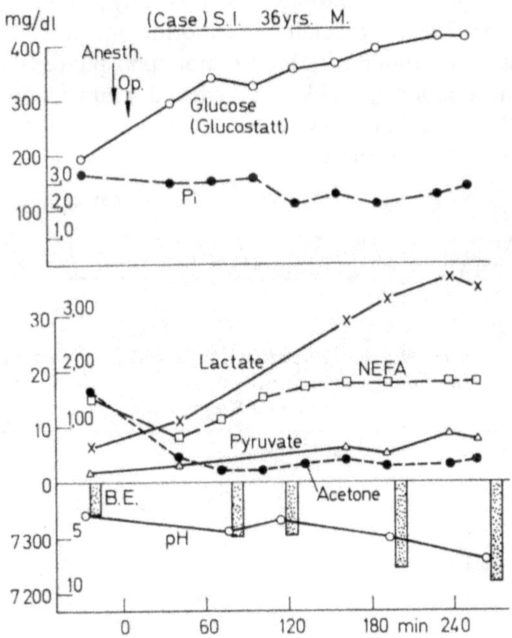

Fig. 6. Metabolic changes during anesthesia in a case of diabetes mellitus (Brittle type). (Kyushu Univ.)

The Effect of Xylitol on Cardiac Arrhythmia induced by Intermittent Injection of Succinylcholine Chloride under Halothane Anesthesia in Man. It is well known that the intermittent injection of succinylcholine has a possibility of causing arrhythmia. Therefore next experiments were performed in order to investigate the ECG change following intermittent injection of succinylcholine chloride and to study whether xylitol can protect it under halothane anesthesia. The patients were anesthetized with halothane-oxygen-nitrous oxide (GOF) with a premedication of pethidine hydrochloride and atropin sulfate. The patients were without any cardiac disorders on ECG. Initially 40 mg of succinylcholine chloride was injected rapidly. Thereafter, 20 mg of succinylcholine chloride was injected every 10 minutes during anesthesia, and ECG was recorded. As shown in Table II, cardiac arrhythmia was recognized in 26.2% of the control group (5% glucose infusion). When 15% xylitol was infused instead of glucose, the rate of cardiac arrhythmia was decreased to 4.3% (Table III), and by hypertonic glucose infusion (12—15%) the incidence was 15% (Table IV). The patterns of ECG changes following succinylcholine chloride injection are shown in Table V. The protective action was also observed by injection of ascorbic acid and glutathione, but the latter was less effective.

At present it is difficult to interpret the xylitol effect on cardiac arrhythmia induced by succinylcholine chloride injection. In pace-maker cells, existence of an abundant Hexose Monophosphate Shunt is speculated [14]. Since ascorbic acid and reduced glutathione prevent arrhythmia, xylitol infusion might have activated the metabolic pattern in the tissue according to the following scheme.

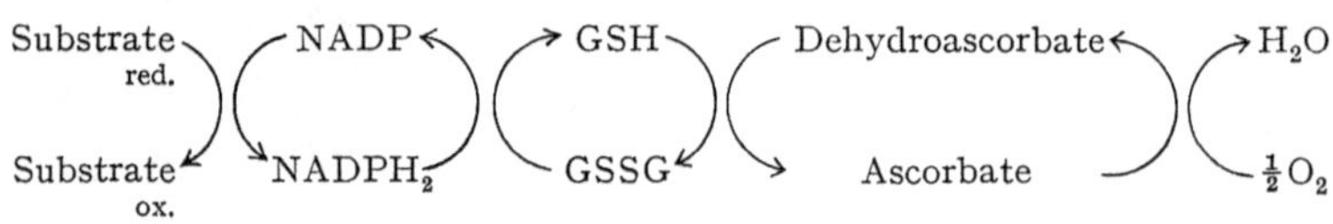

Table II. *Cardiac arrhythmia induced by intermittent injection of succinylcholine chloride under halothane-oxygen-nitrous oxide (GOF) anesthesia with 5% glucose infusion (Kagoshima Univ.)*

	1	2	3	4	5	6	7	8	9	10	Total
Total cases	27	27	27	27	27	25	26	21	13	13	233
Arrhythmia (%)	2 7.4	7 25.5	5 18.5	7 25.5	5 18.5	3 12.0	3 11.5	0 0	1 7.7	0 0	33 14.2
P-wave change	1	3	4	4	5	5	4	2	0	0	28
Total (%)	3 11.1	10 37.0	9 33.3	11 40.8	10 37.0	8 32.0	7 27.0	2 9.5	1 7.7	0 0	61 26.2

Table III. *The effect of hypertonic xylitol (15%) infusion on cardiac arrhythmia (Kagoshima Univ.)*

	1	2	3	4	5	6	7	8	9	10	Total
Total cases	20	20	20	20	20	18	14	13	10	8	163
Arrhythmia	1	2	0	0	1	0	0	0	0	0	4
(%)	5.0	10.0	0	0	5.0	0	0	0	0	0	2.5
P-wave change	0	0	1	1	1	0	0	0	0	0	3
Total	1	2	1	1	2	0	0	0	0	0	7
(%)	5.0	10.0	5.0	5.0	10.0	0	0	0	0	0	4.3

Table IV. *The effect of hypertonic glucose (12—15%) infusion on cardiac arrhytmia (Kagoshima Univ.)*

	1	2	3	4	5	6	7	8	9	10	Total
Total cases	10	10	10	10	10	8	6	7	6	3	80
Arrhythmia	0	2	2	3	2	0	0	0	0	0	9
(%)	0	20.0	20.0	30.0	20.0	0	0	0	0	0	11.3
P-wave change	0	1	0	0	0	2	0	0	0	0	3
Total	0	3	2	3	2	2	0	0	0	0	12
(%)	0	30.0	20.0	30.0	20.0	25.0	0	0	0	0	15.0

Table V. *ECG changes after succinylcholine chloride injection in the control group (5% glucose infusion) (Kagoshima Univ.)*

	1	2	3	4	5	6	7	8	9	10	Total
Total cases	27	27	27	27	27	25	26	21	13	13	233
Arrhythmias											
Card. arrest	0	3	3	2	2	2	1	0	0	0	13
Nodal rhythm.	0	2	1	1	3	1	1	0	1	0	10
A-V block	0	0	0	0	0	0	1	0	0	0	1
A-V dissociat.	0	1	1	4	0	0	0	0	0	0	6
Ventricular extra systole	0	1	0	0	0	0	0	0	0	0	1
Sinus arrhythmia	2	0	0	0	0	0	0	0	0	0	2
Arrhythmia total	2	7	5	7	5	3	3	0	1	0	33
P-wave changes	1	3	4	4	5	5	4	2	0	0	28
Total (arrhythmia + P-wave (changes)	3	10	9	11	10	8	7	2	1	0	61

Effect of Xylitol on Enzyme Activities after Hemorrhagic Shock in Dogs. Enzymatic characterization of hemorrhagic shock has not been established conclusively. The main factor which affects enzyme activity after hemorrhagic shock is obviously anoxia due to the incompleteness of effective blood flow. It has been reported that succinic dehydrogenase in heart muscle is inhibited by shock [15]. In the present experiment

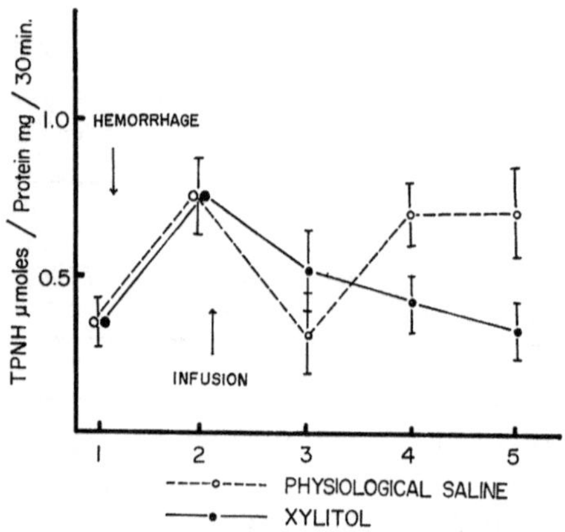

Fig. 7. The change of malic enzyme activity of muscle after hemorrhagic shock. Second arrow indicates the infusion of xylitol (5% solution) or physiological saline. The condition of the shock was as described in the text. (Osaka Univ.)

the activities of malic enzyme and succinic dehydrogenase of skeletal muscle and liver were determined. Shock was induced by withdrawing blood, and blood pressure was maintained at 40 mm Hg for one hour until 5% of the carbohydrate solution was infused (0.5 g/kg body weight). Muscle, liver and blood were sampled at 0, 30, 60 and 90 minutes after infusion of fluid had started. Before the blood withdrawal, those tissues were sampled as the control. Blood gas was analyzed polarographically, and acid-base balance by glass electrode at each period. The tissues were homogenized in 0.14 M KCl with a Potter-Elvehjem homogenizer, and the homogenate was centrifuged at 2,000 $\times$g for 10 minutes. The supernatant was again centrifuged at 20,000 $\times$g for 20 minutes. The precipitate was used for determination of succinic dehydrogenase activity, and the supernatant for the assay of malic enzyme activity [16].

As shown in Fig. 7, the activity of malic enzyme before shock was presented as 0.5 μmoles NADPH/mg protein/30 minutes. Marked elevation of the activity was recognized one hour after shock. The highest value was 1.5 μmoles NADPH/mg protein/30 minutes. The activity was decreased when saline or xylitol was infused. Thirty minutes after

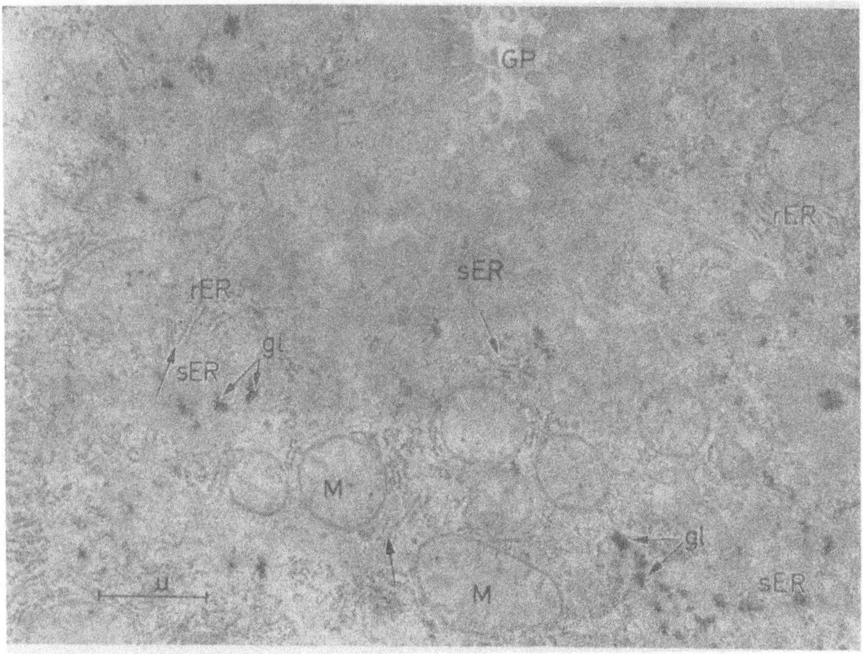

Figs. 8 and 9. Electron micrograph of rabbit liver without the administration of xylitol (Fig. 8) and after the administration of xylitol (2.5 g/kg/day) for one week (Fig. 9). (Kyoto Univ.) Magnitude ×20,000 (Fig. 8), ×15,000

infusion, the mean activity of malic enzyme was 0.8 with xylitol, but 0.5 with physiological saline. However, a large difference between saline and xylitol was recognized thereafter. Activity was not decreased in the saline group, but it was decreased in the xylitol group. During shock the content of potassium in blood did not change markedly, but arterial blood pH decreased. A rise of lactate, pyruvate and glucose in blood during shock was recognized. However differences of those changes between saline and xylitol infusion were not observed.

Activity of succinic dehydrogenase in muscle was slightly decreased during shock, whereas in liver mitochondria elevated by shock. No marked change of its activity in muscle or liver mitochondria was observed with infusion of saline or xylitol.

The metabolic changes of shock have been reviewed by Eagle [17]. Hyperglycemia, and increases of pyruvate and lactate in blood after shock were recognized again in the present experiments.

It has been reported that hyperglycemia is due to reflex stimulation of the adrenal gland [18]. Hyperglycemia after shock was attributed to glucose output from glycogen in liver [19].

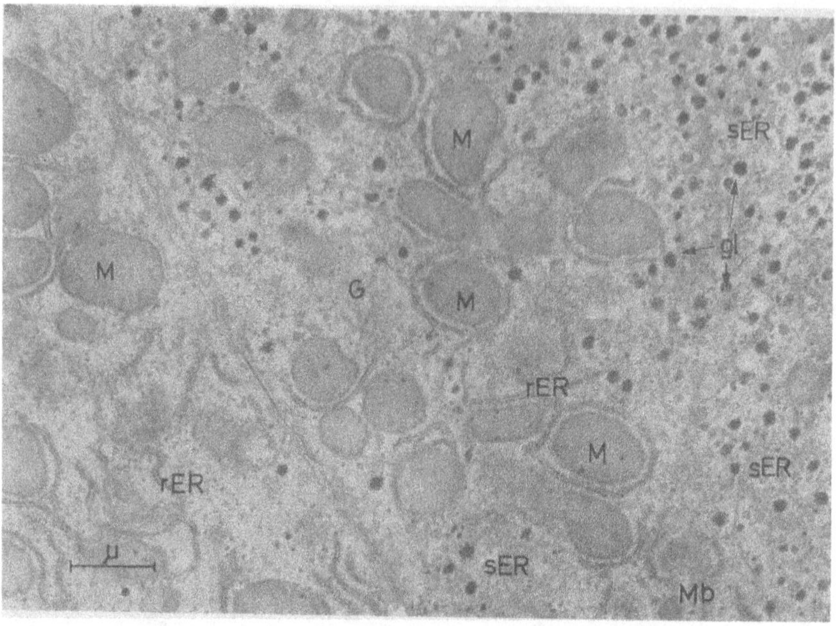

(Fig. 9); *sER* smooth-surfaced endoplasmic reticulum; *rER* rough-surfaced endoplasmic reticulum; *gl* glycogengranules; *M* mitochondria; *G* Golgi apparatus; *Mb* micro body

Next experiment showed that under normal condition glycogen is deposited in the liver after xylitol administration to rabbit. Fig. 8 is an electron micrograph of normal rabbit liver without the administration of xylitol. After xylitol administration (2.5 g/kg/day) for one week, glycogen granules appeared, as indicated in Fig. 9. However, under our experimental conditions, the rise in blood glucose was not observed by infusion of xylitol. Therefore it seems likely that the rate of gluconeogenesis was too low to induce the output of glucose from the tissue in this experiment.

Malic enzyme catalyzes the following reaction and is widely distributed.

$$\text{Pyruvate} + CO_2 + \text{NADPH} + H^+ \rightleftharpoons \text{Malate} + \text{NADP}$$

The reaction tends to favor synthesis of malate, rather than its breakdown. The hemorrhagic shock causes oxygen deficiency, and accumulation of reduced NADP will occur. Even in a normal state malic enzyme tends to move the reaction to the right, owing to the high ratio of NADPH/NADP in cells. After shock, pCO_2 is decreased.

Therefore, CO_2 tension in tissues might be decreased. During shock, muscle will be influenced previous to liver. We believe that the drop in tissue pCO_2 and the increase in the ratio of NADPH/NADP may cause the elevation of malic enzyme activity.

References

1. Burns, J. J., C. Evans, and N. Trousof: Stimulatory effect of barbital on urinary excretion of L-ascorbic acid and non-conjugated D-glucuronic acid. J. biol. Chem. **227**, 785 (1957).
2. Momose, T., A. Inaba, Y. Mukai, and S. Watanabe: Organic analysis. 23. Determination of blood sugar and urine sugar. Talanta **4**, 33 (1960).
3. Friedmann, T. E., and G. E. Haugen: Pyruvic acid. II. The determination of keto acids in blood and urine. J. biol. Chem. **147**, 415 (1943).
4. Barker, S. B., and W. H. Summerson: The colorimetric determination of lactic acid in biological material. J. biol. Chem. **138**, 535 (1941).
5. Tauskky, H. H., and E. Shorr: A microcolorimetric method for the determination of inorganic phosphorus. J. biol. Chem. **202**, 675 (1953).
6. Kvam, D. C., J. G. Schmidt, D. A. Riggiol, and D. G. Gallo: Colorimetric determination of plasma free fatty acids. J. Pharmacol. Sci. **53**, 988 (1964).
7. Ross, E. L., and L. H. Davis: The role of the pancreas in hyperglycemia from ether. Amer. J. Physiol. **53**, 391 (1920).
8. Drucker, W. R., C. Costley, R. Stults, W. D. Holden, J. Craig, M. Miller, N. Hoffman, and H. Woodward: Studies of carbohydrates metabolism during ether anesthesia: effect of ether on glucose and fructose metabolism. Metabolism **8**, 828 (1959).
9. Mann, F. C.: The effects of complete and of partial removal of the liver. Medicine (Baltimore) **6**, 419 (1927).
10. Major, S. G., and J. L. Bollman: Effect of ether and iso-amylethylbarbiturate (Amytal) anesthesia on the glycogen content of skeletal muscle. Proc. Soc. exp. Biol. Med. (N.Y.) **29**, 1109 (1931/32).
11. Annamunthodo, H., V. J. Keating, and S. J. Patrick: Liver glycogen alteration in anesthesia and surgery. Anaesthesia **13**, 429 (1958).
12. Haydon, R. K.: The antiketogenic effects of polyhydric alcohols in rat liver slices. Biochim. biophys. Acta (Amst.) **46**, 598 (1961).
13. Bässler, K. H., u. G. Dreiss: Antiketogene Wirkung von Xylit bei alloxandiabetischen Ratten. Klin. Wschr. **41**, 593 (1963).
14. Laborit, H.: De la prophylaxie à la thérapeutique de choc dit «irreversible». Presse méd. **74**, 2079 (1966).
15. Burdette, W. J.: Oxygen consumption of cardiac muscle during shock. Amer. J. Physiol. **168**, 575 (1952).
16. Ochoa, S., A. H. Mehler, and A. Kornberg: Reversible oxidative decarboxylation of malic acids. J. biol. Chem. **167**, 871 (1947).
17. Eagle, F.: The significance of the metabolic changes during shock. Ann. N.Y. Acad. Sci. **55**, 381 (1952).
18. Ellis, S.: The metabolic effects of epinephrine and related amines. Pharmacol. Rev. **8**, 485 (1956).
19. Drucker, W., and J. De Kiewiet: Glucose uptake by diaphragms from rats subjected to hemorrhagic shock. Amer. J. Physiol. **206**, 317 (1964).

Discussion

Dr. NINOMIYA: Did you measure DPNH-DPN and TPNH-TPN ratios?

Dr. YOSHIKAWA: I did not measure the ratios.

Dr. HALMÁGYI: Have you controlled the pCO_2 values in your investigations with xylitol under the condition of anaesthesia or the pH-values only?

Dr. YOSHIKAWA: Arterial pCO_2, pO_2 and base excess were determined. The acidosis was found to be metabolic.

Dr. FURUKAWA: We checked pH and pCO_2 in every case, and tried to maintain pCO_2 at around 40 mm Hg. So cases with high or low pCO_2 were excluded from this paper.

Dr. ASAKURA: Do you have any idea for the explanation of decreased Pi level in blood after administration of xylitol?

Dr. YOSHIKAWA: No.

Dr. FEKL: Is there any explanation for the very interesting observation on the high rise of lactate after infusion of fructose in contrast to infusion of sorbitol and xylitol? Perhaps Dr. Bässler has an explanation for this.

Dr. BÄSSLER: For an explanation of the difference in blood lactate levels during fructose infusion on one hand and sorbitol and xylitol infusions on the other hand, I refer to our paper „Biochemische Grundlagen für Wirkungsunterschiede zwischen Sorbit und Fructose" [Hoppe-Seylers Z. physiol. Chem. **348**, 533 (1967)].

Observations on the Anticatabolic Effect of Xylitol in the Posttraumatic Phase

K. SCHULTIS and C. A. GESER

Chirurgische Klinik of the Giessen University,
Giessen, West Germany
and Medizinische Poliklinik of the Munich University,
München, West Germany

At the beginning of the century, Abderhalden sen. was of the opinion that an intravenous nutrition might be developed with the basic elements of different substances. Since, during the last 20 years, this prediction has been fulfilled — as far as the metabolism of proteins is concerned — by amino acid solutions of a relatively ideal composition [3, 17—19], we were confronted with the task of finding new ways allowing the parenteral administration of immediately utilizable metabolites of the energy metabolism in sufficient quantities. In practice we have to choose between glucose, fructose, sorbitol and xylitol. There are frequently hyperglycemia and glucosuria in the posttraumatic phase indicating a disturbed utilization of glucose. By means of an intravenous administration of glucose according to Conard [5] we were able to find a disturbed glucose assimilation in more than 20 patients after laparotomy or bone fractures by calculating the elimination constant K_G. The value of glucose infusions during catabolism caused by stress is doubtful. The administration of fructose, sorbitol and xylitol seems to be more advantageous.

Since, during the last few years several teams [1, 10, 11, 20] had repeatedly been able to demonstrate that xylitol had an especially intense antiketogenic effect, we thought it worthwhile to examine this sugar alcohol on its ability to influence the posttraumatic changes in metabolism.

Methods

The first informative tests were carried out in patients (men) after laparotomy or with fractures between the age of 20 to 50 years who had been fasting for 12 hours. In three successive days they were given 1.5 g/kg body weight of sorbitol, xylitol or fructose, respectively, during one hour. Fructose was assessed according to Schmidt [15][1].

1. We thank Prof. Bässler [2] for the determinations of sorbitol and xylitol.

In 10 metabolically normal patients at the age of 20 to 40 who had been operated because of benign stomach and biliary diseases, the glucose utilization was examined preoperatively according to Conard. Samples of venous blood were drawn in fasting condition, 30 to 60 minutes after the end of the glucose injection. Glucose was assessed enzymatically according to Schmidt [15].

The NEFA analysis was made according to Mosinger [12], β-hydroxy-butyrate and aceto-acetate was assessed enzymatically according to Bergmeyer *et al.* [4]. In the 24-hour-urine, the total nitrogen was measured according to Kjeldahl and sodium and potassium were measured flame-photometrically. All the values from urine analyses are mean values of two 24-hour-quantities. We proceeded in the same way during the first three postoperative days. The patients had been nourished normally up to the day before the operation.

The surgical patients were divided into two groups. Four patients (group I) received exclusively water and electrolytes (on an average 2.5 l per day) beginning with the end of the operation, whereas the other 6 test persons (group II) received 100 g/day of xylitol in 5% solution intravenously. The missing volume was substituted during the night in form of Ringer solution.

Results

Fig. 1 presents the results of the administered quantities. Although in this experiment the applied quantity of carbohydrates is extremely high per time unit, the concentrations for the polyols increase only insignificantly — sorbitol from zero to 18 mg-% and xylitol from zero to 9 mg-%. As for sorbitol, 1 hour and as for xylitol, 2 hours after the end of the infusion there is no longer any concentration in the blood to be measured. This corresponds to the rapid turnover known for these substances. We want to draw your attention to the fact that sorbitol can be assessed by polyoldehydrogenase only up to 10 mg-% and xylitol by the same enzyme only up to 2 mg-%. When fructose is infused, the actual increase of the concentration of the supplied substances from 6 to 116 mg-% is by many times higher than when polyols are administered. The increase of the glucose concentration is always the same during the infusion of the three sugars. The average excretion of glucose in the 24-hour-urine is the lowest after xylitol. We consider this as an indication for its good utilization which can be explained by the fact that xylitol enters the metabolism via the pentose-phosphate-cycle. Hence, there is no influence on the concentration of fructose in the blood and on the excretion of fructose via the kidney during or after infusion of xylitol. As expected, both the concentration and the

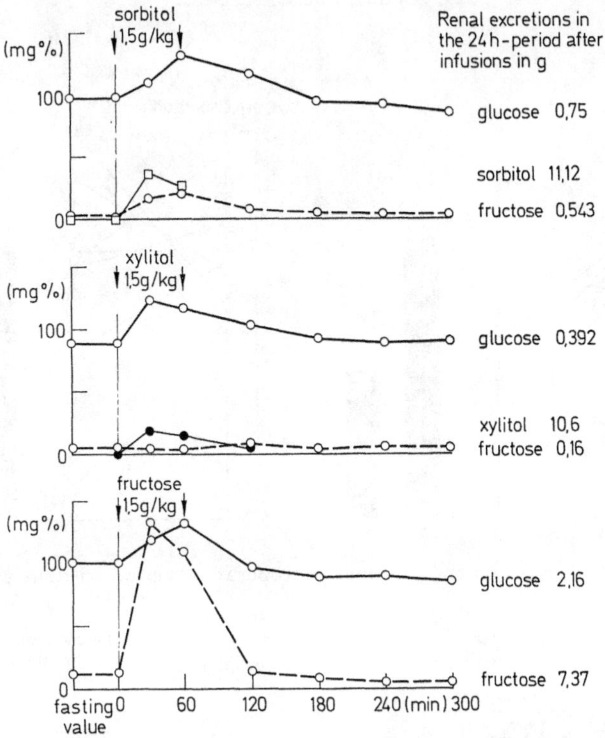

Fig. 1. Values of glucose, fructose, sorbitol and xylitol (concentrations in blood) during and after infusions of sorbitol, xylitol and fructose (5 men 1—3 days after operation)

excretion increase, however, during the supply of sorbitol, since the introduction of sorbitol occurs via fructose.

Because of these results which favour xylitol, we investigated different phases of the intermediary metabolism when xylitol was given in the posttraumatic phase.

In the following we shall report on the surgical patients of group I and II:

Fig. 2 presents the elimination constants K_G for both groups and the mean values in the form of columns for pre- and postoperative phases. Though the K_G-values show an individually different reaction, their mean values and their standard deviation made clear the remarkable difference between these groups. The utilization of glucose is generally not influenced postoperatively, when xylitol was given.

Fig. 3 shows the difference between the fasting blood glucose and the nitrogen losses (expressed as absolute differences) in the pre- and postoperative periods for both groups. In group II — i.e. when

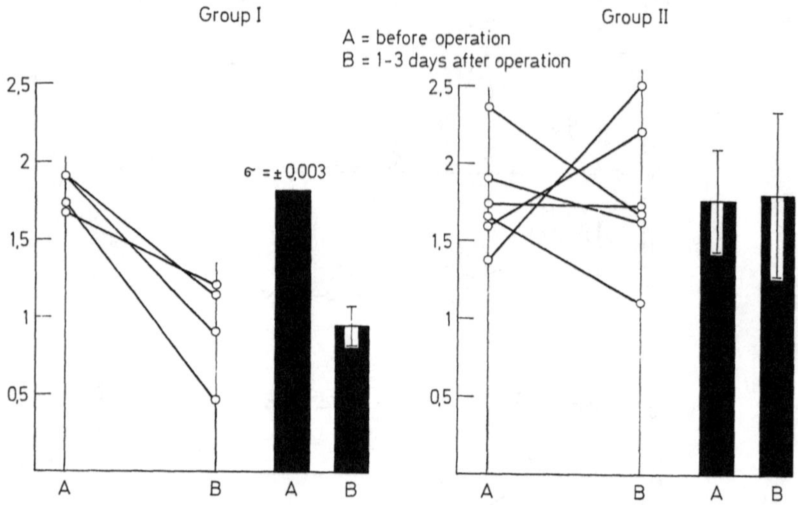

Fig. 2. Ratio of K_G in the pre- and postoperative period without and with xylitol

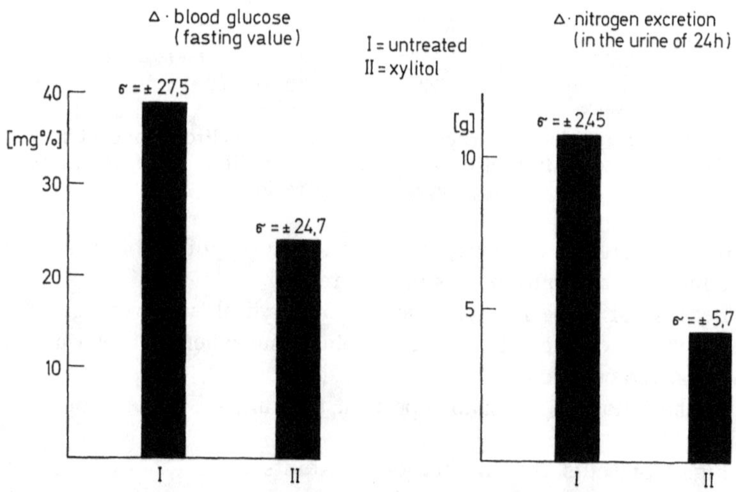

Fig. 3. Fasting value of blood glucose and nitrogen excretion without and with xylitol in comparison with the preoperative period

xylitol was given — a reduced increase of the fasting blood glucose in the postoperative period by approximately 15 mg-% on the average is observed, a standard deviation being found (group I: $s = \pm 27.5$; group II: $s = \pm 24.7$). Hyperglycemia in the postoperative days is, however, less prominent when xylitol is administered; this corresponds also to the reaction of the K_G-values already shown.

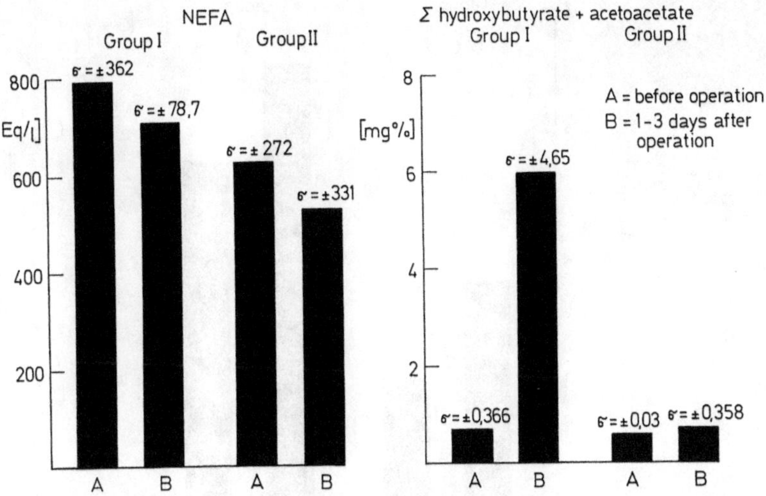

Fig. 4. NEFA and ketone bodies ($\Sigma\beta$-hydroxybutyrate + acetoacetate) in the pre- and postoperative period without and with xylitol

While the nitrogen loss of group I in 24 hours is in the postoperative period by 10.7 g higher than in the preoperative period, this increase amounts to only 4.3 g in group II where xylitol was used. This marked nitrogen-saving effect which is evident in spite of the high standard deviation (see above) must be evaluated as a reduction of the gluconeogenesis from endogenic amino acids.

In the postoperative period, an increased turnover of endogenic fatty acids takes place [14] which results in ketone bodies when no carbohydrates are offered and the ability of utilizing glucose is decreased.

Fig. 4 shows postoperatively no increase of NEFA levels. The development of the ketone bodies is obvious in group I. The figure presents them as the sum of β-hydroxybutyrate and aceto-acetate in blood in mg-%. These increased ketone bodies are no longer found when xylitol has been used. This effect is so prominent that postoperatively the standard deviation of the mean values — indicated as "s" above the columns — does not show in intersection.

Another symptom of the posttraumatic catabolism is the loss of potassium during retention of sodium. It is not yet clear to what extent this phenomenon which leads to a decrease of the sodium/potassium-ratio under 1, is exclusively a result of the posttraumatic temporary secondary hyperaldosteronism, whether the disturbances in intermediary metabolism with an increased catabolism of protein and glycogen and a decrease of the syntheses play the most important part.

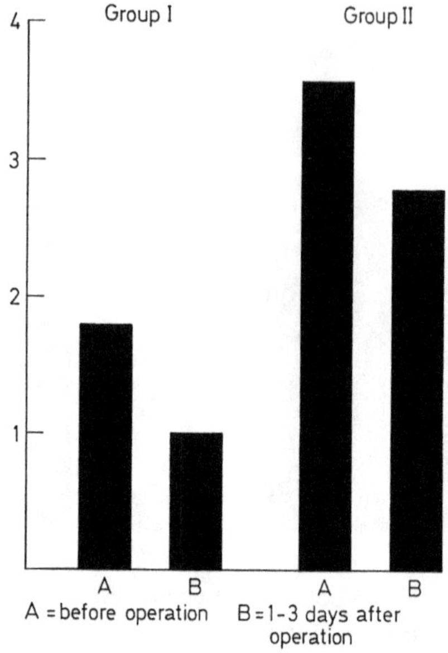

Fig. 5. Na^+/K^+-quotient in the pre- and postoperative period without and with xylitol

In Fig. 5 we can show that the decrease of the sodium/potassium-ratio under 1 does not occur in xylitol administration.

The reasons for our investigations as to possible changes of insulin concentration in serum when xylitol was given were the following:

The favourable effect of xylitol in influencing individual metabolic phases in posttraumatic catabolism, the results of the experiments of the groups of Hirata [8] and Kuzuya [9], according to which xylitol causes an increased liberation of insulin in dogs and the possibility of an absolute or relative lack of insulin — either by lower incretion or by its inhibited effectiveness [13] — being essential for an impairment of glucose utilization and the development of posttraumatic catabolism [16].

0.8 g of xylitol/kg body weight were infused in 15 minutes in 10 healthy subjects of an average age of 24 years [6]. Serum insulin [7], glucose and NEFA were assessed before the infusion, immediately after the end of the infusion and 5, 15, 30, 60 and 120 minutes after the end of the infusion.

Fig. 6 comprises the mean values of the results. The levels of blood glucose rise slightly after xylitol infusion — 15 minutes after the infusion by 5 mg-% — whereas 120 minutes after the infusion they show a level of 14 mg-% below the initial level. The levels of NEFA fall from an initial value of 0.410 meq/l to 0.220 meq/l after 60 minutes. The serum insulin values show a slight increase from 51 μU/ml to

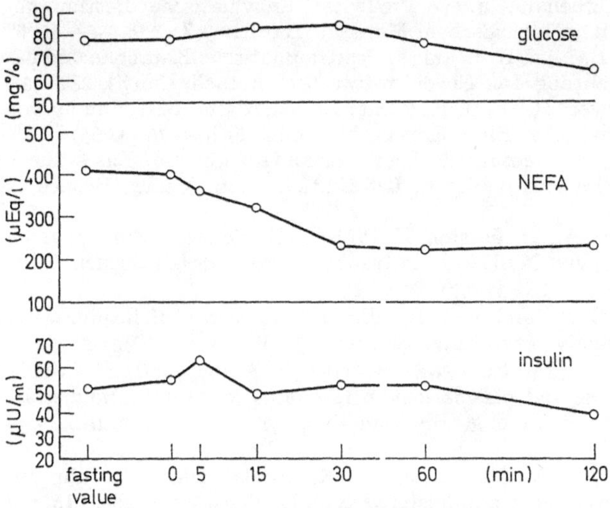

Fig. 6. Blood glucose, NEFA and serum insulin after administration of 0.8 g/kg body weight of xylitol

63 μU/ml after 5 minutes, but this remains within the error of the method and did not prove statistically significant.

According to our results an improvement is to be observed in the same way by postoperative application of xylitol as it is known by the administration of other carbohydrates. This favourable effect is due to the special position of xylitol in the intermediary metabolism and gives further evidence for the good utilization of the sugar alcohol. The other possibility, i.e. that the anabolic effect of xylitol infusions might solely be a result of an increase of insulin secretion, was eliminated by our investigation.

Summary

By giving xylitol to operated patients, we can observe a normalisation of the disturbances of glucose utilization, and a reduction of the formation of ketone bodies as well as an obviously reduced loss of nitrogen. The hyperglycemic blood glucose levels and a deviation of the sodium/potassium-ratio are influenced towards normal. Compared with sorbitol and fructose, xylitol utilization shows some advantages. The serum insulin values in healthy persons are not influenced by xylitol.

References

1. Bässler, K. H., W. Prellwitz, V. Unbehaun u. K. Lang: Xylitstoffwechsel beim Menschen. Zur Frage der Eignung von Xylit als Zucker-Ersatz beim Diabetiker. Klin. Wschr. **40**, 791—793 (1962).

2. — V. Unbehaun u. W. Prellwitz: Enzymatische Bestimmungen von Xylit in biologischem Material. Biochem. Z. **336**, 35—40 (1962).
3. Bauer, B. L., u. K. Schultis: Posttraumatische Katabolie und parenterale Ernährung. Der Eiweißstoffwechsel. Actuelle Chir. **1**, 281—286 (1967).
4. Bergmeyer, H. U., u. E. Bernt: Enzymatische Bestimmung von Keton-Körpern im Blut. Enzym. biol. clin. **5**, 65—76 (1965).
5. Conard, V.: Mesure de l'assimilation du glucose. Bases theoriques et applications cliniques. Les Editions Acta Medica Belgica, Bruxelles 1955.
6. Geser, C. A., H. Förster, H. Pröls u. H. Mehnert: Zur Frage einer Wirkung von Xylit auf die Insulinsekretion des Menschen. Klin. Wschr. **45**, 851—852 (1967).
7. Hales, C. N., and P. J. Randle: Immunoassay of insulin with insulin-antibody precipitate. Biochem. J. **88**, 137 (1963).
8. Hirata, Y., M. Fujisawa, H. Sato, T. Asano, and S. Katsuki: Blood glucose and plasma insulin responses to xylitol administrated intravenously in dog. Biochem. biophys. Res. Commun. **24**, 471—475 (1966).
9. Kuzuya, T., Y. Kanazawa, and K. Kosaka: Plasma insulin response to intravenously administered xylitol in dogs. Metabolism **15**, 1149—1152 (1966).
10. Lang, K.: Xylit als Nahrungskohlenhydrat. Med. u. Ernähr. **4**, 45—50 (1963).
11. Mehnert, H., J. D. Summa u. H. Förster: Untersuchungen zum Xylitstoffwechsel bei gesunden, leberkranken und diabetischen Personen. Klin. Wschr. **42**, 382—387 (1964).
12. Mosinger, F.: Photometric adaptation of Dole's microdetermination of free fatty acids. J. Lipid Res. **6**, 157—159 (1965).
13. Randle, P. J.: The glucose fatty acid cycle, its role in insulin sensitivity and the metabolic disturbances of diabetes mellitus. Lancet **1963**, 785—789.
14. Rodewald, G., u. H. Harms: Beziehungen zwischen P_{O_2} und P_{CO_2} in Normo- und Hypothermie. Anaesthesist **13**, 113—117 (1964).
15. Schmidt, F. H.: Die enzymatische Bestimmung von Glucose und Fructose nebeneinander. Klin. Wschr. **39**, 1244—1247 (1961).
16. Schultis, K.: Utilisation i.v. zugeführter Aminosäuren unter Insulin. Symposion Int. Soc. of Parenteral Nutrition Hamburg 5./6. 8. 1966. Lochham b. München: Pallas Verlag 1967.
17. — Erfahrungen bei der parenteralen Ernährung chirurgischer Patienten. In: Parenterale Ernährung, hrsg. von K. Lang, R. Frey u. M. Halmágyi in Anaesthesiologie und Wiederbelebung, Bd. 6, S. 80—86, hrsg. von R. Frey, F. Kern u. O. Mayerhofer. Berlin-Heidelberg-New York: Springer 1966.
18. —, u. B. L. Bauer: Grundlagen und praktische Anwendung der parenteralen Ernährung in der postoperativen Phase. Z. prakt. Anästh. Wiederbeleb. **2**, 228—239 (1967).
19. —, u. H. L'Allemand: Zur Ernährung Tetanuskranker. Anaesthesist **17**, 196—201 (1968).
20. Toussaint, W., K. Roggenkamp u. K. H. Bässler: Behandlung der Ketonämie im Kindesalter mit Xylit. Z. Kinderheilk. **98**, 146—154 (1967).

Discussion

Dr. HALMÁGYI: Did you calculate the nitrogen-potassium ratio in the urinari output in your experiments.

Dr. SCHULTIS: No, we did not.

Effect of Xylitol Administration on Levels of Blood Sugar and Plasma Free Amino Acids Following Abdominal Surgery

MINORU UEKUSA, TOICHI TOSEN, MASATOMI OHBA, TAKASHI AKAMATSU, NORIYUKI NARAHARA, and SUSUMU KODAIRA

Department of Surgery, Keio University School of Medicine, Tokyo, Japan

It is important to supply carbohydrate to patients who have undergone major abdominal surgery. In order to evaluate xylitol as a postoperative carbohydrate source, the effect of xylitol administration on blood sugar, plasma free amino acids and urea was compared with that of glucose administration following some abdominal surgery.

76 patients, aged 23 to 72, were selected for this study. The majority of them had a partial gastrectomy for peptic ulcer or gastric cancer. They were within 10% of their standard weight and none had evidence of any endocrine disease, renal disease or hypertension. Fluid and electrolyte requirements were given post-operatively. The operations were carried out in the morning with pentothal induction, nitrous oxide, oxygen and fluothane inhalation as anesthesic; it lasted between 1.5 and 4 hours. They were regarded as abdominal surgery of moderate severity. All measurements were performed on blood samples taken from a forearm vein of patients.

Changes in post-operative blood sugar level and glucose tolerance in gastric cancers and other benign diseases are indicated in Fig. 1. Blood sugar levels were most conspicuously raised immediately after completion of surgery, still were elevated after 24 hours, and then recovered to initial levels on the 3rd to 5th post-operative day. The glucose tolerance was markedly delayed after 24 hours, and in some case it was not recovered even on the 3rd post-operative day. These changes were relatively large and tended to exhibit a retarded recovery in most gastric cancer cases. Thus it is evident that carbohydrate utilization is depressed during the first few days after abdominal surgery.

It has been proposed that the levels of plasma free amino acids are very closely related to the nutritional condition of individuals and that they are reasonable indicator of the protein metabolism reflecting surgical stress [1—3]. In our study, eighteen individual plasma free amino acids were determined by ion exchange chromatography on columns of Amberlite resins, using an automatic amino acid analyzer, as described by

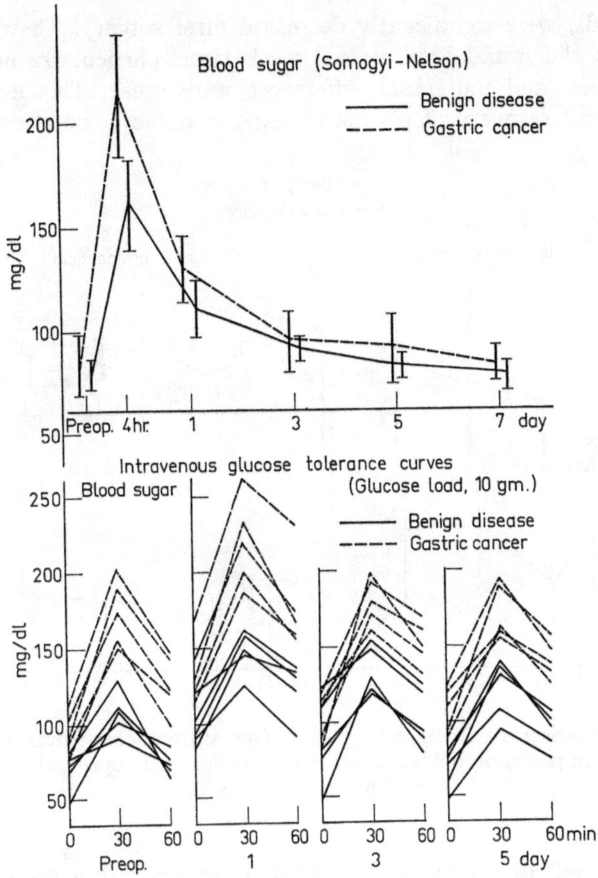

Fig. 1. Changes of blood sugar and glucose tolerance in patients undergone surgery

Spackman, Stein and Moore. Although all measured amino acids decreased immediately after completion of surgery, their subsequent fluctuations were characteristic of each individual amino acid.

The ketogenic and other amino acids decreased immediately after surgery recovered to normal levels at the 1st or 3rd post-operative day or to higher than the preoperative levels. However the extent of these differences was generally little, and presented a small proportion of the changes in total plasma free amino acids. These changes are shown in Fig. 2 by mean value and range of values obtained in gastric cancers and other benign diseases.

Glucogenic amino acids, including all non-essential amino acids except threonine and valine, and about 70% of the total plasma free amino acids, were significantly decreased after surgery. They did not recover to the initial level even a week later, particularly in gastric cancer cases, and individual differences were great. This decreasing tendence and recovery of plasma glucogenic amino acids after surgery

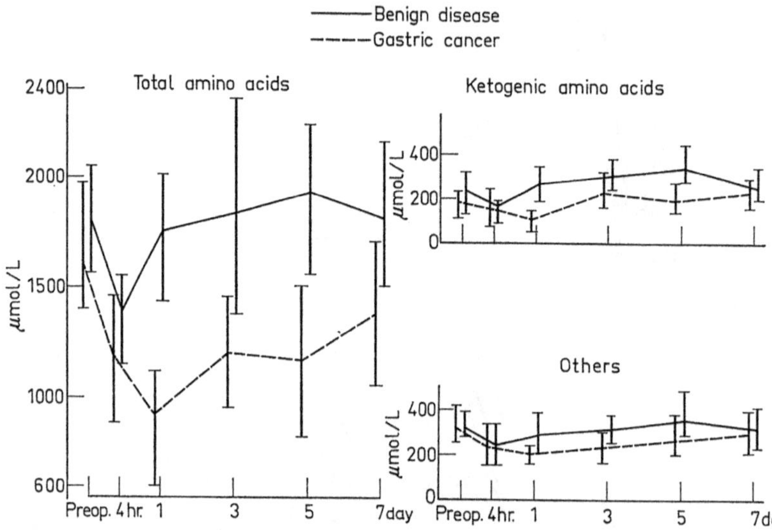

Fig. 2. Comparison of changes on plasma free amino acids (total, ketogenic and others) in patients undergone surgery. Vertical line represents maximum and minimum ranges

responded well to the change of blood sugar levels (Fig. 3). Although the effect of starvation and the post-operative change in blood sugar levels are not caused unitarily, one possibility might be that the glucogenic amino acids are utilized for post-operative energy metabolism.

It has been well established that glucose causes protein sparing and a decrease of plasma free amino acids [4, 5]. The changes in blood sugar and plasma glucogenic amino acids levels in the peptic ulcer cases which have had an infusion of 75 g of xylitol or glucose daily lasting for 6 days from the day of surgery are illustrated in Fig. 4.

No remarkable difference was observed between administration of xylitol and glucose in regard to the fasting blood sugar level, whereas plasma glucogenic amino acids levels on the first few post-operative days revealed a relatively smaller decrease in the xylitol-administered cases than in the glucose cases. However, it is impossible to supply required calories only by post-operative parenteral administration of carbohydrate.

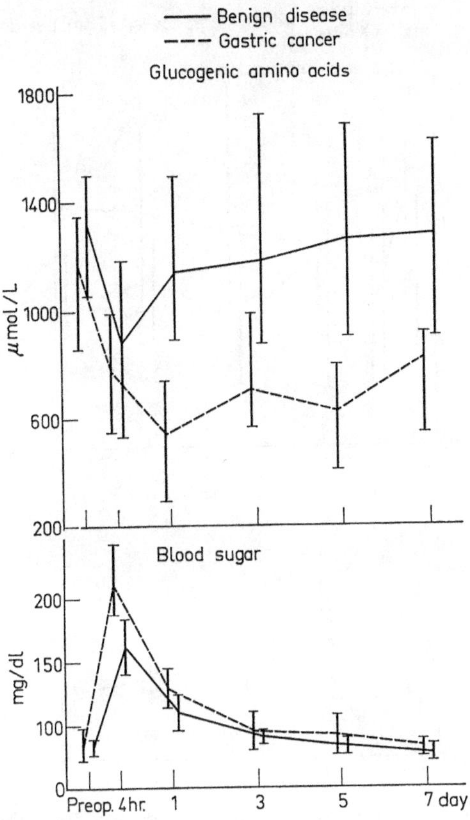

Fig. 3. Correlation of changes between levels of plasma free glucogenic amino acids and blood sugar. Vertical line on upper figure represents maximum and minimum ranges. Vertical line on lower figure represents ± S.D.

In order to clarify this disparity, then, the patients subjected to gastrectomy received intravenous loading of 500 ml of either 10% xylitol or glucose for 120 minutes at 24 hours after surgery and in the morning of the 3rd post-operative day. Amino acid levels at preloading and 30 minutes after termination of loading, when changes are most significant, were determined. The results were compared with those obtained at preoperative day by the same method (Fig. 5). The decrease of all amino acids was observed by preoperative xylitol or glucose loading except for a slight elevation of glycine by xylitol loading. The extent of change was different for each individual amino acid, and the decrease of glucogenic amino acids was remarkable especially in glucose loading. A similar effect was obtained by loading on the 3rd post-operative day. The level of plasma glucogenic amino acids, however, by glucose loading

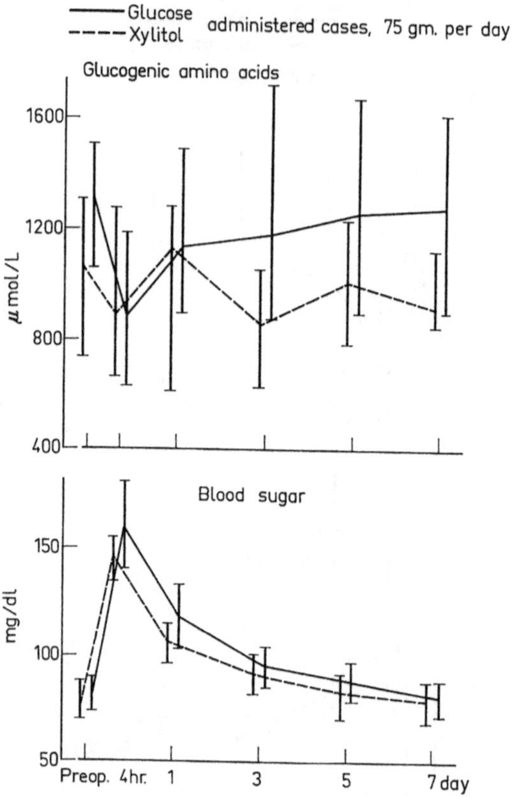

Fig. 4. Comparison of changes on levels of plasma free glucogenic amino acids and blood sugar between glucose and xylitol administered cases after surgery. Vertical line on upper figure represents maximum and minimum ranges. Vertical line on lower figure represents ± S.D.

24 hours after surgery displayed either no decrease in most of them or a notable elevation in some cancer cases.

A main factor in these changes of plasma glucogenic amino acids was the change of alanine, glycine and proline (Fig. 6).

By xylitol loading 24 hours after surgery, however, all plasma glucogenic amino acids except glycine decreased, and it resulted in a slight decrease in total glucogenic amino acids. The disparity of effect on the levels of plasma free amino acids by glucose or xylitol loading 24 hours after surgery was remarkable in glucogenic amino acids. No marked difference was noted among ketogenic and other amino acids.

According to the previous report by Albanese [6], the changes induced by the administration of carbohydrate in the blood amino nitrogen levels of fasting subjects might serve as a dynamic index of its protein-sparing

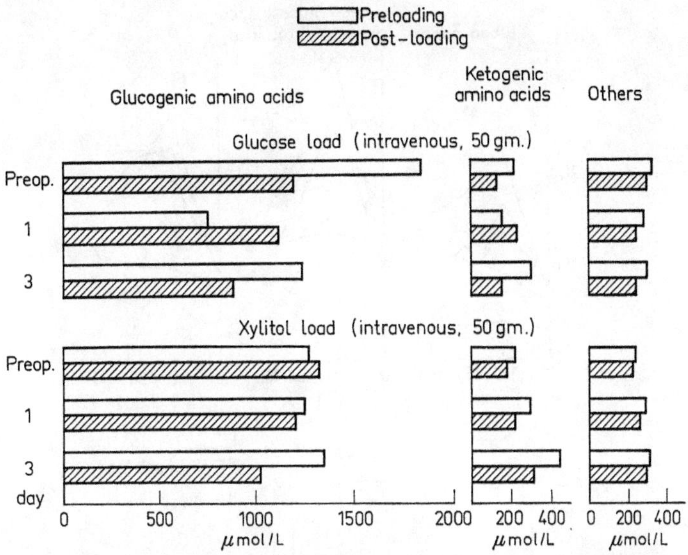

Fig. 5. Effect on plasma free amino acids (glucogenic, ketogenic and others) after glucose or xylitol loading in patients undergone surgery

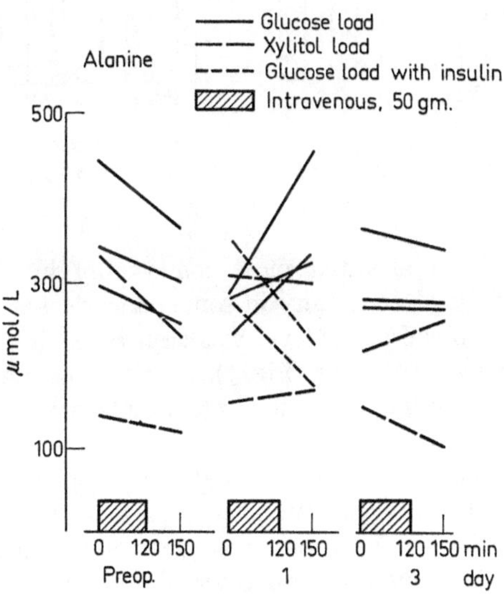

Fig. 6. Changes on plasma alanine level after carbohydrates loadings in patients undergone surgery

activity. Thus glucose administered during this period is likely to have less protein-sparing effect than xylitol.

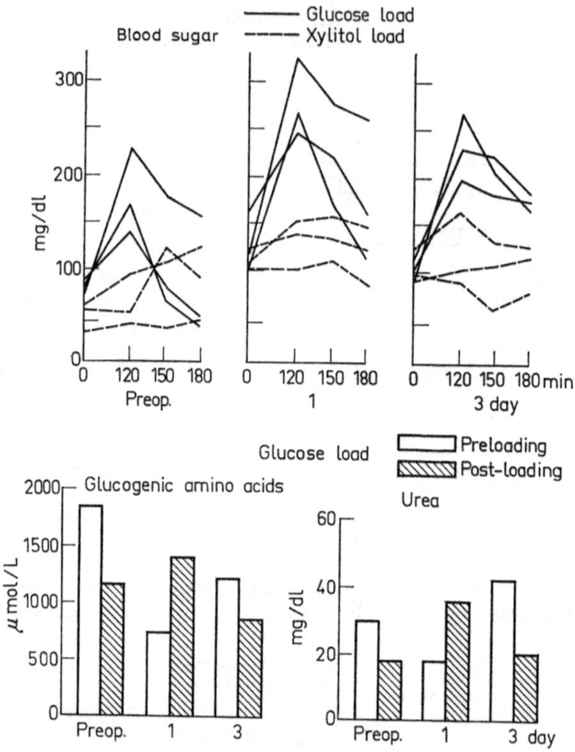

Fig. 7. Changes of blood sugar, plasma glucogenic amino acids, and urea after 50 g of intravenous carbohydrates loadings for 120 minutes in patients undergone surgery

Blood sugar level elevated most remarkably by glucose loading 24 hours and 3 days after surgery in some cancer cases. However, only slight elevation in blood sugar level was seen by xylitol loading during pre- and post-operative periods (Fig. 7). Also the plasma urea level was elevated in corresponding with increase in plasma glucogenic amino acids after post-operative glucose loading.

Repeatedly it has been demonstrated that exogenous insulin causes not only a fall in blood sugar but also in plasma free amino acids [7]. When 10 units of regular insulin was given subcutaneously prior to glucose loading, blood sugar elevation was definitely restrained, even on the first post-operative day (Fig. 8). Also seen was the depressed increase in plasma glucogenic amino acids, particularly alanine and glycine, and that in plasma urea concentration. Thus exogenous insulin seems to enhance glucose utilization, and results in a decrease of plasma free amino acids.

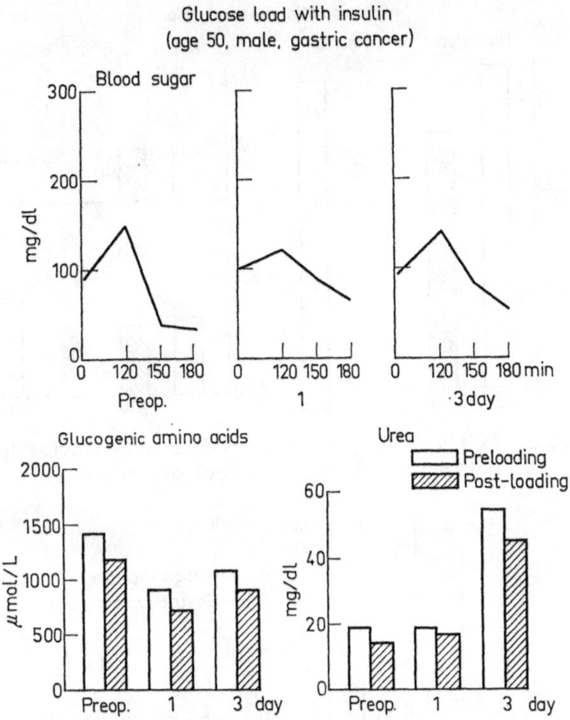

Fig. 8. Changes on blood sugar, glucogenic amino acids, and urea after 50 g of intravenous glucose loading with simultaneous administration of insulin to the patient with gastric cancer undergone surgery

When 25 units of ACTH dissolved in 500 ml of normal saline was loaded intravenously for 6 hours in preoperative patient [8], the level of plasma glucogenic amino acids decreased at the completion of loading (Fig. 9). Many experimental investigations [9, 10] revealed that the increase in activity of serum transaminase and gluconeogenesis, and the elevation of blood sugar results from administration of glucocorticoids. Therefore, surgical stress stimulates the release of cortisol, which in turn may lead to enhanced transamination of glucogenic amino acids and gluconeogenesis. It also is well known that cortisol is a powerful insulin antagonist [11].

The plasma insulin, which was estimated by the double radio-immunoassay technique, was moderately elevated immediately after surgery for 24 hours, and returned to the preoperative level after 5 days (Fig. 10). Nevertheless, the results of glucose loading with or without exogenous insulin, glucose tolerance, plasma free amino acids levels and

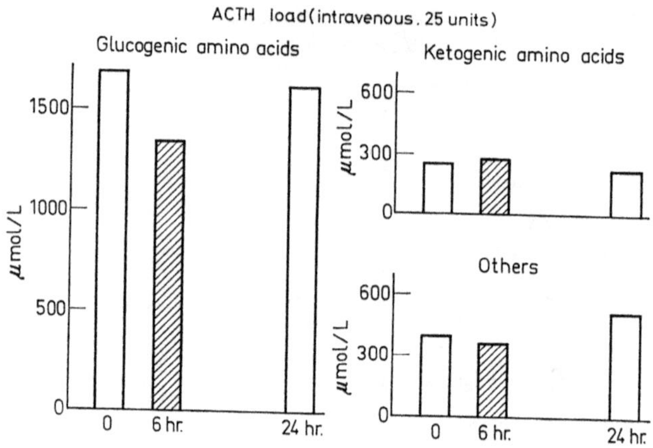

Fig. 9. Effect of ACTH loading on plasma free amino acids (glucogenic, ketogenic and others) in control studies

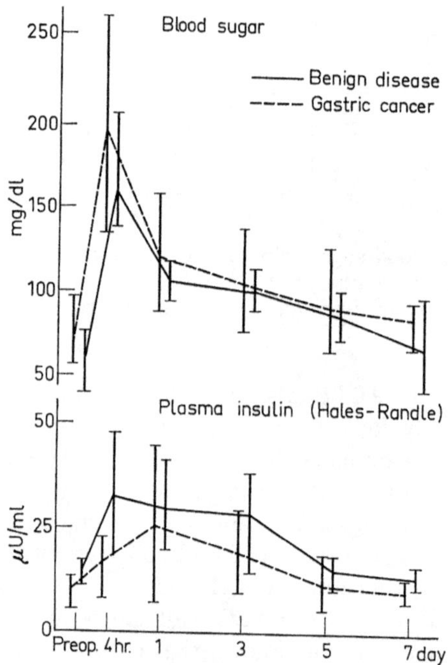

Fig. 10. Correlation of changes between levels of blood sugar and plasma insulin (I.R.I.) in patients undergone surgery. Vertical line represents maximum and minimum ranges

urea concentrations indicated that the diabetogenic condition appeared during the early post-operative period. The insulin antagonist which

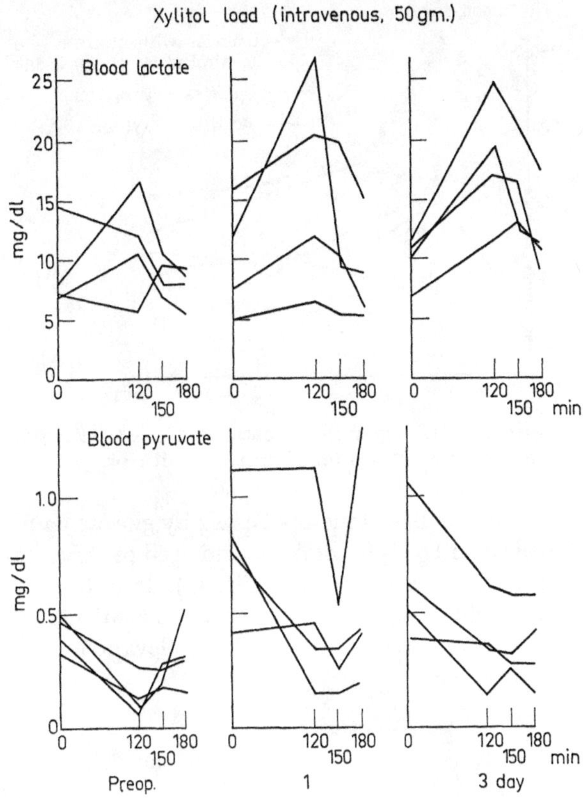

Fig. 11. Effect of pre- and post-operative xylitol loadings on blood lactate and pyruvate in patients undergone surgery

increases post-operatively in blood, is not only a cortisol [12]. Therefore it is thought that surgical stress induces comprehensive endocrine responses which act diabetogenically, resulting in the impairment of glucose utilization and then, interferring with amino acid metabolism.

Pyruvate and lactate in blood were elevated by post-operative glucose loading. Meanwhile, the decrease of pyruvate and increase of lactate were observed by xylitol loading during the same period (Fig. 11).

As xylitol utilization without insulin action [13] shows a protein-sparing effect and does not lead to a marked increase of blood sugar even when administered during the early post-operative period, xylitol may be advantageous for the post-operative parenteral use of glucose.

Many reports assumed the effect of xylitol administration on diabetic patients [14]. Surgical stress always aggravates the diabetic condition,

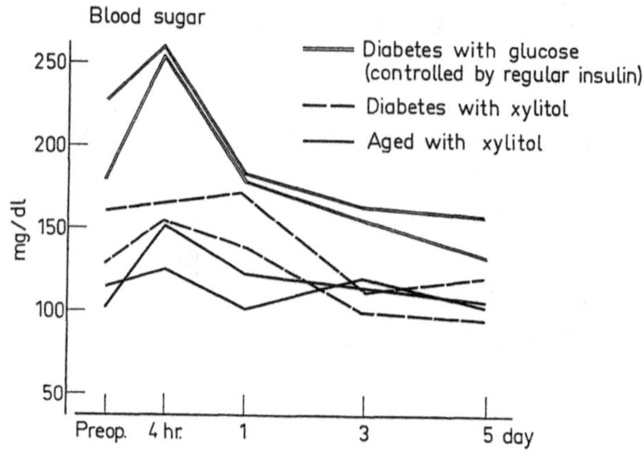

Fig. 12. Changes on blood sugar in diabetic and aged patients with post-operative glucose or xylitol administrations with or without insulin

and there is some possibility of metabolic risk by glucose administration. Changes in blood sugar levels for diabetic and aged patients who received xylitol post-operatively were plotted (Fig. 12). It is thought that the administration of xylitol to post-operative diabetic patients as the caloric source is preferable on account of its slight elevation of blood sugar and antiketogenic action.

References

1. Man, E. B., P. G. Bettcher, C. M. Cameron, and J. P. Peters: J. clin. Invest. **25**, 701 (1946).
2. Everson, T. C., and M. J. Fritschel: Surgery **30**, 931 (1951).
3. Albanese, A. A., L. A. Orto, and D. N. Zavattaro: Metabolism **7**, 256 (1958).
4. Munro, H. N., and W. S. T. Thomson: Metabolism **2**, 354 (1953).
5. Crofford, O. B., P. W. Felts, and W. W. Lacy: Proc. Soc. Exp. Biol. (N.Y.) **117**, 11 (1964).
6. Albanese, A. A., L. Orto, J. Rossy, R. Dilallo, and A. Belmont: Metabolism **4**, 160 (1955).
7. Lotspeich, W. D.: J. biol. Chem. **179**, 175 (1949).
8. Eik-Nes, K., A. A. Sandberg, D. H. Nelson, F. H. Tyler, and L. T. Samuel: J. clin. Invest. **33**, 1502 (1954).
9. Rosen, F., N. R. Roberts, L. E. Bundwick, and C. A. Nichol: Endocrinology **65**, 256 (1959).
10. Segal, H. L., R. G. Rosso, A. Hopper, and M. M. Weber: J. biol. Chem. **237**, 3303 (1962).
11. De Boto, R. C., and N. Altzulaer: Physiol. Rev. **38**, 389 (1958).
12. Ross, H., I. D. A. Johnson, T. A. Welborn, and A. D. Wright: Lancet **1966 II**, 563.
13. Prellwitz, W., u. K. H. Bässler: Klin. Wschr. **41**, 196 (1963).
14. Bässler, K. H., W. Prellwitz, V. Unbehaun u. K. Lang: Klin. Wschr. **40**, 791 (1962).

Closing Remarks

Now, we are approaching the end of our Symposium. It was the first of this kind in the world. We have discussed in detail the biochemistry, physiology, and clinical use of pentoses and pentitols. Most of the experimental and clinical work has been done with xylitol, and this substance proved to be a very interesting one with a broad spectrum of useful therapeutical properties. The high capacity of the body to utilize xylitol is interesting if not surprising. The independence of xylitol metabolism of insulin, its strong antiketogenic action, its depressing effect on free fatty acid level, and its ability to activate many metabolic pathways by increasing pentose phosphate cycle, are also important. Apart from this, xylitol does not only act as a useful substrate for correcting disturbances in the intermediary metabolism, especially in diabetic subjects, but it seems to me well-established that xylitol prevents a pathological activity decrease of the adrenal cortex.

Xylitol enhances the production and delivery of corticosteroids in blood. We understand therefore that xylitol may exert anabolic effects and has proved to be very useful in the treatment of the catabolic phases after hemorrhagic shock or surgical interventions.

Many papers presented during our Symposium gave some evidence that xylitol may have also other effects in the endocrinological system.

There is no doubt that xylitol because of its complex activity mechanism is not comparable to similar substances like sorbitol or fructose which act only as substrates and only in some restricted metabolic pathways.

I believe that this Symposium gave us first preliminary results but they are very interesting and encouraging.

We are not only impressed by the high quality of the Symposium but also by the friendly atmosphere in which the discussions took place. I hope that this Symposium may induce and activate new research work in this field.

Finally, I have to thank both companies, *Eisai Co.* and *J. Pfrimmer and Co.*, who rendered possible this Symposium. Especially, the sponsorship and wonderful organisation of *Eisai Co.* are remarkable and worth mentioning. The Committee responsible for the organisation arranged everything in such a perfect manner that we are strongly impressed and will surely never forget this Symposium.

Last not least, we have to thank *Sony Corporation*. Due to their new equipment, it was possible to carry out image transcription and sound recording simultaneously.

The fact that this Symposium was held in Japan gave it a special charm, not only because a great part of the research work on the problems we discussed here has been done with great success in this country, but also because the stay in Japan will remain an unforgettable event for the guests from abroad. I believe that all of us will leave this country with the hope to have once more in their life the opportunity to see this country with these beautiful landscapes, wonderful buildings and old culture so impressive for us.

The Symposium is closed.

K. Lang

Participants at the Symposium

MASANOBU AKAGI	IInd Department of Surgery School of Medicine Kumamoto University Kumamoto, Japan
TAKASHI AKAMATSU	Department of Surgery School of Medicine Keio University Tokyo, Japan
KAZUYA AONO	Department of Anesthesiology School of Medicine Kyushu University Fukuoka, Japan
TOSHIO ASAKURA	Department of Physiological Chemistry and Nutrition Faculty of Medicine University of Tokyo Tokyo, Japan
K. H. BAESSLER	Physiologisch-Chemisches Institut University of Mainz Mainz, West Germany
WILLIAM S. BECK	Department of Medicine Harvard Medical School and Massachusetts General Hospital Boston, Massachusetts, U.S.A.
HEINER BEISBARTH	J. Pfrimmer & Co. Erlangen Erlangen, West Germany
HANNE BICKEL	J. Pfrimmer & Co. Erlangen Erlangen, West Germany
K. H. DEHMEL	IIIrd Medical Department Städtisches Krankenhaus München-Schwabing München, West Germany
G. ERDMANN	Universitäts-Kinderklinik Mainz, West Germany

WERNER FEKL	J. Pfrimmer & Co. Erlangen Erlangen, West Germany
KIYOSHI FUJISAWA	Ist Department of Internal Medicine School of Medicine Jikeikai University Tokyo, Japan
SADAMITSU FUKUI	Ist Department of Internal Medicine Nara Medical College Nara, Japan
TETSUJI FURUKAWA	Department of Anesthesiology School of Medicine Kyushu University Fukuoka, Japan
C. A. GESER	Medizinische Poliklinik University of München München, West Germany
YOSHIO GOTO	Department of Internal Medicine School of Medicine Tohoku University Sendai, Japan
YUICHIRO GOTO	Department of Internal Medicine School of Medicine Keio University Tokyo, Japan
M. P. B. HALMAGYI	Institut für Anaesthesiologie Mainz, West Germany
KIYOMICHI HANABUSA	Department of Biochemistry School of Medicine Keio University Tokyo, Japan
HIROSHI HAYASAKA	Ist Department of Surgery Sapporo Medical College Sapporo, Hokkaido, Japan
RUTH VAN HEYNINGEN	Nuffield Laboratory of Ophthalmology University of Oxford Oxford, England
YUKIMASA HIRATA	IInd Department of Internal Medicine School of Medicine Kyushu University Fukuoka, Japan
SIEGFRIED HOLLMANN	Physiologisch-Chemisches Institut University of Düsseldorf Düsseldorf, West Germany

B. L. HORECKER	Department of Molecular Biology Albert Einstein College of Medicine Bronx, New York, U.S.A.
NORIMASA HOSOYA	Nutritional Laboratory School of Health Sciences Faculty of Medicine University of Tokyo Tokyo, Japan
YOSHIO IKEDA	IVth Department of Internal Medicine School of Medicine Jikeikai University Tokyo, Japan
AKIRA INAMOTO	Department of Anesthesiology School of Medicine Kyoto University Kyoto, Japan
BUNMEI INATSUKI	IInd Department of Internal Medicine School of Medicine Tokyo Medical and Dental University Tokyo, Japan
HIROMASA ISHII	Department of Internal Medicine School of Medicine Keio University Tokyo, Japan
YASUNORI KANAZAWA	IIIrd Department of Internal Medicine Faculty of Medicine University of Tokyo Tokyo, Japan
TOSHIO KANEKO	Ist Department of Internal Medicine Faculty of Medicine University of Tokyo Tokyo, Japan
KUNIZO KATAOKA	Department of Internal Medicine School of Medicine Keio University Tokyo, Japan
HIDEO KATO	Department of Pediatrics School of Medicine Juntendo University Tokyo, Japan

KAZUHIRO KODEKI — Department of Surgery, School of Medicine, Juntendo University, Tokyo, Japan

TOJI KONISHI — Eisai Co., Ltd., Tokyo, Japan

KINORI KOSAKA — IInd Department of Internal Medicine, Tokyo Women's Medical College, Tokyo, Japan

YAHITO KOTAKE — IInd Department of Biochemistry, School of Medicine, Nagoya University, Nagoya, Japan

MITIO KUMAGAI — Department of Pediatrics, School of Medicine, Keio University, Tokyo, Japan

KAKUGEN KUZUYA — IIIrd Department of Internal Medicine, School of Medicine, Kyoto Prefectural University, Kyoto, Japan

TAKESHI KUZUYA — IIIrd Department of Internal Medicine, Faculty of Medicine, University of Tokyo, Tokyo, Japan

KONRAD LANG — Physiologisch-Chemisches Institut, University of Mainz, Mainz, West Germany

AGNE LARSSON — Karolinska Institutet, Kemiska Institutionen II, Stockholm, Sweden

DIETER LEIS — J. Pfrimmer Co. Erlangen, Erlangen, West Germany

I. MACDONALD — Department of Physiology, Guy's Hospital Medical School, London, England

YOSHISUKE MARUHAMA — Department of Internal Medicine, School of Medicine, Tohoku University, Sendai, Japan

MASASUKE MASUDA	IIIrd Department of Internal Medicine School of Medicine Kyoto Prefectural University Kyoto, Japan
SHIGEKI MINAKAMI	Department of Physiological Chemistry and Nutrition Faculty of Medicine University of Tokyo Tokyo, Japan
KATSUMI MORI	IIIrd Department of Internal Medicine School of Medicine Kyoto Prefectural University Kyoto, Japan
SHIRO NAGANO	Department of Internal Medicine School of Medicine Keio University Tokyo, Japan
KOJI NAITO	Eisai Co., Ltd. Tokyo, Japan
HARUMI NAKAJIMA	Department of Pediatrics Kosei Hospital Tokyo, Japan
HARUO NAKAMURA	Department of Internal Medicine School of Medicine Keio University Tokyo, Japan
HARUTADA NINOMIYA	Department of Physiological Chemistry and Nutrition Faculty of Medicine University of Tokyo Tokyo, Japan
SHIGEYOSHI OBUCHI	IInd Department of Internal Medicine School of Medicine Tokyo Medical and Dental University Tokyo, Japan
TOSHITSUGU ODA	Ist Department of Internal Medicine Faculty of Medicine University of Tokyo Tokyo, Japan

MINORU OHNUKI
IInd Department of Internal Medicine
School of Medicine
Tokyo Medical and Dental University
Tokyo, Japan

YASUYUKI OHTA
Ist Department of Internal Medicine
School of Medicine
Okayama University
Okayama, Japan

HIROSHI OKA
Ist Department of Internal Medicine
Faculty of Medicine
University of Tokyo
Tokyo, Japan

KAZUO OKADA
Department of Anesthesiology
Faculty of Medicine
University of Tokyo
Tokyo, Japan

K. OPITZ
Pharmakologisches Institut
Westf. Wilhelms-Universität
Münster, West Germany

BURTON M. POGELL
Department of Biochemistry
Albany Medical College
Albany, New York, U.S.A.

LENNART RODÉN
Departments of Pediatrics and Biochemistry and the La Rabida-University of Chicago Institute
University of Chicago
Chicago, Illinois, U.S.A.

KEN SAMBE
Department of Internal Medicine
School of Medicine
Keio University
Tokyo, Japan

KIYOO SATOH
Eisai Co., Ltd.
Tokyo, Japan

MITSUO SATOH
Department of Anesthesiology
School of Medicine
Juntendo University
Tokyo, Japan

RYOZABURO SENAGA
Department of Pediatrics
School of Medicine
Keio University
Tokyo, Japan

NORIO SHIMAZONO	Department of Biochemistry Tokyo Medical College Tokyo, Japan
KEIJIRO SURUGA	Department of Surgery School of Medicine Juntendo University Tokyo, Japan
YASUYUKI TAKAGI	Department of Biochemistry School of Medicine Kyushu University Fukuoka, Japan
TADAO TAKAHASHI	Ist Department of Internal Medicine School of Medicine Jikeikai University Tokyo, Japan
KAZUHISA TAKEDA	Ist Department of Internal Medicine School of Medicine Okayama University Okayama, Japan
TERUJI TANAKA	Ist Department of Internal Medicine School of Medicine Jikeikai University Tokyo, Japan
SEIICHIRO TARUI	Central Laboratory for Clinical Investigation School of Medicine Osaka, University Osaka, Japan
TOICHI TOSEN	Department of Surgery School of Medicine Keio University Tokyo, Japan
OSCAR TOUSTER	Department of Molecular Biology Vanderbilt University Nashville, Tennessee, U.S.A.
TADASU TSUJII	Ist Department of Internal Medicine Nara Medical College Nara, Japan
MINORU UEKUSA	Department of Surgery School of Medicine Keio University Tokyo, Japan

ALBERT I. WINEGRAD George S. Cox Medical Research Institute
Department of Medicine
Hospital of the University of Pennsylvania
Philadelphia, Pennsylvania, U.S.A.

HIDEO YAMAMURA Department of Anesthesiology
Faculty of Medicine
University of Tokyo
Tokyo, Japan

AKIRA YOKOTA Department of Internal Medicine
National Hospital of Ohkura
Tokyo, Japan

HARUHISA YOSHIKAWA Department of Physiological Chemistry and Nutrition
Faculty of Medicine
University of Tokyo
Tokyo, Japan

KIYOSHI YOSHIKAWA Department of Anesthesiology
School of Medicine
Osaka University
Osaka, Japan

TOSHIO YOSHIKAWA Department of Pediatrics
School of Medicine
Juntendo University
Tokyo, Japan

JUNICHI YOSHITAKE Central Operation Room
School of Medicine
Kagoshima University
Kagoshima, Japan

YAWARA YOSHITOSHI Ist Department of Internal Medicine
Faculty of Medicine
University of Tokyo
Tokyo, Japan

HITOSHI YAMAMOTO Eisai Co., Ltd.
Tokyo, Japan

Universitätsdruckerei H. Stürtz AG Würzburg

www.ingramcontent.com/pod-product-compliance
Lightning Source LLC
Chambersburg PA
CBHW071411230926
51846CB00001B/61
* 9 7 8 3 6 4 2 4 6 1 9 3 4 *